Protein Pharmacokinetics
and Metabolism

Pharmaceutical Biotechnology

Series Editor: Ronald T. Borchardt
The University of Kansas
Lawrence, Kansas

Protein Pharmacokinetics and Metabolism

Edited by

Bobbe L. Ferraiolo
R. W. Johnson Pharmaceutical Research Institute
Spring House, Pennsylvania

Marjorie A. Mohler
Genentech, Inc.
South San Francisco, California

and

Carol A. Gloff
Alkermes, Inc.
Cambridge, Massachusetts

Plenum Press • New York and London

Library of Congress Cataloging-in-Publication Data

Protein pharmacokinetics and metabolism / edited by Bobbe L.
  Ferraiolo, Marjorie A. Mohler, and Carol A. Gloff.
        p.   cm. -- (Pharmaceutical biotechnology ; v. 1)
     Includes bibliographical references and index.
     ISBN 0-306-44151-9
     1. Proteins--Metabolism. 2. Protein drugs--Pharmacokinetics.
  I. Ferraiolo, Bobbe L. II. Mohler, Marjorie A. III. Gloff, Carol
  A. IV. Series.
     [DNLM: 1. Proteins--metabolism. 2. Proteins--pharmacokinetics.
  QU 55 P9674]
  QP551.P69747 1992
  615'.7--dc20
  DNLM/DLC
  for Library of Congress   92-49392
                                                                 CIP

QP551
.P69747
1992

ISBN 0-306-44151-9

© 1992 Plenum Press, New York
A Division of Plenum Publishing Corporation
233 Spring Street, New York, N.Y. 10013

Printed in the United States of America

Contributors

Robert A. Baughman, Jr. • Research and Development, Emisphere Technologies, Hawthorne, New York 10532

Gerhard Baumann • Center for Endocrinology, Metabolism, and Nutrition, Department of Medicine, Northwestern University Medical School, Chicago, Illinois 60611

Jennifer E. Cook • Department of Safety Evaluation, Genentech, Inc., South San Francisco, California 94080

Paul A. Cossum • Department of Preclinical Development, Isis Pharmaceuticals, Carlsbad, California 92008

Bobbe L. Ferraiolo • R. W. Johnson Pharmaceutical Research Institute, Spring House, Pennsylvania 19477

Carol A. Gloff • Alkermes, Inc., Cambridge, Massachusetts 02139

Marjorie A. Mohler • Department of Safety Evaluation, Genentech, Inc., South San Francisco, California 94080

Jerome A. Moore • Department of Safety Evaluation, Genentech, Inc., South San Francisco, California 94080; *present address:* Celtrix Pharmaceuticals, Santa Clara, California 95052

John B. Stoudemire • Cytel Corporation, San Diego, California 92121

John M. Trang • Department of Pharmaceutical Sciences, Samford University Pharmacokinetics Center, School of Pharmacy, Samford University, Birmingham, Alabama 35229

v

Robert J. Wills • R. W. Johnson Pharmaceutical Research Institute, Raritan, New Jersey 08869

Peter K. Working • Department of Pharmacology and Toxicology, Liposome Technology, Inc., Menlo Park, California 94025

Victor J. Wroblewski • Lilly Research Laboratories, Lilly Corporate Center, Indianapolis, Indiana 46285

Preface to the Series

A major challenge confronting pharmaceutical scientists in the future will be to design successful dosage forms for the next generation of drugs. Many of these drugs will be complex polymers of amino acids (e.g., peptides, proteins), nucleosides (e.g., antisense molecules), carbohydrates (e.g., polysaccharides), or complex lipids.

Through rational drug design, synthetic medicinal chemists are preparing very potent and very specific peptides and antisense drug candidates. These molecules are being developed with molecular characteristics that permit optimal interaction with the specific macromolecules (e.g., receptors, enzymes, RNA, DNA) that mediate their therapeutic effects. However, rational drug design does not necessarily mean rational drug delivery, which strives to incorporate into a molecule the molecular properties necessary for optimal transfer between the point of administration and the pharmacological target site in the body.

Like rational drug design, molecular biology is having a significant impact on the pharmaceutical industry. For the first time, it is possible to produce large quantities of highly pure proteins, polysaccharides, and lipids for possible pharmaceutical applications. The design of successful dosage forms for these complex biotechnology products represents a major challenge to pharmaceutical scientists.

Development of an acceptable drug dosage form is a complex process requiring strong interactions between scientists from many different divisions in a pharmaceutical company, including discovery, development, and manufacturing. The series editor, the editors of the individual volumes, and the publisher hope that this new series will be particularly helpful to scientists in the development areas of a pharmaceutical company (e.g., drug metabolism, toxicology, pharmacokinetics and pharmacodynamics, drug delivery, preformulation, formulation, and physical and analytical chemistry). In addition, we hope this series will help to build bridges between the development scientists and scientists in discovery (e.g., medicinal chemistry, pharmacology, immunology, cell biology, molecular biology) and in manufacturing (e.g., process chemistry, engineering). The design of successful dosage forms for the next generation of drugs will require not only a high level of expertise by individual scientists, but also a high degree of interaction between scientists in these different divisions of a pharmaceutical company.

Finally, everyone involved with this series hopes that these volumes will also be useful to the educators who are training the next generation of pharmaceutical scientists. In addition to having a high level of expertise in their respective disciplines, these young scientists will need to have the scientific skills necessary to communicate with their peers in other scientific disciplines.

RONALD T. BORCHARDT
Series Editor

Preface

Investigation of the pharmacokinetics and metabolism of human proteins has escalated over the last two decades because of the use of recombinant human proteins as therapeutic agents. In addition, the development and improvement of analytical techniques enabling the detection of minute quantities of proteins in biological matrices have aided this process.

In assembling this volume, we sought to provide a state-of-the-art assessment of the pharmacokinetics and metabolism of protein therapeutics through complete reviews of selected examples. A comprehensive review of all protein therapeutics was not attempted; the majority of the therapeutic protein classes and crucial scientific issues have been addressed, however. Therefore, we are confident that this volume will provide a useful reference for scientists in this field.

The volume has been divided into two general parts. The first part (Chapters 1–3) is composed of general reviews of topics of importance in pharmacokinetic/metabolism studies of proteins: goals and analytical methodologies, effects of binding proteins, and effects of antibody induction, respectively. The second part (Chapters 4–8) consists of specific, detailed reviews by therapeutic protein class: growth factors and hormones, cytokines, cardiovascular proteins, hematopoietic proteins, and antibodies, respectively.

The editors are grateful to the contributors for the patience, personal sacrifice and perseverance required to complete this volume.

BOBBE L. FERRAIOLO
MARJORIE A. MOHLER
CAROL A. GLOFF

Contents

Chapter 3

Potential Effects of Antibody Induction by Protein Drugs
Peter K. Working

Chapter 4

Pharmacokinetics and Metabolism of Protein Hormones
Jerome A. Moore and Victor J. Wroblewski

Chapter 5

Pharmacokinetics and Metabolism of Therapeutic Cytokines
Carol A. Gloff and Robert J. Wills

Chapter 6

Pharmacokinetics and Metabolism of Cardiovascular Therapeutic Proteins
Paul A. Cossum and Robert A. Baughman, Jr.

Chapter 7

Pharmacokinetics and Metabolism of Hematopoietic Proteins
John B. Stoudemire

Chapter 8

**Pharmacokinetics and Metabolism of Therapeutic and
Diagnostic Antibodies**
John M. Trang

Chapter 1

Goals and Analytical Methodologies for Protein Disposition Studies

Bobbe L. Ferraiolo and Marjorie A. Mohler

1. INTRODUCTION

This chapter will present an overview of the goals of protein disposition studies and some of the analytical tools used to achieve those goals. The examples cited in the goals portion of this chapter will be briefly presented since these will be explored in greater depth in the ensuing chapters. Analytical issues are of major concern in protein disposition studies since many of the standard protein analysis methods do not individually provide a positive identification of the analyte. The question of which existing methodologies and analytical technologies are most productive will be addressed.

2. GOALS OF PROTEIN DISPOSITION STUDIES

The questions that need to be addressed in protein disposition studies are the same as those that are addressed for conventional drugs (Table I).

Bobbe L. Ferraiolo • R. W. Johnson Pharmaceutical Research Institute, Spring House, Pennsylvania 19477. *Marjorie A. Mohler* • Department of Safety Evaluation, Genentech, Inc., South San Francisco, California 94080.

Protein Pharmacokinetics and Metabolism, edited by Bobbe L. Ferraiolo *et al.* Plenum Press, New York, 1992.

1

Table I
Protein Disposition Study Goals

Fate of intact drug
Correlation of concentrations and effects
Binding protein effects
Mechanisms of clearance; sites of catabolism, excretion
Active and inactive metabolites

2.1. Fate of Intact Drug

In common with other drugs, protein therapeutics must be present in adequate concentrations at the site(s) of action to produce an effect. Since it is usually not possible to sample these (often unknown) sites, it is generally assumed that the concentration of the protein therapeutic at a readily accessible site such as blood is related to its concentration at its site of action (Benet *et al.*, 1990). As will be discussed below, many protein analysis techniques may not provide useful information on the disposition of unchanged active parent because they do not discriminate between the parent and its processed forms.

2.2. Pharmacokinetic/Pharmacodynamic Correlations

An important early goal of protein disposition studies should be to correlate the time course of the product-related materials in the body with pharmacologic (or toxicologic) effects. This may aid in the selection of new drug candidates based on their disposition properties. In addition, these correlations help identify optimum routes of administration and regimens. Analytical issues are pertinent since meaningful results may depend on an unequivocal knowledge of the identity of the analyte.

Correlations of effects with concentrations have been limited for protein therapeutics, perhaps because their mechanisms and sites of action are often unknown (Gloff and Benet, 1990). However, the relationship between the hypoglycemic effects of insulin and its plasma concentrations has been intensively investigated (Sorensen *et al.*, 1982; Bergman *et al.*, 1985). Similar correlations have been made between the hypoglycemia observed after intravenous administration of insulin-like growth factor (IGF-I) and the plasma concentrations of free IGF-I (Maneval *et al.*, 1990).

2.3. Regimen-Dependent Effects

The regimen employed for protein therapeutics may influence the pharmacodynamic effects observed. For example, one dose of growth hormone (GH) per day in hypophysectomized rats caused significant gains in body weight and bone growth rate; however, the same dose of GH in nine pulses per day doubled the previous weight gain and increased the bone growth rate by over 270% (Cronin *et al.*, 1991; Clark *et al.*, 1985). Continuous intravenous delivery of the same daily GH dose was less efficacious in increasing body weight than pulsatile delivery. Other investigators have confirmed and extended these observations (Isgaard *et al.*, 1988; Maiter *et al.*, 1988). Similar results were achieved with pulsatile administration of an amidated fragment of human GH releasing factor (Clark and Robinson, 1985). These preclinical studies have been supported by clinical investigations showing that daily GH administration promotes a greater growth rate in GH-deficient children than three times a week administration of the same total dose (Sherman *et al.*, 1988; Tuvemo, 1989; Rosenbloom *et al.*, 1990). For parathyroid hormone and its fragments, intermittent doses appear to be more effective than continuous infusion in promoting anabolic skeletal effects (Tam *et al.*, 1982; Podbesek *et al.*, 1983). Regimen-dependent effects have also been noted in preclinical studies with relaxin (Ferraiolo *et al.*, 1989), tissue plasminogen activator (Klabunde *et al.*, 1990), and clinical studies with insulin (Ruotolo *et al.*, 1990; Selam, 1990). It has also been shown that pulsatile delivery of gonadotropin-releasing hormone is optimal for stimulating gonadotropin release; continuous input leads to suppression of secretion (Mazer, 1990).

2.4. Binding Proteins

Pharmacologic effects are usually attributed to unbound drug, although the relationship is less clear for protein therapeutics. An agent that is extensively and strongly bound to plasma proteins has limited access to cellular sites of action (Benet *et al.*, 1990); binding protein interactions may also affect the protein's pharmacokinetics and elimination. The effects of binding proteins on protein pharmacokinetics and metabolism will be reviewed in detail in a separate chapter; the analytical difficulties that binding proteins present are discussed in the analytical section of this chapter. Several well-known examples of binding proteins for protein therapeutics are presented briefly here as illustrations.

2.4.1. Insulinlike Growth Factor-I

IGF-I is extensively bound in plasma to multiple plasma carrier proteins (Ooi, 1990; Zapf *et al.*, 1990a). The clearances of free IGF-I and IGF-I bound in the binding protein complexes differ; free IGF-I is cleared very rapidly compared to the bound forms (Cook *et al.*, 1989). These binding proteins may modulate IGF-I actions by either inhibiting or potentiating its effects at the cellular level (Ooi, 1990; Clemmons, 1989; Zapf *et al.*, 1990a,b).

2.4.2. Tissue Plasminogen Activator (t-PA)

t-PA circulates unbound and in complexes with fast-acting plasminogen activator inhibitor (PAI-1) and other protease inhibitors such as α_2-antiplasmin, C_1-esterase inhibitor, α_2-macroglobulin, and plasminogen inhibitors 1 and 2 (Higgins and Bennett, 1990; Lucore and Sobel, 1988; Sprengers and Kluft, 1987; Haggroth *et al.*, 1984). The relative importance of the various inhibitors may be dose, species, or disease state specific (Mohler *et al.*, 1988; Higgins and Bennett, 1990). While unbound t-PA is cleared rapidly, t-PA in protease inhibitor complexes persists in the circulation (Lucore and Sobel, 1988). Elevated PAI-1 activity associated with myocardial infarction could attenuate the fibrinolytic activity of t-PA postinfusion and contribute to the risk of reocclusion (Lucore and Sobel, 1988).

2.4.3. Human Growth Hormone

Human GH circulates in plasma in part bound to at least one plasma binding protein (Baumann *et al.*, 1986; Baumann and Shaw, 1990; Herington *et al.*, 1986). *In vitro* studies have shown that GH binding protein inhibits the biological effects and receptor binding of GH (Lim *et al.*, 1990). Recent studies indicate that coinjection of human GH and the GH binding protein alters GH's half-life and clearance in rats (Moore *et al.*, 1988, 1989; Baumann *et al.*, 1987, 1989).

2.4.4. Miscellaneous Binding Proteins/Inhibitors

Binding protein/inhibitor phenomena have recently been described for other proteins including nerve growth factor/α_2-macroglobulin (Koo and

Stach, 1989); transforming growth factor-β/α_2-macroglobulin (LaMarre *et al.*, 1990; Danielpour and Sporn, 1990); interleukin-4 (Fernandez-Botran and Vitetta, 1990); interleukin-2 (Lelchuk and Playfair, 1985); and the interferons (Havredaki and Barona, 1985).

2.5. Catabolism of Protein Therapeutics

Catabolism of therapeutic proteins occurs, presumably, by proteolysis; the potential number of sites of catabolism is vast due to the ubiquitous nature of proteolytic enzymes (Lee, 1988; Bocci, 1987, 1990). Preproteolysis modifications of proteins may render them much more susceptible to degradation (Daggett, 1987). For example, the role of carbohydrates in regulating the *in vivo* plasma survival time of glycoproteins is well established (Ashwell and Harford, 1982). Removal of terminal sialic acid residues exposing specific carbohydrate determinants for hepatic recognition is one mechanism proposed to explain this phenomenon. Other alterations of carbohydrate structure may have similar effects (Hotchkiss *et al.*, 1988; Barbeau, 1990). The importance of carbohydrate moieties in modulating plasma clearance has been demonstrated for therapeutic proteins such as t-PA (Lucore *et al.*, 1988; Hotchkiss *et al.*, 1988), interferon-β (Bocci *et al.*, 1982), and erythropoietin (Spivak and Hogans, 1989; Fukuda *et al.*, 1989). Other determinants of *in vivo* protein catabolism that have been suggested include the amino-terminal residue (the N-end rule) (Bachmair *et al.*, 1986), primary structure (the PEST hypothesis) (Rogers *et al.*, 1986; Lombardo *et al.*, 1988), and tertiary structure (Kossiakoff, 1988).

2.5.1. Sites of Catabolism

2.5.1a. Liver. The liver has been shown to contribute significantly to the metabolism of protein therapeutics. Receptor-mediated endocytosis followed by degradation in lysosomes appears to apply to nerve growth factor, chorionic gonadotropin, prolactin, and thyrotropin releasing hormone (Desbuquois and Postel-Vinay, 1980), t-PA (Kuiper *et al.*, 1988; Bakhit *et al.*, 1987, 1988; Marks *et al.*, 1990), transforming growth factor-β (Coffey *et al.*, 1987), and epidermal growth factor (Kim *et al.*, 1989; St. Hilaire *et al.*, 1983; Sugiyama and Hanano, 1989). For insulin and glucagon, both lysosomal and extralysosomal hepatic catabolic mechanisms have been proposed (Hagopian and Tager, 1987; Zingg *et al.*, 1987; Surmacz *et al.*, 1988). The role of the liver in the metabolism of interferons has been reviewed elsewhere (Wills, 1990; Bocci et al., 1982).

Biliary excretion of therapeutic proteins such as insulin (LaRusso, 1984; Zingg et al., 1987) and epidermal growth factor (LaRusso, 1984; St. Hilaire et al., 1983) has been reported. These proteins may undergo partial proteolysis prior to biliary excretion.

2.5.1b. Kidney.   The kidney appears to play a dominant role in the catabolism of many proteins (Maack et al., 1979, 1985; Strober and Waldmann, 1974). After filtration at the glomerulus, proteins such as interferon (Wills, 1990; Bocci, 1987, 1990), insulin, GH, and parathyroid hormone are reabsorbed at the proximal tubule by specific or nonspecific luminal endocytosis (Maack et al., 1979, 1985; Carone et al., 1982; Christensen et al., 1988; Rabkin et al., 1984; Nielsen and Christensen, 1989). Specific kidney clearance receptors for atrial natriuretic factor have also recently been described (Maack et al., 1988). Small peptides may be hydrolyzed by brush border membrane proteases (Carone and Peterson, 1980). Reabsorbed proteins may be proteolytically degraded in lysosomes, although cytosolic protein processing or processing in other intracellular compartments may participate as well (Thomas et al., 1987). Peritubular extraction and catabolism may also contribute for proteins such as insulin and parathyroid hormone (Maack et al., 1979, 1985; Hellfritzsch et al., 1988; Rabkin et al., 1984). Urinary clearance of intact therapeutic proteins in the absence of renal insufficiency appears to be negligible (Bocci, 1990; Maack et al., 1979; Strober and Waldmann, 1974; Katz and Emmanouel, 1978), with a few exceptions [e.g., erythropoietin in anemia (Miyake et al., 1977)].

2.5.1c. Catabolism at Extravascular Sites of Administration.   Proteins may also be degraded at extravascular sites of injection; this may contribute to incomplete bioavailability (Lee, 1988). Catabolism of protein therapeutics at extravascular sites of administration has been described for insulin (Okumura et al., 1985; Berger et al., 1979; Davies et al., 1980; Lee, 1988); parathyroid hormone and calcitonin (Parsons et al., 1979); and interferon-β (Bocci et al., 1986a). Coadministration of various protease inhibitors or alternate substrates for proteases has been shown to prevent local degradation of therapeutic proteins including: collagen (Hori et al., 1989); benzyloxycarbonyl-Gly-Pro-Leu-Gly (Hori et al., 1983); bacitracin, leupeptin, and phosphoramidon (Komada et al., 1985); the enkephalinase inhibitor, thiorphan (Chipkin et al., 1984); ophthalmic acid and a bovine pancreatic proteinase inhibitor (Offord *et al.*, 1979); albumin (Bocci *et al.*, 1986a); gelatin (Bocci *et al.*, 1986b); and aprotinin (Berger *et al.*, 1980, 1982; Freidenberg *et al.*, 1981). Explanations for the effects of these compounds on the absorption of therapeutic proteins other than inhibition of proteolysis have also been suggested (Williams *et al.*, 1983).

2.5.1d. Active and Inactive Metabolites. Investigating drug metabolites is significantly more difficult for protein therapeutics than for conventional drugs, due, in part, to the abundance of potential sites of metabolism, the potency and physiochemistry of most protein therapeutics, and the nondiscriminating nature of many of the commonly used analytical techniques. Notwithstanding, metabolites of protein therapeutics have been identified or partially characterized for insulin (Lee, 1988; Thomas *et al.*, 1983; Clot *et al.*, 1990), glucagon (Hagopian and Tager, 1987), atrial natriuretic factor (Bertrand and Doble, 1988; Koehn *et al.*, 1987), human relaxin (Ferraiolo *et al.*, 1991), and human interferon-γ (Ferraiolo *et al.*, 1990b). Active metabolites of protein therapeutics have been identified or proposed (see Section 3.1.3b).

2.6. Drug–Drug Interactions

Proteins may exert effects on the elimination of concurrently administered therapeutic compounds (Gloff and Benet, 1990). For example, interferons have been shown to alter the pharmacologic effects of other drugs (Moore *et al.*, 1983; Taylor *et al.*, 1985) and to depress cytochrome P450 and to alter drug metabolism activities (Singh *et al.*, 1982; Rustgi *et al.*, 1987; Franklin and Finkle, 1985, 1986; Williams and Farrell, 1986; Parkinson *et al.*, 1982). Clinically, interferons have been shown to alter the pharmacokinetics and/or metabolism of other drugs (Nokta *et al.*, 1989; Jonkman *et al.*, 1989) and interferon-α has been shown to alter the disposition and distribution of a radiolabeled antibody (Rosenblum *et al.*, 1988). Other authors have been unable to demonstrate significant effects of interferons on pharmacokinetic and metabolic parameters of other drugs (Secor and Schenker, 1984; Echizen *et al.*, 1990). Treatment of mice with tumor necrosis factor (TNF-α) caused a marked depression in cytochrome P450 and ethoxycoumarin deethylase and arylhydrocarbon hydroxylase activities (Ghezzi *et al.*, 1986). GH is postulated to modulate the development and/or expression of the liver mixed-function oxidase system (Wilson, 1970; Wilson and Frohman, 1974; Yamazoe *et al.*, 1986). Interleukin-I has been reported to depress cytochrome P450 levels and drug metabolizing activities in mice (Shedlofsky *et al.,* 1987).

2.7. Interspecies Scaling

For many protein therapeutics, the clearance in laboratory animals is highly predictive of the human pharmacokinetics when analyzed as a function of body weight or surface area (Mordenti *et al.,* 1991; Young *et al.,* 1990). The proteins that have been studied cover a large range of molecular sizes (6–150 kDa) and are eliminated by a variety of mechanisms. This predictive information during the drug development process provides a rational basis for preclinical and clinical dose selection (Mordenti and Green, 1991; Mordenti *et al.,* 1992).

3. METHODS

The methods highlighted in this chapter are those used in the majority of current published protein disposition studies—namely, immunoassays, bioassays, and the use of radiolabeled proteins. While the potential drawbacks of relying on any one method will be emphasized, the authors do not mean to imply that these methods or their results are not valuable. Our intent, instead, is to suggest that in many cases, the validation of these methods may sometimes be inadequate, and to caution the user to avoid drawing unwarranted conclusions from the resultant data.

Immunoassays have been the method of choice for quantitating proteins in pharmacokinetic studies since these methods are rapid, sensitive, economical, and suitable for batch processing (Chen *et al.,* 1991). In addition, for many clinically important proteins, no other viable alternatives exist (Gosling, 1990). Of the immunoassays, the most frequently used are enzyme-linked immunosorbent assays (ELISA), radioimmunoassays (RIA), and immunoradiometric assays (IRMA). *In vitro* or *in vivo* bioassays may be used separately or in conjunction with immunoassays. Proteins may also be radiolabeled with a variety of isotopes for disposition studies. Whatever the method used, it must be sufficiently sensitive so that samples may be taken over a long enough period of time to adequately characterize the disposition. In addition, it should be clear that none of these methods may be absolutely reliable in isolation; it is advisable to supplement them with other methods such as HPLC, electrophoresis, and/or mass spectrometry.

3.1. Immunoassays

It is beyond the scope of this chapter to review the immunoassay literature. The authors refer the interested reader to several reviews and texts describing immunoassay methods in detail (Chan and Perlstein, 1987; Maggio, 1980; Gosling, 1990; Kemeny and Chantler, 1988). Immunoassays are relatively specific and easy to perform compared to bioassays, and they have an objective endpoint. If a bioassay is available, it is desirable to correlate the immuno- and bioactivities as part of the immunoassay validation.

3.1.1. ELISA

The most commonly used immunoassay method is the ELISA, primarily because it is sensitive, nonradioactive, maximally automated, has a long reagent shelf-life, can be done in a batch process, and has the minimum requirements for difficult-to-obtain reagents such as antibodies. These double-antibody sandwich assays may employ polyclonal or monoclonal antibodies, or both. An antianalyte antibody is bound to a solid support (coat antibody). The matrix containing the analyte is added, followed by a second antianalyte antibody (conjugated antibody) that provides the means for detection (Maggio, 1980; Kemeny and Chantler, 1988; Gosling, 1990).

3.1.2. Other Immunoassays

RIAs are also frequently employed in protein disposition studies (Parker, 1976). In these immunoassays, the quantitation is through a radioactive label. RIAs are competitive assays; the unlabeled antigen competes with the labeled antigen for a limited number of binding sites on the antibody. The results obtained with RIAs, like other immunoassays (see below), may be highly dependent on the choice of antibodies (Venturini et al., 1990). RIAs may present special problems in that their use may not permit assay of radiolabeled proteins. RIAs may also be subject to endogenous interferences which may compromise their specificity (Temple et al., 1990). Binding proteins can interfere with RIA results; the binding protein may bind tracer and artifactually increase or decrease the final result depending on the separation method. Some of the disadvantages of RIAs relative to ELISAs include short reagent lifetimes, less convenient protocols, less automation, potential health hazards necessitating radiological protection, and generation of radioactive waste and the consequent disposal problems (Gosling, 1990).

3.1.3. Immunoassay Disadvantages

One disadvantage of immunoassays is that positive identification (e.g., exact biochemical composition, sequence) of the analyte is not possible. Immunoassays may not distinguish bioactive from inactive forms of the protein. Partial degradation of the protein may alter or eliminate its interaction with the antibodies used in the assay, resulting in incomplete characterization of its disposition. There may be interference by a variety of endogenous or exogenous materials (Table II). In addition, one can only measure one analyte at a time; the protein therapeutic and its metabolites cannot, generally, be measured simultaneously. Immunoassay results must be evaluated very critically; a reasonable, reproducible standard curve is no assurance of meaningful experimental results. *In vitro* spikes cannot always substitute for actual samples in testing the assay's performance. Finally, immunoassays may have relatively large coefficients of variation (CV); even a good immunoassay may vary ±15–20% (Chen *et al.*, 1991).

3.1.3a. Binding Protein Interferences.   The presence of binding proteins for protein therapeutics can present serious analytical problems. Interference by binding proteins may make a direct immunoassay useless; however, low-affinity interferences may be diluted out (Chen *et al.*, 1991), although this solution effectively reduces the assay sensitivity.

Specific binding proteins may result in poor or variable recoveries in direct immunoassays, and dilutions may not be able to overcome high-affinity interferences; extraction methods may be necessary to eliminate these effects (Chen *et al.*, 1991). IGF-I provides an example of how binding proteins may complicate the ability to quantitate therapeutic proteins using immunoassays. In order to quantitate total and free IGF-I in the presence of multiple plasma binding proteins, multistep sample processing is necessary. Total IGF-I may be obtained by acidification and subsequent neutralization of each plasma sample prior to immunoassay (Holland *et al.*, 1988). To measure free IGF-I, each sample may be subjected to size exclusion HPLC,

Table II
Potential Immunoassay Interferences

Binding proteins/inhibitors
Metabolites
Antibodies
Matrix

and the presumed appropriate free IGF-I fractions may be collected blind and then assayed by RIA (e.g., Celniker *et al.,* 1990).

3.1.3b. Metabolite Interferences.

Partially degraded forms of protein therapeutics may or may not be immunoreactive in the parent assay; immunoassays may be relatively insensitive to small changes in primary and secondary structure. Therefore, it may be impossible to quantitatively or qualitatively detect metabolites. Yet, it is particularly important to detect and quantitate biologically active partially degraded forms of protein therapeutics. Controlled *in vitro* digestion of human GH yielded supra-active products (lacking residues 138–147) that were indistinguishable from intact GH by RIA (Lewis *et al.,* 1975). In cases where limited proteolysis results in recoverable and/or active metabolites, there is a significant risk that the polyclonal-based immunoassays or C-terminal/N-terminal specific immunoassays may yield misleading information about protein disposition (Hagopian and Tager, 1987).

Recombinant human interferon-γ (rIFN-γ) provides an example of the problems that can arise due to reliance on immunoassay data alone for protein disposition studies. After IV and SC administration of rIFN-γ (dimer) in rhesus monkeys, the SC bioavailability as determined by ELISA was 200% (Ferraiolo *et al.,* 1990b); this was a reproducible finding (Ferraiolo *et al.,* 1988a), and similar observations have been made in human bioavailability studies (Kurzrock *et al.,* 1985; Reed *et al.,* 1990; Chen *et al.,* 1990). [^{125}I]-rIFN-γ was administered IV and SC to rhesus monkeys to investigate the source of the nonphysiological bioavailability. SDS–PAGE autoradiography of plasma samples showed, for both routes of administration, protein bands with apparent molecular weights of approximately 41–42 kDa (presumed nondissociated dimer), 15–16 kDa, and 14 kDa, but in route-dependent proportions. After SC administration, more of the detectable radioactivity was of lower molecular weight than after IV administration (Ferraiolo *et al.,* 1990b). Since the carboxy terminus of rIFN-γ contains many basic residues and is highly susceptible to the action of trypsinlike proteases (Gray *et al.,* 1982; Rinderknecht and Burton, 1985; Rinderknecht *et al.,* 1984; Finbloom, 1988; Arakawa *et al.,* 1986; Honda *et al.,* 1987; Gray and Goeddel, 1987; Pan *et al.,* 1987), and since endogenous IFN-γ has multiple carboxy termini (Rinderknecht and Burton, 1985; Rinderknecht *et al.,* 1984; Pan *et al.,* 1987), it was reasonable to suspect that slightly degraded forms of exogenous rIFN-γ may circulate. The rate and extent of production of these metabolites may be route dependent and the degraded forms may have different disposition properties than the parent protein. If degraded forms of rIFN-γ cross-react in the parent immunoassay, they could significantly interfere with the determination of bioavailability.

3.1.3c. Antibody Interferences.   A form of endogenous interference peculiar to therapeutic proteins results from antibody formation. Antibodies are particularly troublesome in preclinical studies where the administered human protein is clearly foreign, although their interference in clinical studies with GH, insulin, and interferons has been reported (Pringle *et al.*, 1989; Giannarelli *et al.*, 1988; Steis *et al.*, 1988; van Haeften, 1989; Bolli, 1989; Gray *et al.*, 1985). These antibodies may be neutralizing or nonneutralizing (Gloff and Benet, 1990). They may affect clearance (van Haeften, 1989; Bolli, 1989; Arquilla *et al.*, 1987, 1989; Gray *et al.*, 1985; Rosenblum *et al.*, 1985) or the pharmacologic effect (Steis *et al.*, 1988), and they may have a major impact on quantitation of the administered protein (Pringle *et al.*, 1989). For example, Table III shows the results of a study using recombinant human tissue factor (rhTF) in dogs. Animals received daily IV doses of rhTF for 28 days and rhTF plasma concentrations were measured weekly by ELISA. In the presence of anti-rhTF antibodies (week 3 and 4 samples), the rhTF plasma concentrations were anomalously low compared to week 2 and 3 levels in the absence of antibodies.

Table III
rhTF Antigen and Anti-rhTF Antibody Levels in Dogs after Daily IV Doses
(25 μg/kg) for 28 Days[a]

Predose ng/ml	Week 2		Week 3		Week 4	
	Ab titer	ng/ml	Ab titer	ng/ml	Ab titer	ng/ml
4.1[b]	−	490[c]	4.1	17.9[c,d]	4.4	LTS[c,d]
LTS	−	468	3.4	LTS[d]	4.1	LTS[d]
LTS	−	357	3.1	LTS[d]	3.9	LTS[d]
LTS	−	541	−	434	2.8	125[d]
LTS	−	381	−	262	3.6	1.2[d]
LTS	−	453	−	377	3.2	LTS[d]
LTS	−	285	−	274	3.6	LTS[d]
LTS	−	607	−	562	3.5	LTS[d]
LTS	−	422	−	174	2.4	19.0[d]
LTS	−	367	−	306	−	5.5
LTS	−	425	−	389	−	352
LTS	−	471	−	329	−	1.6

[a] Data courtesy of P. A. Cossum, M. Lewandowski, and B. Hutchins, Genentech, Inc.
[b] Value reflects nonspecific interaction.
[c] Corrected for nonspecific interaction.
[d] Value suspect when Ab titer positive; (−) = negative; LTS = below assay limit.

Table IV

rhTF Antigen and Anti-rhTF Antibody Levels in Cynomolgous Monkeys after Daily IV Doses (25 μg/kg) for 28 Days[a]

Predose ng/ml	Week 1		Week 2		Week 4	
	Ab titer	ng/ml	Ab titer	ng/ml	Ab titer	ng/ml
LTS	−	596	−	620	−	164
LTS	−	814	−	813	−	LTS
LTS	−	606	−	220	−	55.5
LTS	−	655	−	674	−	125
LTS	−	647	−	222	−	ND
LTS	−	790	−	262	−	10.3
LTS	−	858	−	295	−	25.5
LTS	−	692	−	763	−	258
LTS	−	600	−	592	2.9	2.5[b]
LTS	−	933	−	714	−	136
LTS	−	590	−	647	−	454
284	−	LTS	−	397	−	LTS

[a] Data courtesy of P. A. Cossum, M. Lewandowski, and B. Hutchins, Genentech, Inc.

[b] Value suspect when Ab titer positive; (−) = negative; LTS = below assay limit; ND = not determined.

Conversely, the continued presence of the administered protein may adversely affect the ability to quantitate the antibody. In a similar study in cynomolgous monkeys (Table IV), the extreme reduction in the rhTF concentrations at 4 weeks strongly suggests the presence of anti-rhTF antibodies, although only one animal shows measurable titers. It is likely that the continued presence of rhTF in plasma during this multiple dose study interfered with the quantitation of the anti-rhTF antibodies by the ELISA method, since the ELISA captures the antibodies in plasma with rhTF bound to a solid support.

3.1.3d. Matrix Interferences. Immunoassays are matrix specific; they must be qualified individually for each matrix, as well as species, sometimes sex. Biological fluids tend to modify the reactivity of the protein analyte compared to its reactivity in buffer solutions (Chen *et al.*, 1991). Some matrix effects may be dealt with by adding immunoglobulins to serum samples prior to assay (Lucas *et al.*, 1989). The assay standard curve may have to be developed in the relevant biological fluid. Another way to overcome matrix effects is to apply a minimum dilution prior to assay (Chen *et al.*, 1991).

3.1.3e. Miscellaneous Interferences. Other problems may be peculiar to specific proteins. The endogenous protein itself may be troublesome if it exists in high concentrations naturally (Chen *et al.*, 1991). In the case of IGF-I, the high endogenous concentrations make it difficult to visualize changes in total IGF-I after administration of low pharmacological doses. Pulsatile secretion of endogenous hormones like GH may present similar problems. Cross-reacting endogenous proteins will have a similar effect. Heterophilic antibodies are nonanalyte substances (antibodies to nonhuman immunoglobulins) that bind to antibodies multivalently, leading to erroneous analyte quantitation in two-site immunoassays and RIAs (Boscato and Stuart, 1988). These are antibodies that can bind immunoglobulins from the species used to generate the reagents used in the immunoassays (Gosling, 1990; Dahlmann and Bidlingmaier, 1989).

3.2. Bioassays

Traditionally, bioassays are *in vivo* or *in vitro* tissue- or cell-based assays (Quiron, 1982). *In vivo* bioassays are expensive and time-consuming; they may take days or weeks to perform; they may lack specificity and sensitivity, have subjective endpoints and high variability. In addition, they require animals and often surgical procedures. Cell-based bioassays may be less expensive and time-consuming, but they may still be difficult to perform and they are subject to environmental and supply variables. If transformed cells are used, they may not be representative of normal cells. These assays may be based on, e.g., proliferation, differentiation, or cytotoxicity. In general, bioassays are also subject to interferences similar to those that pertain for immunoassays. Like immunoassays, bioassays may provide no information about degraded products. Species specificity of biological effects may also limit the usefulness of bioassays in particular species (Gloff and Benet, 1990). Other *in vitro* assays that have been proposed as substitutes for *in vivo* and cell-based bioassays include receptor binding assays (and HPLC-based receptor binding assays), so-called bio-ELISAs (which combine aspects of bioassays and immunoassays), and biomimetic assays.

3.3. Radiolabels

An alternative to relying exclusively on immuno- or bioassays in protein disposition studies involves the use of radiolabeled proteins. While radiolabeled protein studies provide advantages with respect to detection, numerous pitfalls must be avoided.

3.3.1. Radiolabeling Methods

3.3.1a. External Labels. If the protein contains a suitable amino acid such as tyrosine or lysine, an external label such as ^{125}I may be used. The label may be chemically coupled to the intact protein; the interested reader is referred to previous reviews that describe these methods in detail (Regoeczi, 1984; Bolton, 1985; Woltanski et al., 1990). Once the iodination method has been selected (often through trial and error), it must be optimized. Labeled protein and free iodine may be separated by gel filtration chromatography or HPLC. A high percentage of the radioactivity should be precipitable with acid (or immunoprecipitable), as an indication that the label is associated with protein. Iodine has several advantages as a protein label: high specific activity, relative simplicity of preparation of the labeled material, and the choice of several isotopes with relatively short radioactive half-lives.

3.3.1b. Internal Labels. If external labeling is not possible, the protein may be internally labeled by growing the production cell line in the presence of amino acids labeled with, e.g., ^3H, ^{14}C, or ^{35}S (Ferraiolo et al., 1988b; Cossum et al., 1992). Alternate methods of labeling proteins with these isotopes are also available (Halban and Offord, 1975).

3.3.1c. Characterization of Radiolabeled Proteins. Regardless of the method used, the labeled product must be thoroughly characterized prior to dosing. The labeled molecule should be shown to be essentially identical in physiochemical and biological properties to the unlabeled material (Bennett and McMartin, 1979). Its state of aggregation may be determined by size exclusion HPLC or gel filtration chromatography. Its purity may also be verified by gel filtration chromatography, HPLC (reversed phase, ion exchange, size exclusion), or electrophoresis. Functional aspects of the protein may be evaluated by bioassay or receptor binding assays. Obviously, the yield and specific activity must be determined (measured, not assumed). The extent of labeling should be determined, as it may be important that the labeling be 1:1; if labeled molecules are a small proportion of the total population, their deviant behavior may not be recognized (e.g., in bioassays with large coefficients of variation).

3.3.2. Dosing Strategies for Radiolabeled Proteins

The choice of whether to administer tracer alone or tracer plus excess unlabeled protein depends on the goals of the disposition study. Use of a pharmacological dose (achieved by addition of cold carrier) may defeat the purpose of some disposition studies by displacing the radiolabel from specific (saturable) binding sites. For example, in the presence of excess unlabeled IGF-I, binding of $[^{125}I]$-IGF-I to its multiple plasma binding proteins (determined by size exclusion HPLC) was completely eliminated (Cook *et al.*, 1989). Determination of recovery of the labeled protein from the dosing solution is important, especially if no carrier unlabeled protein is used and the total protein concentration is low (less than 100 μg/ml), since proteins may adhere to surfaces during transfers. This characteristic may require the use of adsorption retardants such as gelatin or albumin. Because of the high potency of protein therapeutics, large doses of radioactivity (100–500 μCi/kg) may be required to ensure that sufficient radioactivity will be present in biological matrices for analysis. This requirement may be reduced if concentration techniques such as immunoprecipitation can be employed.

3.3.3. Disadvantages of Radiolabeled Proteins

There are numerous pitfalls to avoid in the use of radiolabeled proteins in disposition studies. It is imperative to establish whether radioactivity in a given sample represents intact labeled protein, radiolabeled degradation products, or liberated label (Bennett and McMartin, 1979). *In vivo* loss of the label is a problem for both externally and internally labeled proteins. For iodine-labeled proteins, dehalogenation may occur, liberating free radioactive iodine. Administration of iodine-labeled human GH in clinical studies has shown that free iodine appears in plasma within minutes postdose, and by 60–90 min accounts for the majority of the total plasma radioactivity (Cameron *et al.*, 1969; Parker *et al.*, 1962). Acid precipitation may distinguish grossly between protein-associated label (>1–3 kDa) and small fragments or free label. Size exclusion and other HPLC modalities may also aid in distinguishing among intact protein, degraded protein, and free label. However, with a nonuniformly labeled protein, any degradation product that does not contain the labeled amino acid will be undetectable. Proteolysis will liberate labeled amino acids which may circulate, localize in tissues, or may be reutilized and incorporated into endogenous proteins (Regoeczi, 1984, 1987a; Schwenk *et al.*, 1985). For example, after administration of internally labeled ([^{3}H]leucine) recombinant human GH in rats, it was found that radiolabel either was distributed to the tissues in a pattern similar to that observed for free [^{3}H]leucine (J. A. Moore, personal communication) or was reincorporated into endogenous proteins (see below). Because of these problems, it is foolhardy to rely on the results of total radioactivity without characterization of the radioactive species; the results may be meaningless or misleading (Jansen, 1979; Bier, 1989; Regoeczi, 1987b). Internally labeled proteins may be less desirable than iodinated proteins because of the serious potential for reutilization. Reutilization of iodoamino acids has been reported to occur; however, this subject is controversial (Regoeczi, 1987b).

An illustration of this reutilization phenomenon has recently been presented (Ferraiolo *et al.,* 1990a). After IV administration of internally labeled human [^{3}H]-GH to rats, the initial (1–10 min) plasma time course of radioactivity was similar to that determined by ELISA. However, within 30 min postdose, the plasma ELISA reactivity decline exceeded the plasma radioactivity decline. Analysis of the plasma samples by size exclusion HPLC revealed that between 1 and 20 min postdose, the radioactivity was contained in either GH or small peptides (1–2 amino acids, suggested by reversed-phase HPLC). As early as approximately 60 min postdose, when there was very little intact radioactive GH detectable in plasma, other high-molecular-weight radioactive proteins were observed. One of these proteins was determined to be rat serum albumin (RSA) by its specific precipitation with anti-RSA antibodies and its apparent molecular weight. Apparently, following degradation of [^{3}H]-GH, the liberated amino acids were rapidly incorporated into newly synthesized albumin.

Studies in rats using semisynthetic [^{3}H]insulin (prepared by replacement of the B chain N-terminal phenylalanine with [^{3}H]phenylalanine) showed that the plasma radioactivity was only partially associated with tritiated insulin. Three peaks of plasma radioactivity were observed after gel filtration: one with the same apparent molecular weight as insulin, one at the total column volume (coeluted with unlabeled phenylalanine), and a high-molecular-weight peak at the column void volume. Subtilisin digestion of the high-molecular-weight peak suggested that [^{3}H]phenylalanine was present in sequences other than the N-terminal sequence of the B chain of insulin (Halban *et al.,* 1979; Davies *et al.,* 1980; Berger *et al.,* 1978).

The literature contains numerous other examples of the problems associated with disposition studies employing radiolabeled proteins. A study using ^{125}I- and ^{3}H-labeled recombinant human tumor necrosis factor-α (rhTNF) showed that while the labeled and unlabeled materials had identical immunoreactive pharmacokinetic characteristics after IV administration in mice (Ferraiolo *et al.,* 1988b), total radioactivity in serum for both labels plateaued after approximately 50 min and did not accurately reflect the true time course of intact rhTNF. A similar plateau in total plasma radioactivity was noted for murine TNF in mice (Beutler *et al.,* 1985); the radioactivity in the plateau phase was not associated with intact TNF as determined by SDS–PAGE.

In the dual label study described above, the tissue distribution of [^{125}I]- and [^{3}H]-rhTNF was also determined in mice. Liver and kidney appeared to be the major organs of accumulation of the ^{125}I radiolabel. [Since the kidney is the primary iodide-eliminating organ (Regoeczi, 1987b), total radioactivity was expected to be high in kidney, bladder, and urine.] At later time points, however, the stomach also accumulated a significant amount of ^{125}I radioactivity. This was troublesome considering the known cachectic effect of TNF-α. The expected accumulation of radiolabel was observed in liver and kidney after [^{3}H]-rhTNF administration, but there was no accumulation of ^{3}H radioactivity in the stomach. Therefore, with the iodinated protein, the gastric accumulation was probably secondary to dehalogenation and release of free iodide which is known to accumulate in the thyroid, gastrointestinal tract, and skin (Regoeczi, 1987b). The gastrointestinal iodide cycle is always a consideration in distribution studies with radioiodide. Iodide is secreted and concentrated by the salivary glands and stomach, and reabsorbed into the circulation from the intestines (Regoeczi, 1987a). The magnitude of potential misunderstandings is illustrated by reports that the gastrointestinal tract contribution to [^{131}I]albumin catabolism was in excess of 50%, even though the rate of albumin degradation was unaffected by intestinal resection (Regoeczi, 1987a). The disposition of radioiodide may affect the interpretation of both plasma and total body radioactivities, since the gastrointestinal tract may represent a delay compartment (Regoeczi, 1987b). Radioactive iodine concentration in the skin further confounds studies based on total radioactivity measurements. It has been shown that accumulation of radioiodide in skin may account for as much as 25% of a dose of Na[131]I (Regoeczi, 1987b). Accrual of label derived from iodoproteins in skin and hair may prolong the apparent residence time of labeled proteins.

3.3.4. Clarification of Results with Radiolabeled Proteins

A summary of some of the controls or ancillary studies that may help clarify the results of radiolabeled protein disposition studies is listed in Table V. Pretreatment with sodium iodide may work well for blocking specific free

Table V
Clarification of Results of Protein Disposition Studies
with Radiolabeled Proteins

Pretreatment with sodium iodide
Pretreatment with dehalogenase inhibitors
Distribution of sodium iodide or labeled amino acid
Chase with excess unlabeled protein
Repeat with second label
TCA-precipitable radioactivity
Immunoprecipitation/immunoassay
SDS–PAGE, HPLC, bioassay

iodide uptake (e.g., thyroid), but may not have much impact on uptake in skin, stomach, and intestines. Dehalogenase inhibitors may prevent the liberation of free labeled iodide; however, protein degradation will yield iodine-labeled tyrosine. Administration of the label alone (sodium iodide or the internally labeled amino acid) may aid in determining specific distribution by difference. The labeled protein may also be administered as a tracer dose and as a tracer with excess unlabeled protein to distinguish specific from nonspecific uptake or binding.

Use of internally and externally labeled proteins in separate experiments may also help to clarify results (see above). TCA-precipitable radioactivity is often used as an indicator of intact protein, but this may be compromised if the label is reutilized and incorporated into other proteins, or if large-molecular-mass (>1–3 kDa) fragments or metabolites are generated. The nature of the radioactivity in tissues or blood/plasma/serum should be further characterized using PAGE, HPLC, immuno- or bioassays.

3.4. Other Analytical Methods

The previously described limitations of immunoassays, bioassays, and radioactivity techniques apply as well to other analytical techniques such as chromatography and electrophoresis. The authors suggest that the results derived from the former analytical methods may be improved by supplementing them with other techniques (Bennett and McMartin, 1979; Walker, 1984; Deutscher, 1990; Villafranca, 1990; Hugli, 1989). The same scrutiny of the resultant data and consideration of the limitations of the techniques should be applied as has been suggested in the preceding sections.

Chromatography may offer poorer recoveries and less sensitivity than immunoassays; however, good specificity and quantitation may be provided, as well as allowance for measurement of multiple analytes simultaneously. Immunoaffinity chromatography may be useful for isolation of protein products from dilute mixtures containing numerous contaminants that are often present in much higher concentrations than the desired product (Grandics et al., 1990; Philips, 1989). Mass spectrometry, when practical, can significantly aid in identification of the protein analyte. However, laborious sample preparation and small amounts of the analyte in biological matrices may limit the usefulness and general applicability of this method (Ferraiolo et al., 1991). Other methodologies that may aid in the identification of protein therapeutic-related materials in vivo include immunoprecipitation (Firestone and Winguth, 1990), the use of N- and C-terminal specific immunological techniques, sequencing (Bhown, 1988; Matsudaira, 1989), electrophoresis (MacNamara and Whicher, 1990), and immunoblotting techniques (Timmons and Dunbar, 1990; Hossenlopp et al., 1986).

The application of size exclusion HPLC to the identification of protein binding/inhibitor phenomena for protein therapeutics has been described for IGF-I (Cook *et al.*, 1989) and t-PA (Harris *et al.*, 1988). A similar approach has been used to characterize the intactness of GH *in vivo* (Ferraiolo *et al.*, 1990a). Reversed-phase HPLC has also been used to characterize radioiodinated insulin in disposition studies (Sato *et al.*, 1990).

Mass spectrometry has recently been used to characterize the disposition of relaxin (Ferraiolo *et al.*, 1991). Relaxin consists of two chains (A and B, in analogy to insulin). Two forms of relaxin (hRlx-2 and hRlx) have been extensively characterized. hRlx-2 consists of an A chain 24 amino acids in length and a 33-amino-acid B chain. hRlx differs from hRlx-2 in the absence of 4 amino acids from the C terminus of the B chain (A24B29). These two relaxins were indistinguishable by immunoassay (ELISA). Relaxin was isolated from rhesus monkey plasma by monoclonal antibody affinity purification after administration of a large IV dose (0.5–1.0 mg/kg) of hRlx-2. Reversed-phase HPLC separated hRlx-2 from hRlx and other products; this method resolved single amino acid differences in the B chain. The column fractions were submitted to fast atom bombardment mass spectroscopy for identification. Sixty minutes after IV administration, the majority of the recoverable relaxin was in the form of hRlx-2 and its oxidation products. Significant (12–32%) conversion to hRlx was observed; other products (A24B32, A24B27) also appeared in small amounts.

While this mass spectroscopy approach aided in the identification of the analyte, recoveries in the multistep purification procedure were variable and raised concerns about selective purification or artifact production. In addition, the purification procedure was labor-intensive and large volumes of biological matrix containing high concentrations of the test protein were required.

4. FUTURE DIRECTIONS

The procedures used for the characterization of protein therapeutics *in vivo* are not cut-and-dried; the extent of characterization necessary appears to be decided on a case-by-case basis. In addition, key analytical approaches are just beginning to be explored, developed, and/or improved. The applicable methods depend on the physiochemical properties of the protein and its biological properties, including potency. Protein size alone may exclude some techniques from consideration. The high potency of many protein therapeutics, resulting in very low therapeutic concentrations (ng–pg/ml or less), may also limit the choice of applicable methodologies, particularly in clinical studies where supratherapeutic doses cannot be justified.

The rationale for doing these studies from a regulatory/safety point of view may be debatable in cases where the protein therapeutic is identical to the endogenous protein and is being used as replacement therapy. Conventional metabolism studies may be of greater concern for second-generation molecules that have been substantially altered or endogenous proteins used in nonphysiological doses.

In any case, the goals and standards that apply to conventional drug disposition studies should also apply to protein disposition studies. It should be clear that a battery of analytical approaches are required; a single approach may not provide an unambiguous answer.

REFERENCES

Arakawa, T., Hsu, Y.-R., Parker, C. G., and Lai, P.-H., 1986, Role of the polycationic C-terminal portion in the structure and activity of recombinant human interferon-gamma, *J. Biol. Chem.* **261**:8534–8539.

Arquilla, E. R., Stenger, D., McDougall, B., and Ulich, T. R., 1987, Effect of IgG subclasses of the *in vivo* bioavailability and metabolic fate of immune-complexed insulin in Lewis rats, *Diabetes* **36**:144–151.

Arquilla, E. R., McDougall, B. R., and Stenger, D. P., 1989, Effect of isologous and autologous insulin antibodies on the *in vivo* bioavailability and metabolic fate of immune-complexed insulin in Lou/M rats, *Diabetes* **38**:343–349.

Ashwell, G., and Harford, J., 1982, Carbohydrate-specific receptors of the liver, *Annu. Rev. Biochem.* **51**:531–554.

Bachmair, A., Finley, D., and Varshavsky, A., 1986, *In vivo* half life of a protein is a function of its amino terminal residue, *Science* **234**:179–186.

Bakhit, C., Lewis, D., Billings, R., and Malfroy, B., 1987, Cellular catabolism of recombinant tissue-type plasminogen activator, *J. Biol. Chem.* **262**:8716–8720.

Bakhit, C., Lewis, D., Busch, U., Tanswell, P., and Mohler, M., 1988, Biodisposition and catabolism of tissue-type plasminogen activator in rats and rabbits, *Fibrinolysis* **2**:31–36.

Barbeau, D., 1990, Protein glycosylation and pharmacokinetics, Controlled Release Society Newsletter, July, pp. 6–10.

Boscato, L. M., and Stuart, M. C., 1988, Heterophilic antibodies: A problem for all immunoassays, *Clin. Chem.* **34**:27–33.

Baumann, G., and Shaw, M. A., 1990, A second lower affinity growth hormone-binding protein in human plasma, *J. Clin. Endocrinol. Metab.* **70**:680–686.

Baumann, G., Stolar, M. W., Amburn, K., Barsano, C. P., and DeVries, B. C., 1986, A specific growth hormone binding protein in human plasma: Initial characterization, *J. Clin. Endocrinol. Metab.* **62**:134–141.

Baumann, G., Amburn, K. D., and Buchanan, T. A., 1987, The effect of circulating growth hormone-binding protein on metabolic clearance, distribution and degradation of human growth hormone, *J. Clin. Endocrinol. Metab.* **64**:657–660.

Baumann, G., Shaw, M. A., and Buchanan, T. A., 1989, *In vivo* kinetics of a covalent growth hormone-binding protein complex, *Metabolism* **38**:330–333.

Benet, L. Z., Mitchell, J. R., and Sheiner, L. B., 1990, Pharmacokinetics: The dynamics of drug absorption, distribution and elimination, in: *The Pharmacological Basis of Therapeutics* (A. G. Gilman, T. W. Rall, A. S. Nies, and P. Taylor, eds.), Pergamon Press, Elmsford, N.Y., pp. 3–32.

Bennett, H. P. J., and McMartin, C., 1979, Peptide hormones and their analogues: Distribution, clearance from the circulation, and inactivation *in vivo, Pharmacol. Reviews* **30**:247–292.

Berger, M., Halban, P. A., Muller, W. A., Offord, R. E., Renold, A. E., and Vranic, M., 1978, Mobilization of subcutaneously injected tritiated insulin in rats: Effects of muscular exercise, *Diabetologia* **15**:133–140.

Berger, M., Halban, P. A., Girardier, L., Seydoux, J., Offord, R. E., and Renold, A. E., 1979, Absorption kinetics of subcutaneously injected insulin, *Diabetologia* **17**:97–99.

Berger, M., Cuppers, H. J., Halban, P. A., and Offord, R. E., 1980, The effect of aprotinin on the absorption of subcutaneously injected regular insulin in normal subjects, *Diabetes* **29**:81–83.

Berger, M., Cuppers, H. J., Hegner, H., Jorgens, V., and Berchtold, P., 1982, Absorption kinetics and biological effects of subcutaneously injected insulin preparations, *Diabetes Care* **5**:77–91.

Bergman, R. N., Finegood, D. T., and Ader, M., 1985, Assessment of insulin sensitivity in vivo, *Endocrine Rev.* **6**:45–86.

Bertrand, P., and Doble, A., 1988, Degradation of atrial natriuretic peptides by an enzyme in rat kidney resembling neutral endopeptidase 24.11, *Biochem. Pharmacol.* **37**:3817–3821.

Beutler, B. A., Milsark, I. W., and Cerami, A., 1985, Cachectin/tumor necrosis factor. Production, distribution and metabolic fate *in vivo, J. Immunol.* **135**: 3972–3977.

Bhown, A. S., (ed.), 1988, *Protein/Peptide Sequence Analysis: Current Methodologies,* CRC Press, Boca Raton, Fla.

Bier, D. M., 1989, Intrinsically difficult problems: The kinetics of body proteins and amino acids in man, *Diab. Metab. Rev.* **5**:111–132.

Bocci, V., 1987, Metabolism of protein anticancer agents, *Pharmacol. Ther.* **34**:1–49.

Bocci, V., 1990, Catabolism of therapeutic proteins and peptides with implications for drug delivery, *Adv. Drug Del. Rev.* **4**:149–169.

Bocci, V., Pacini, A., Bandinelli, L., Pessina, G. P., Muscettola, M., and Paulesu, L., 1982, The role of the liver in the catabolism of human alpha- and beta-interferons, *J. Gen. Virol.* **60**:397–400.

Bocci, V., Muscettola, M., and Naldini, A., 1986a, The lymphatic route—III. Pharmacokinetics of human natural interferon-beta injected with albumin as a retarder in rabbits, *Gen. Pharmacol.* **17**:445–448.

Bocci, V., Pacini, A., Maioli, E., Muscettola, M., and Paulesu, L., 1986b, Prolonged interferon plasma levels after administration of interferon with retarders, *IRCS Med. Sci.* **14**:360–361.

Bolli, G. B., 1989, The pharmacokinetic basis of insulin therapy in diabetes mellitus, *Diabetes Res. Clin. Pract.* **6**:S3–S16.

Bolton, A. E., 1985, *Radioiodination Techniques,* Review 18, Amersham, Arlington Heights, pp. 1–88.

Cameron, D. P., Burger, H. G., Catt, K. J., and Dong, A., 1969, Metabolic clearance rate of radioiodinated human growth hormone in man, *J. Clin. Invest.* **48**:1600–1608.

Carone, F. A., and Peterson, D. R., 1980, Hydrolysis and transport of small peptides by the proximal tubule, *Am. J. Physiol.* **238**:F151–F158.

Carone, F. A., Peterson, D. R., and Flouret, G., 1982, Renal tubular processing of small peptide hormones, *J. Lab. Clin. Med.* **100**:1–14.

Celniker, A. C., Chen, S., Spanski, N., Pocekay, J., and Perlman, A. J., 1990, IGF-I: Assay methods and pharmacokinetics of free IGF-I in the plasma of normal human subjects following intravenous administration, Proc. U.S. Endocrine Soc., Vol. 72, Program and Abstracts, p. 310, Abstract No. 1144.

Chan, D. W., and Perlstein, M. T., (eds.), 1987, *Immunoassay—A Practical Guide,* Academic Press, New York.

Chen, A. B., Baker, D. L., and Ferraiolo, B. L., 1991, Points to consider in correlating bioassays and immunoassays in the quantitation of peptides and proteins, in: *Peptides, Peptoids and Proteins.* (P. D. Garzone, W. A. Colburn, and M. Mokotoff, eds.) Harvey Whitney, Inc., Cincinnati, pp. 53–71.

Chen, S. A., Izu, A. E., Baughman, R. A., Ferraiolo, B. L., Mordenti, J., Reed, B. R., and Jaffee, H. S., 1990, Pharmacokinetic disposition of recombinant interferon-gamma following intravenous and subcutaneous administration in normal volunteers, Annual Meeting of the International Society for Interferon Research, San Francisco.

Chipkin, R. E., Kreutner, W., and Billard, W., 1984, Potentiation of the hypoglycemic effect of insulin by thiorphan, an enkephalinase inhibitor, *Eur. J. Pharmacol.* **102**:151–154.

Christensen, E. I., Nielsen, S., Hellfritzsch, M., and Nielsen, J. T., 1988, Luminal uptake of insulin in renal proximal tubules, *Contrib. Nephrol.* **68**:78–85.

Clark, R. G., and Robinson, I. C. A. F., 1985, Growth induced by pulsatile infusion of an amidated fragment of human growth hormone releasing factor in normal and GHRF-deficient rats, *Nature* **314**:281–283.

Clark, R. G., Jansson, J.-O., Isaksson, O., and Robinson, I. C. A. F., 1985, Intravenous growth hormone: Growth responses to patterned infusions in hypophysectomized rats, *J. Endocrinol.* **104**:53–61.

Clemmons, D. R., 1989, The role of insulin-like growth factor binding proteins in controlling the expression of IGF actions, in: *Molecular and Cellular Biology of the Insulin-Like Growth Factors and Their Receptors* (D. LeRoith and M. K. Raizada, eds.), Plenum Press, New York, pp. 381–394.

Clot, J. P., Janicot, M., Desbuquois, B., Fouque, F., Haumont, P. Y., and Lederer, F., 1990, Characterization of insulin degradation products generated in liver endosomes: *In vivo* and *in vitro* studies, *FASEB J.* **4**:A2115, Abstract 2445.

Coffey, R. J., Jr., Kost, L. J., Lyons, R. M., Moses, H. L., and LaRusso, N. F., 1987, Hepatic processing of transforming growth factor beta in the rat, *J. Clin. Invest.* **80:**750–757.

Cook, J. E., Ferraiolo, B. L., and Mohler, M. A., 1989, The role of binding proteins in the metabolism of IGF-I, *Pharm. Res.* **6:**S30, Abstract BT219.

Cossum, P. A., Dwyer, K. A., Roth, M., Chen, S. A., Moffat, B., Vandlen, R., and Ferraiolo, B. L., 1992, The disposition of a human relaxin (hRlx-2) in pregnant and nonpregnant rats, *Pharm. Res.* **9:**415–420.

Cronin, M. J., Ferraiolo, B. L., and Moore, J. A., 1991, Contemporary issues involving the activities of recombinant human hormones, in: *Peptides, Peptoids and Proteins,* (P. D. Garzone, W. A. Colburn, and M. Motokoff, eds.) Harvey Whitney, Inc., Cincinnati, pp. 138–146.

Daggett, V., 1987, Protein degradation: The role of mixed-function oxidases, *Pharm. Res.* **4:**278–284.

Dahlmann, N., and Bidlingmaier, F., 1989, Circulating antibodies to mouse monoclonal immunoglobulins caused false-positive results in a two-site assay for alpha-fetoprotein, *Clin. Chem.* **35:**2339.

Danielpour, D., and Sporn, M. B., 1990, Differential inhibition of transforming growth factor beta1 and beta2 activity by alpha2-macroglobulin, *J. Biol. Chem.* **265:**6973–6977.

Davies, J. G., Offord, R. E., Halban, P. A., and Berger, M., 1980, The chemical characterization of the products of the processing of subcutaneously injected insulin, in: *Insulin: Chemistry, Structure and Function of Insulin and Related Hormones* (D. Brandenburg and A. Wollmer, eds.), de Gruyter, Berlin, pp. 517–523.

Desbuquois, B., and Postel-Vinay, M.-C., 1980, Receptor-mediated internalization of insulin, glucagon and growth hormone in intact rat liver. A biochemical study, in: *Insulin: Chemistry, Structure and Function of Insulin and Related Hormones* (D. Brandenburg and A. Wollmer, eds.), de Gruyter, Berlin, pp. 285–292.

Deutscher, M. P., (ed.), 1990, *Guide to Protein Purification,* Academic Press, New York.

Echizen, H., Ohta, Y., Shirataki, H., Tsukamoto, K., Umeda, N., Oda, T., and Ishizaki, T., 1990, Effects of subchronic treatment with natural human interferons on antipyrine clearance and liver function in patients with chronic hepatitis, *J. Clin. Pharmacol.* **30:**562–567.

Fernandez-Botran, R., and Vitetta, E. S., 1990, A soluble, high-affinity, interleukin-4-binding protein is present in the biological fluids of mice, *Proc. Natl. Acad. Sci. USA* **87:**4202–4206.

Ferraiolo, B. L., Fuller, G. B., Burnett, B., and Chan, E., 1988a, Pharmacokinetics of recombinant human interferon-gamma in the rhesus monkey after intravenous and subcutaneous administration, *J. Biol. Resp. Mod.* **7:**115–122.

Ferraiolo, B. L., Moore, J. A., Crase, D., Gribling, P., Wilking, H., and Baughman, R. A., 1988b, Pharmacokinetics and tissue distribution of recombinant human tumor necrosis factor-alpha in mice, *Drug Metab. Dispos.* **16:**270–275.

Ferraiolo, B. L., Cronin, M., Bakhit, C., Chestnut, M., Lyon, R., and Roth, M., 1989, Pharmacokinetics and pharmacodynamics of relaxin in mice, *Endocrinology* **125**:2922–2926.

Ferraiolo, B. L., Mohler, M. A., Cossum, P. A., Moore, J. A., Reed, B., and Vandlen, D., 1990a, Characterization of therapeutic proteins in disposition studies, Society of Toxicology Annual Meeting, Miami.

Ferraiolo, B. L., Mohler, M., Cook, J., Chen, A., Reed, B., O'Connor, J., and Keck, R., 1990b, The metabolism of recombinant human interferon-gamma (rIFN-gamma) in rhesus monkeys, *Pharm. Res.* **7**:S-46 (Abstract BIOTEC 2028).

Ferraiolo, B. L., Winslow, J., Laramee, G., Celniker, A., and Johnston, P., 1991, Pharmacokinetics and metabolism of human relaxins in rhesus monkeys, *Pharm. Res.* **8**:1032–1038.

Finbloom, D. S., 1988, Internalization and degradation of human recombinant interferon-gamma in the human histiocytic lymphoma cell line, U937: Relationship to Fc receptor enhancement and antiproliferation, *Clin. Immunol. Immunopathol.* **47**:93–105.

Firestone, G. L., and Winguth, S. D., 1990, Immunoprecipitation of proteins, *Methods Enzymol.* **182**:688–699.

Franklin, M. R., and Finkle, B. S., 1985, Effect of murine gamma-interferon on the mouse liver and its drug metabolizing enzymes: Comparison with human hybrid alpha interferon, *J. Interferon Res.* **5**:265–272.

Franklin, M. R., and Finkle, B. S., 1986, The influence of recombinant DNA-derived human and murine gamma interferons on mouse hepatic drug metabolism, *Fundam. Appl. Toxicol.* **7**:165–169.

Freidenberg, G. R., White, N., Cataland, S., O'Dorisio, T. M., Sotos, J. F., and Santiago, J. V., 1981, Diabetes responsive to intravenous but not subcutaneous insulin: Effectiveness of aprotinin, *N. Engl. J. Med.* **305**:363–368.

Fukuda, M. N., Sasaki, H., Lopez, L., and Fukuda, M., 1989, Survival of recombinant erythropoietin in the circulation: The role of carbohydrates, *Blood* **73**:84–89.

Ghezzi, P., Saccardo, B., and Bianchi, M., 1986, Recombinant tumor necrosis factor depresses cytochrome P450-dependent microsomal drug metabolism in mice, *Biochem. Biophys. Res. Commun.* **136**:316–321.

Giannarelli, R., Marchetti, P., Giannecchini, M., Di Cianni, G., Cecchetti, P., Masoni, A., and Navalesi, R., 1988, Free insulin concentrations in immediately extracted plasma samples and their relationships to clinical and metabolic parameters in insulin-treated diabetic patients, *Acta Diabetol. Lat.* **25**:257–262.

Gloff, C. A., and Benet, L. Z., 1990, Pharmacokinetics and protein therapeutics, *Adv. Drug Del. Rev.* **4**:359–386.

Gosling, J. P., 1990, A decade of development of immunoassay methodology, *Clin. Chem.* **36**:1408–1427.

Grandics, P., Szathmary, Z., and Szathmary, S., 1990, A novel immunoaffinity system for the purification of therapeutic proteins, *Ann. N.Y. Acad. Sci.* **589**:148–156.

Gray, P. W., and Goeddel, D. V., 1987, Molecular biology of interferon-gamma, *Lymphokines* **13**:151–162.

Gray, P. W., Leung, D. W., Pennica, D., Yelverton, E., Najarian, R., Simonsen, C. C., Derynck, R., Sherwood, P. J., Wallace, D. M., Berger, S. L., Levinson, A. D., and Goeddel, D. V., 1982, Expression of human immune interferon cDNA in *E. coli* and monkey cells, *Nature* **295**:503–508.

Gray, R. S., Cowan, P., di Mario, U., Elton, R. A., Clarke, B. F., and Duncan, L. J. P., 1985, Influence of insulin antibodies on pharmacokinetics and bioavailability of recombinant human and highly purified beef insulins in insulin dependent diabetics, *Br. Med. J.* **290**:1687–1691.

Haggroth, L., Mattsson, C., and Friberg, J., 1984, Inhibition of the human tissue plasminogen activator in plasma from different species, *Thromb. Res.* **33**:583–594.

Hagopian, W. A., and Tager, H. S., 1987, Hepatic glucagon metabolism, *J. Clin. Invest.* **79**:409–417.

Halban, P. A., and Offord, R. E., 1975, The preparation of a semisynthetic tritiated insulin with a specific radioactivity of up to 20 curies per millimole, *Biochem. J.* **151**:219–225.

Halban, P. A., Berger, M., and Offord, R. E., 1979, Distribution and metabolism of intravenously injected tritiated insulin in rats, *Metabolism* **28**:1097–1104.

Harris, R., Frade, L. G., Creighton, L. J., Gascoine, P. S., Alexandroni, M. M., Poole, S., and Gaffney, P. J., 1988, Investigation by HPLC of the catabolism of recombinant tissue plasminogen activator in the rat, *Thromb. Haemostas.* **60**:107–112.

Havredaki, M., and Barona, F., 1985, Variations in interferon inactivators and/or inhibitors in human serum and their relationship to interferon therapy, *Jpn. J. Med. Sci. Biol.* **38**:107–111.

Hellfritzsch, M., Nielsen, S., Christensen, E. I., and Nielsen, J. T., 1988, Basolateral tubular handling of insulin in the kidney, *Contrib. Nephrol.* **68**:86–91.

Herington, A. C., Ymer, S., and Stevenson, J., 1986, Identification and characterization of specific binding proteins for growth hormone in normal human serum, *J. Clin. Invest.* **77**:1817–1823.

Higgins, D. L., and Bennett, W. F., 1990, Tissue plasminogen activator: The biochemistry and pharmacology of variants produced by mutagenesis, *Annu. Rev. Pharmacol. Toxicol.* **30**:91–121.

Holland, M. D., Hossner, K. L., Niswender, G. D., Elsasser, T. H., and Odde, K. G., 1988, Validation of a heterologous radioimmunoassay for insulin-like growth factor-I in bovine serum, *J. Endocrinol.* **119**:281–285.

Honda, S., Asano, T., Kajio, T., Nakagawa, S., Ikeyama, S., Ichimori, Y., Sugino, H., Nara, K., Kakinuma, A., and Kung, H.-F., 1987, Differential purification by immunoaffinity chromatography of two carboxy-terminal portion-deleted derivatives of recombinant human interferon-gamma from *Escherichia coli, J. Interferon Res.* **7**:145–154.

Hori, R., Komada, F., and Okumura, K., 1983, Pharmaceutical approach to subcutaneous dosage forms of insulin, *J. Pharm. Sci.* **72**:435–439.

Hori, R., Komada, F., Iwakawa, S., Seino, Y., and Okumura, K., 1989, Enhanced bioavailability of subcutaneously injected insulin coadministered with collagen in rats and humans, *Pharm. Res.* **6**:813–816.

Hossenlopp, P., Seurin, D., Segovia-Quinson, B., Hardouin, S., and Binoux, M., 1986, Analysis of serum insulin-like growth factor binding proteins using Western blotting: Use of the method for titration of the binding proteins and competitive binding studies, *Anal. Biochem.* **154**:138–143.

Hotchkiss, A., Refino, C. J., Leonard, C. K., O'Connor, J. V., Crowley, C., McCabe, J., Tate, K., Nakamura, G., Powers, D., Levinson, A., Mohler, M., and Spellman, M. W., 1988, The influence of carbohydrate structure on the clearance of recombinant tissue-type plasminogen activator, *Thromb. Haemost.* **60**: 255–261.

Hugli, T. E., (ed.), 1989, *Techniques in Protein Chemistry,* Academic Press, New York.

Isgaard, J., Carlsson, L., Isaksson, O. G. P., and Jansson, J.-O., 1988, Pulsatile intravenous growth hormone (GH) infusion to hypophysectomized rats increases insulin-like growth factor I messenger ribonucleic acid in skeletal tissues more effectively than continuous GH infusion, *Endocrinology* **123**:2605–2610.

Jansen, A. B. A., 1979, Total radioactivity half-lives, *Drug Metab. Dispos.* **7**:350.

Jonkman, J. H. G., Nicholson, K. G., Farrow, P. R., Eckert, M., Grassmijer, G., Oosterhuis, B., De Noord, O. E., and Guentert, T. W., 1989, Effects of alpha-interferon on theophylline pharmacokinetics and metabolism, *Br. J. Clin. Pharmacol.* **27**:795–802.

Katz, A. I., and Emmanouel, D. S., 1978, Metabolism of polypeptide hormones by the normal kidney and in uremia, *Nephron* **22**:69–80.

Kemeny, D. M., and Chantler, S., 1988, An introduction to ELISA, in: *ELISA and Other Solid Phase Immunoassays* (D. M. Kemeny and S. J. Challacombe, eds.), Wiley, New York, pp. 1–29.

Kim, D. C., Sugiyama, Y., Fuwa, T., Sakamoto, S., Iga, T., and Hanano, M., 1989, Kinetic analysis of the elimination process of human epidermal growth factor (hEGF) in rats, *Biochem. Pharmacol.* **38**:241–249.

Klabunde, R. E., Burke, S. E., and Henkin, J., 1990, Enhanced lytic efficacy of multiple bolus injections of tissue plasminogen activator in dogs, *Thromb. Res.* **58**:511–517.

Koehn, J. A., Norman, J. A., Jones, B. N., LeSueur, L., Sakane, Y., and Ghai, R. D., 1987, Degradation of atrial natriuretic factor by kidney cortex membranes: Isolation and characterization of the primary proteolytic product, *J. Biol. Chem.* **262**:11623–11627.

Komada, F., Okumura, K., and Hori, R., 1985, Fate of porcine and human insulin at the subcutaneous injection site. II. In vitro degradation of insulins in the subcutaneous tissue of the rat, *J. Pharmacobio-Dyn.* **8**:33–40.

Koo, P. H., and Stach, R. W., 1989, Interaction of nerve growth factor with murine alpha-macroglobulin, *J. Neurosci. Res.* **22**:247–261.

Kossiakoff, A. A., 1988, Tertiary structure is a principal determinant of protein deamidation, *Science* **240**:191–194.

Kuiper, J., Otter, M., Rijken, D. C., and van Berkel, T. J. C., 1988, Characterization of the interaction *in vivo* of tissue type plasminogen activator with liver cells, *J. Biol. Chem.* **263**:18220–18224.

Kurzrock, R., Rosenblum, M. G., Sherwin, S. A., Rios, A., Talpaz, M., Quesada, J. R., and Gutterman, J. U., 1985, Pharmacokinetics, single-dose tolerance and biological activity of recombinant gamma-interferon in cancer patients, *Cancer Res.* **45**:2866–2872.

LaMarre, J., Wollenberg, G. K., Gauldie, J., and Hayes, M. A., 1990, alpha2-macro-globulin and serum preferentially counteract the mitoinhibitory effect of trans-forming growth factor-beta2 on rat hepatocytes, *Lab. Invest.* **62**:545–551.

LaRusso, N. F., 1984, Proteins in the bile: How they get there and what they do, *Am. J. Physiol.* **247**:G199–G205.

Lee, V. H. L., 1988, Enzymatic barriers to peptide and protein absorption, *Crit. Rev. Ther. Drug Carrier Syst.* **5**:69–97.

Lelchuk, R., and Playfair, J. H. L., 1985, Serum IL-2 inhibitor in mice. I. Increase during infection, *Immunology* **56**:113–118.

Lewis, U. J., Pence, S. J., Singh, R. N. P., and VanderLaan, W. P., 1975, Enhance-ment of growth promoting activity of human growth hormone, *Biochem. Biophys. Res. Commun.* **67**:617–624.

Lim, L., Spencer, S. A., McKay, P., and Waters, M. J., 1990, Regulation of growth hormone (GH) bioactivity by a recombinant human GH-binding protein, *Endo-crinology* **127**:1287–1291.

Lombardo, Y. B., Morse, E. L., and Adibi, S. A., 1988, Specificity and mechanism of influence of amino acid residues on hepatic clearance of oligopeptides, *J. Biol. Chem.* **263**:12920–12926.

Lucas, C., Bald, L. N., Martin, M. C., Jaffe, R. B., Drolet, D. W., Mora-Worms, M., Bennett, G., Chen, A. B., and Johnston, P. D., 1989, An enzyme-linked immuno-sorbent assay to study human relaxin in human pregnancy and in pregnant rhesus monkeys, *J. Endocrinol.* **120**:449–457.

Lucore, C. L., and Sobel, B. E., 1988, Interactions of tissue-type plasminogen activa-tor with plasma inhibitors and their pharmacological implications, *Circulation* **77**:660–669.

Lucore, C. L., Fry, E. T. A., Nachowiak, D. A., and Sobel, B. E., 1988, Biochemical determinants of clearance of tissue-type plasminogen activator from the circula-tion, *Circulation* **77**:906–914.

Maack, T., Johnson, V., Kau, S. T., Figueiredo, J., and Sigulem, D., 1979, Renal filtration, transport and metabolism of low-molecular weight proteins: A review, *Kidney Int.* **16**:251–270.

Maack, T., Park, C. H., and Camargo, M. J. F., 1985, Renal filtration, transport and metabolism of proteins, in: *The Kidney: Physiology and Pathophysiology* (D. W. Seldin and G. Giebisch, eds.), Raven Press, New York, pp. 1773–1803.

Maack, T., Almeida, F. A., Suzuki, M., and Nussenzveig, D. R., 1988, Clearance receptors of atrial natriuretic factor, *Contrib. Nephrol.* **68**:58–65.

MacNamara, E. M., and Whicher, J. T., 1990, Electrophoresis and densitometry of serum and urine in the investigation and significance of monoclonal immunoglobulins, *Electrophoresis* **11**:376–381.

Maggio, E. T., (ed.), 1980, *Enzyme-Immunoassay,* CRC Press, Boca Raton.

Maiter, D., Underwood, L. E., Maes, M., Davenport, M. L., and Ketelslegers, J. M., 1988, Different effects of intermittent and continuous growth hormone (GH) administration on serum somatomedin-C/insulin-like growth factor I and liver GH receptors in hypophysectomized rats, *Endocrinology* **123**:1053–1059.

Maneval, D. C., Chen, S. A., Ferraiolo, B. L., Mordenti, J., Clark, R., Cook, J., and Mohler, M. A., 1990, Pharmacokinetic/pharmacodynamic modeling of the hypoglycemic response to recombinant human insulin-like growth factor (rhIGF-I), American Association of Pharmaceutical Scientists 5th Annual National Meeting, Symposium Abstract.

Marks, G. J., Hart, T. K., Rush, G. F., Hoffstein, S. T., Fong, K.-L. L., and Bugelski, P. J., 1990, Internalization of recombinant tissue-type plasminogen activator by isolated rat hepatocytes is via coated pits, *Thromb. Haemost.* **63**:251–258.

Matsudaira, P. T., (ed.), 1989, *A Practical Guide to Protein and Peptide Purification for Microsequencing,* Academic Press, New York.

Mazer, N. A., 1990, Pharmacokinetic and pharmacodynamic aspects of polypeptide delivery, *J. Control. Release* **11**:343–356.

Miyake, T., Kung, C.K.-H., and Goldwasser, E., 1977, Purification of human erythropoietin, *J. Biol. Chem.* **252**:5558–5564.

Mohler, M. A., Tate, K., Bringman, T. S., Fuller, G., Keyt, B., Vehar, G., and Hotchkiss, A. J., 1988, Circulatory metabolism of recombinant tissue type plasminogen activator in monkeys and rabbits, *Fibrinolysis* **2**:17–23.

Moore, J. A., Marafino, B. J., and Stebbing, N., 1983, Influence of various purified interferons on effects of drugs in mice, *Res. Commun. Chem. Pathol. Pharmacol.* **39**:113–125.

Moore, J. A., Vandlen, R., McKay, P., and Spencer, S. A., 1988, Serum clearance of human growth hormone bound to growth hormone binding protein, Proc. U.S. Endocrine Soc., Vol. 70, p. 121, Abstract No. 404.

Moore, J. A., Celniker, A., Fuh, G., Light, D., McKay, P., and Spencer, S., 1989, Cloned human growth hormone binding protein effects on disposition of human growth hormone in rats, Proc. U.S. Endocrine Soc., Vol. 71, p. 435, Abstract No. 1652.

Mordenti, J., and Green, J. D., 1991, The role of pharmacokinetics and pharmacodynamics in the development of therapeutic proteins, in: *New Trends in Pharmacokinetics* (A. Rescigno and A. Thakur, eds.), Plenum Press, New York, pp. 411–424.

Mordenti, J., Chen, S., Moore, J., and Ferraiolo, B., 1991, Interspecies scaling of clearance data for five recombinant proteins, *Pharm. Res.* **8**:1351–1359.

Mordenti, J., Shaieb, D., Chow, P., Cossum, P., Ferraiolo, B., Lewandowski, M., Moore, J., and Green, J. D., 1992, Preclinical safety evaluation strategy for biomacromolecules—A perspective, in: *Safety Assessment for Pharmaceuticals* (S. C. Gad, ed.), Van Nostrand–Reinhold, Princeton, N.J. (in press).

Nielsen, S., and Christensen, E. I., 1989, Insulin absorption in renal proximal tubules: A quantitative immunocytochemical study, *J. Ultrastruct. Mol. Struct.
Res.* **102:**205–220.

Nokta, M., Loh, J. P., Douidar, S. M., Snodgrass, W. R., Ahmed, E. A., and Pollard,
R. B., 1989, Molecular interaction of recombinant beta interferon and zidovu-
dine (AZT): Alteration of AZT pharmacokinetics in HIV infected patients, Fifth
International Conference on AIDS, Montreal, Abstract M.B.P.341, p. 278.

Offord, R. E., Philippe, J., Davis, J. G., Halban, P. A., and Berger, M., 1979, Inhibition of degradation of insulin by ophthalmic acid and a bovine pancreatic proteinase inhibitor, *Biochem. J.* **182:**249–251.

Okumura, K., Komada, F., and Hori, R., 1985, Fate of porcine and human insuin at
the subcutaneous injection site. I. Degradation and absorption of insulins in the
rat, *J. Pharmacobio-Dyn.* **8:**25–32.

Ooi, G. T., 1990, Insulin-like growth factor-binding proteins (IGFBPs): More than
just 1,2,3, *Mol. Cell. Endocrinol.* **71:**C39–C43.

Pan, Y.-C. E., Stern, A. S., Familletti, P. C., Khan, F. R., and Chizzonite, R., 1987,
Structural characterization of human interferon gamma, *Eur. J. Biochem.*
166:145–149.

Parker, C. W., 1976, *Radioimmunoassay of Biologically Active Compounds,* Pren-
tice–Hall, Englewood Cliffs, N.J.

Parker, M. L., Utiger, R. D., and Daughaday, W. H., 1962, Studies on human growth
hormone. II. The physiological disposition and metabolic fate of human growth
hormone in man, *J. Clin. Invest.* **41:**262–268.

Parkinson, A., Lasker, J., Kramer, M. J., Huang, M.-T., Thomas, P. E., Ryan, D. E.,
Reik, L. M., Norman, R. L., Levin, W., and Conney, A. H., 1982, Effects of
three recombinant human leukocyte interferons on drug metabolism in mice,
Drug Metabl. Dispos. **10:**579–585.

Parsons, J. A., Rafferty, B., Stevenson, R. W., and Zanelli, J. M., 1979, Evidence that
protease inhibitors reduce the degradation of parathyroid hormone and calcitonin injected subcutaneously, *Br. J. Pharmacol.* **66:**25–32.

Philips, T. M., 1989, Isolation and recovery of biologically active proteins by high
performance immunoaffinity chromatography, in: *The Use of HPLC in Receptor Biochemistry,* Liss, New York, pp. 129–154.

Podbesek, R., Edouard, C., Meunier, P. J., Parsons, J. A., Reeve, J., Stevenson,
R. W., and Zanelli, J. M., 1983, Effects of two treatment regimes with a synthetic
human parathyroid hormone fragment on bone formation and the tissue balance of trabecular bone in greyhounds, *Endocrinology* **112:**1000–1006.

Pringle, P. J., Hindmarsh, P. C., De Silvio, L., Teale, J. D., Kurtz, A. B., and Brook,
C. G. D., 1989, The measurement and effect of growth hormone in the presence
of growth hormone-binding antibodies, *J. Endocrinol.* **121:**193–199.

Quiron, R., 1982, Bioassays in modern peptide research, *Peptides* **3:**223–230.

Rabkin, R., Ryan, M. P., and Duckworth, W. C., 1984, The renal metabolism of
insulin, *Diabetologia* **27:**351–357.

Reed, B. R., Chen, A. B., Gibson, U. E. M., Chen, S., Baughman, R., Ferraiolo, B. L.,
and Mordenti, J., 1990, Parenteral administration of recombinant interferon-

gamma results in route dependent processing in normal human subjects, Annual Meeting of the International Society for Interferon Research, San Francisco.

Regoeczi, E., 1984, *Iodine-Labeled Plasma Proteins,* Vol. I, CRC Press, Boca Raton, pp. 4–5, 35–102.

Regoeczi, E., 1987a, *Iodine-Labeled Plasma Proteins,* Vol. II, Part A, CRC Press, Boca Raton, pp. 6, 112, 116–117.

Regoeczi, E., 1987b, *Iodine-Labeled Plasma Proteins,* Vol. II, Part B, CRC Press, Boca Raton, pp. 43–62.

Rinderknecht, E., and Burton, L. E., 1985, Biochemical characterization of natural and recombinant IFN-gamma, in: *The Biology of the Interferon System 1984* (H. Kirchner and H. Schellenkens, eds.), Elsevier, Amsterdam, pp. 397–402.

Rinderknecht, E., O'Connor, B. H., and Rodriguez, H., 1984, Natural human interferon-gamma: Complete amino acid sequence and determination of sites of glycosylation, *J. Biol. Chem.* **259**:6790–6797.

Rogers, S., Wells, R., and Rechsteiner, M., 1986, Amino acid sequences common to rapidly degraded proteins: The PEST hypothesis, *Science* **234**:364–368.

Rosenbloom, A. L., Knuth, C., and Shulman, D., 1990, Growth hormone by daily injection in patients previously treated for growth hormone deficiency, *South. Med. J.* **83**:653–655.

Rosenblum, M. G., Unger, B. W., Gutterman, J. U., Hersh, E. M., David, G. S., and Frincke, J. M., 1985, Modification of human leukocyte interferon pharmacology with a monoclonal antibody, *Cancer Res.* **45**:2421–2424.

Rosenblum, M. G., Lamki, L. M., Murray, J. L., Carlo, D. J., and Gutterman, J. U., 1988, Interferon-induced changes in pharmacokinetics and tumor uptake of ^{111}In-labeled antimelanoma antibody 96.5 in melanoma patients, *J. Natl. Cancer Inst.* **80**:160–165.

Ruotolo, G., Miscossi, P., Galimberti, G., Librenti, M. C., Petrella, G., Marcovina, S., Pozza, G., and Howard, B. V., 1990, Effects of intraperitoneal versus subcutaneous insulin administration on lipoprotein metabolism in type I diabetes, *Metabolism* **39**:598–604.

Rustgi, V. K., Suou, T., Jones, D. B., Lisker-Melman, M., Vergalla, J., Jones, E. A., and Hoofnagle, J. H., 1987, The effect of rat gamma interferon on antipyrine metabolism by the isolated perfused rat liver, *Clin. Res.* **35**:414A.

St. Hilaire, R. J., Hradek, G. T., and Jones, A. L., 1983, Hepatic sequestration and biliary secretion of epidermal growth factor: Evidence for a high-capacity uptake system, *Proc. Natl. Acad. Sci. USA* **80**:3797–3801.

Sato, H., Tsuji, A., Hirai, K.-I., and Kang, Y. S., 1990, Application of HPLC in disposition study of A-14-^{125}I-labeled insulin in mice, *Diabetes* **39**:563–569.

Schwenk, W. F., Tsalikain, E., Beaufrere, B., and Haymond, M. W., 1985, Recycling of an amino acid label with prolonged isotope infusion: Implications for kinetic studies, *Am. J. Physiol.* **248**:E482–E487.

Secor, J., and Schenker, S., 1984, Effect of recombinant alpha-interferon on *in vivo* and *in vitro* markers of drug metabolism in mice, *Hepatology* **4**:1081 (Abstract 298).

Selam, J. L., 1990, Insulin therapy of diabetes with implantable infusion pumps: Clinical aspects, *Int. J. Artif. Org.* **13**:261–266.

Shedlofsky, S. I., Swim, A. T., Robinson, J. M., Gallicchio, V. S., Cohen, D. A., and McClain, C. J., 1987, Interleukin-I (IL-I) depresses cytochrome P450 levels and activities in mice, *Life Sci.* **40**:2331–2336.

Sherman, B., Frane, J., and the Genentech Cooperative Group, 1988, Optimizing treatment of growth hormone deficiency (GHD): Influence of growth hormone (GH) schedule and dose, The Endocrine Society 70th Annual Meeting, Program and Abstracts, Abstract 406.

Singh, G., Renton, K. W., and Stebbing, N., 1982, Homogeneous interferon from *E. coli* depresses hepatic cytochrome P450 and drug biotransformation, *Biochem. Biophys. Res. Commun.* **106**:1256–1261.

Sorensen, J. T., Colton, C. K., Hillman, R. S., and Soeldner, J. S., 1982, Use of a physiologic pharmacokinetic model of glucose homeostasis for assessment of performance requirements for improved insulin therapies, *Diabetes Care* **5**:148–157.

Spivak, J. L., and Hogans, B. B., 1989, The *in vivo* metabolism of recombinant human erythropoietin in the rat, *Blood* **73**:90–99.

Sprengers, E. D., and Kluft, C., 1987, Plasminogen activator inhibitors, *Blood* **69**:381–387.

Steis, R. G., Smith, J. W., Urba, W. J., Clark, J. W., Itri, L. M., Evans, L. M., Schoenberger, C., and Longo, D. L., 1988, Resistance to recombinant interferon alpha-2a in hairy-cell leukemia associated with neutralizing anti-interferon antibodies, *N. Engl. J. Med.* **318**:1409–1413.

Strober, W., and Waldmann, T. A., 1974, The role of the kidney in the metabolism of plasma proteins, *Nephron* **13**:35–66.

Sugiyama, Y., and Hanano, M., 1989, Receptor-mediated transport of peptide hormones and its importance in the overall hormone disposition in the body, *Pharm. Res.* **6**:192–202.

Surmacz, C. A., Wert, J. J., Ward, W. F., and Mortimore, G. E., 1988, Uptake and intracellular fate of [^{14}C]sucrose-insulin in perfused rat livers, *Am. J. Physiol.* **255**:C70–C75.

Tam, C. S., Heersche, J. N. M., Murray, T. M., and Parsons, J. A., 1982, Parathyroid hormone stimulates the bone apposition rate independently of its resorptive action: Differential effects of intermittent and continuous administration, *Endocrinology* **110**:506–512.

Taylor, G., Marafino, B. J., Moore, J. A., Gurley, V., and Blaschke, T. F., 1985, Interferon reduces hepatic drug metabolism in vivo in mice, *Drug Metab. Dispos.* **13**:459–463.

Temple, R. C., Clark, P. M. S., Nagi, D. K., Schneider, A. E., Yudkin, J. S., and Hales, C. N., 1990, Radioimmunoassay may overestimate insulin in non-insulin dependent diabetes, *Clin. Endocrinol.* **32**:689–693.

Thomas, J. H., Jenkins, C. D. G., Davey, P. G., and Papachristodoulou, D. K., 1983, The binding and degradation of ^{125}I-labeled insulin by rat kidney brush-border membranes, *Int. J. Biochem.* **15**:329–336.

Thomas, J. H., Corbett, S. A., and Davey, P. G., 1987, Inhibition of insulin degradation in isolated rat kidney tubules, *Biochem. Soc. Trans.* **15**:433–434.

Timmons, T. M., and Dunbar, B. S., 1990, Protein blotting and immunodetection, *Methods Enzymol.* **182:**679–688.

Tuvemo, T., 1989, What is the best mode of growth hormone administration, *Acta Paediatr. Scand. Suppl.* **362:**44–49.

van Haeften, T. W., 1989, Clinical significance of insulin antibodies in insulin-treated diabetic patients, *Diabetes Care* **12:**641–648.

Venturini, P. L., Remorgida, V., Aguggia, V., and De Cecco, L., 1990, Luteinizing hormone determinations obtained with either a monoclonal and a polyclonal antibody radioimmunoassay and their correlations with clinical findings, *J. Endocrinol. Invest.* **13:**227–234.

Villafranca, J. J., (ed.), 1990, *Current Research in Protein Chemistry: Techniques, Structure and Function,* Academic Press, New York.

Walker, J. M., (ed.), 1984, *Proteins,* Humana Press, Clifton, N.J.

Williams, G., Pickup, J. C., Bowcock, S., Cooke, E., and Keen, H., 1983, Subcutaneous aprotinin causes local hyperaemia, *Diabetologia* **24:**91–94.

Williams, S. J., and Farrell, G. C., 1986, Inhibition of antipyrine metabolism by interferon, *Br. J. Clin. Pharmacol.* **22:**610–612.

Wills, R. J., 1990, Clinical pharmacokinetics of interferons, *Clin. Pharmacokinet.* **19:**390–399.

Wilson, J. T., 1970, Alteration of normal development of drug metabolism by injection of growth hormone, *Nature* **225:**861–863.

Wilson, J. T., and Frohman, L. A., 1974, Concomitant association between high plasma levels of growth hormone and low mixed-function oxidase activity in the young rat, *J. Pharmacol. Exp. Ther.* **189:**255–270.

Woltanski, K.-P., Besch, W., Keilacker, H., Ziegler, M., and Kohnert, K.-D., 1990, Radioiodination of peptide hormone and immunoglobulin preparations: Comparison of the chloramine T and Iodogen method, *Exp. Clin. Endocrinol.* **95:**39–46.

Yamazoe, Y., Shimada, M., Kamataki, T., and Kato, R., 1986, Effects of hypophysectomy and growth hormone treatment on sex-specific forms of cytochrome P-450 in relation to drug and steroid metabolism in rat liver microsomes, *Jpn. J. Pharmacol.* **42:**371–382.

Young, J. D., Bell, D. P., Luo, Z. P., Marian, M., and Bauer, R., 1990, Comparative pharmacokinetics of lymphokines, cytokines and antibodies from mouse to man, Drug Information Association Workshop, Nonclinical Development Issues for Biotechnology-Derived Products, San Diego.

Zapf, J., Kiefer, M., Merryweather, J., Masiarz, F., Bauer, D., Born, W., Fischer, J. A., and Froesch, E. R., 1990a, Isolation from adult human serum of four insulin-like growth factor (IGF) binding proteins and molecular cloning of one of them that is increased by IGF I administration and in extrapancreatic tumor hypoglycemia, *J. Biol. Chem.* **265:**14892–14898.

Zapf, J., Schmid, C., Binz, K., Guler, H. P., and Froesch, E. R., 1990b, Regulation and function of carrier proteins for insulin-like growth factors, in: *Growth Factors: From Genes to Clinical Application* (V. R. Sara, ed.), Raven Press, New York, pp. 227–240.

Zingg, W., Rappaport, A. M., and Leibel, B. S., 1987, Transhepatic absorption and biliary excretion of insulin, *Can. J. Physiol. Pharmacol.* **65:**1982–1987.

Chapter 2

Binding Proteins of Protein Therapeutics

Marjorie A. Mohler, Jennifer E. Cook, and Gerhard Baumann

Numerous examples of binding proteins for protein therapeutics have been reported. Binding proteins have been shown to exist for insulinlike growth factors I and II, tissue plasminogen activator, growth hormone, deoxyribonuclease I, tissue factor, nerve growth factor, transforming growth factor-β I and II, as well as others. Binding proteins may have either inhibitory or stimulatory effects, may modulate efficacy at the cellular level, and may also affect the pharmacokinetics and metabolism of protein therapeutics. Furthermore, the relative importance of binding proteins may be species or disease state specific. It is also important to realize that binding proteins may play a different role in the regulation of proteins when physiological concentrations of proteins are involved compared to pharmacological doses of protein therapeutics. The pharmacology, physiology, regulation, and interactions of binding proteins with their target proteins will be discussed. This review chapter will focus on insulinlike growth factor, tissue plasminogen activator, and growth hormone.

1. HUMAN INSULINLIKE GROWTH FACTOR-I

The insulinlike growth factors (IGFs) resemble insulin in their structure and in many of their actions. Originally termed somatomedins because they

Marjorie A. Mohler and Jennifer E. Cook • Department of Safety Evaluation, Genentech, Inc., South San Francisco, California 94080. *Gerhard Baumann* • Center for Endocrinology, Metabolism, and Nutrition, Department of Medicine, Northwestern University Medical School, Chicago, Illinois 60611.

Protein Pharmacokinetics and Metabolism, edited by Bobbe L. Ferraiolo *et al.* Plenum Press, New York, 1992.

were defined as mediators of the somatogenic actions of growth hormone (GH), these peptides are now termed IGF-I and IGF-II. In a classical view, the actions of the IGFs were presumed to be endocrine. However, the IGFs are synthesized and released by many tissues and cell types (Nissley and Rechler, 1984; Daughaday and Rotwein, 1989), and IGF concentrations of physiologically significant levels occur in the tissues, implying possible autocrine and paracrine functions as well (Ooi and Herington, 1990); IGF-I has been shown to mediate the effects of several hormones at the local level (Rutanen and Pekonen, 1990).

1.1. Structure

Human IGF-I is a basic plasma peptide (pI 8.4) with a molecular mass of 7649 Da (Rinderknecht and Humbel, 1978a). IGF-I contains 70 amino acid residues in a single chain with three disulfide bridges (Baxter, 1988). IGF-I is highly homologous to insulin with an amino-terminal region 29 amino acids in length which corresponds to the B chain of insulin. There is a sequence of 12 amino acids (corresponding to but not homologous to the proinsulin C-peptide) which links the amino-terminal region to a region 21 amino acids long which is homologous to the A chain of insulin. IGF-I differs from proinsulin in two ways: its C-peptide region is not cleaved, and it has an 8-amino-acid sequence, termed the D-peptide, at the carboxy terminus of the A-chain region (Rinderknecht and Humbel, 1978a,b).

1.2. Actions

IGF-I actions fall into three classes: metabolic activity, mitogenesis, and differentiation. The metabolic actions are principally anabolic and include insulinlike actions such as stimulation of glucose uptake, glycogen synthesis, amino acid transport, and protein synthesis. The injection of IGF-I into rats or humans elicits a hypoglycemic response similar to that caused by insulin, with about 7.5% of the potency of insulin on a molar basis (Guler *et al.,* 1987).

The mitogenic activity of IGF-I has been well documented in many *in vitro* cell culture systems (Leof *et al.,* 1982). Addition of IGF-I to growth media in these systems stimulates DNA synthesis and cellular proliferation (Riss *et al.,* 1988). *In vivo* studies comparing IGF-I and growth hormone (GH) responses in hypophysectomized rats (Guler *et al.,* 1988) have shown that IGF-I has effects on whole body growth, bone formation, and organ growth, specifically of the thymus, spleen, and kidney.

The role of IGF-I in stimulating cell differentiation and the expression of differentiated functions has only recently been recognized. In granulosa cells, IGF-I acts synergistically with follicle-stimulating hormone to induce receptors for luteinizing hormone, increase cyclic AMP production, and stimulate steroidogenesis (Adashi *et al.*, 1985); comparable effects have also been described in the testis (Lin *et al.*, 1986) and adrenal cortex (Morera *et al.*, 1986). In the thyroid gland, IGF-I synergizes with thyroid-stimulating hormone in stimulating cell proliferation (Tramontano *et al.*, 1986). It may be concluded that full expression of the activity of some pituitary hormones on their target endocrine tissues (gonads, adrenal, thyroid) may require the concomitant action of IGF-I.

1.3. Regulation

One of the definitive characteristics of IGF-I is its regulation by GH. IGF-I transcription is enhanced by the administration of GH, and a concomitant increase in mRNA in most rodent tissues is also observed (Daughaday and Rotwein, 1989). In fact, the close relationship between the secretion of GH and circulating IGF-I levels forms the basis for the use of IGF-I measurement as an indicator of GH secretory status in both GH deficiency and acromegaly (Copeland *et al.*, 1980).

There are a number of other factors which modify IGF-I gene expression in addition to GH. Other hormones, tissue-specific signals, and factors of age and development all influence the synthesis of IGF-I (Daughaday and Rotwein, 1989). When treated with GH and estrogen, ovariectomized hypophysectomized rats showed an increase in mRNA for IGF-I in the uterus, while showing inhibition of GH-stimulated hepatic IGF-I expression (Murphy and Friessen, 1988). IGF-I mRNA levels also vary greatly with tissue-specific factors as well as cell type and regional differences within an organ (Daughaday and Rotwein, 1989).

Age and nutritional status have also been shown to significantly affect IGF-I expression. IGF-I is low in neonates, rises through the prepubertal years to peak during puberty, then slowly declines with increasing age (Copeland *et al.*, 1980). In elderly people, IGF-I levels have been observed to be consistently depressed (Donahue *et al.*, 1990). Evidence is strong that nutritional status is at least as important as hormonal signals in the regulation of IGF-I concentrations. A rapid fall in IGF-I is induced by fasting: in rats fasted for 30 hr, IGF-I mRNA levels in the liver fell to <40% of control values and returned to normal within 6 hr of feeding (Daughaday and Rotwein, 1989). In man and experimental animals, deprivation of dietary protein results in a decrease in serum IGF-I (Phillips and Vassilopoulou-Selin,

1979; Prewitt *et al.*, 1982; Underwood *et al.*, 1986). After 12 and 24 hr of protein restriction in rats, serum IGF-I concentrations were reduced by 58 and 66%, respectively (Maiter *et al.*, 1988).

Three classes of receptors have been shown to mediate IGF action: type I and type II IGF receptors and the insulin receptor. The IGF-I receptor, type I, is similar to the insulin receptor and shows ligand-dependent tyrosine kinase activity and autophosphorylation. The type I IGF receptor binds IGF-I, IGF-II, and insulin, but it has the highest affinity for IGF-I. The second type of receptor binds IGF-II with a higher affinity than IGF-I, does not cross-react with insulin, and therefore has a lesser influence on IGF-I regulation (Ooi, 1990). Both IGF-I and IGF-II can cross-react with the insulin receptor but neither binds as well as insulin (Zapf *et al.*, 1978; Rechler *et al.*, 1980). IGF-I is found in the body fluids associated with binding proteins (Rechler and Nissley, 1985). IGF-I does not appear to be stored within secretory granules in any tissue, so the pool of IGF-I in the blood seems to be a storage form of the peptide and accounts for the high plasma concentration seen in normal individuals. These binding proteins may play an important role in ligand–receptor interactions on target cells (Ooi, 1990).

1.4. Insulinlike Growth Factor Binding Proteins

Numerous proteins with IGF binding characteristics have been studied in various species, body fluids, and cell lines. To date, three distinct classes of IGF binding proteins (IGFBPs) have been well characterized and their amino acid sequences determined. A recent report has attempted to provide terminology based on the source, species, and molecular size of these IGFBPs (Ballard *et al.*, 1990). The proposed designations for the three well-characterized classes of IGFBPs are IGFBP-1, IGFBP-2, and IGFBP-3. The complete amino acid sequence has recently been determined for a fourth binding protein, IGFBP-4 (Mohan *et al.*, 1991). For two other binding proteins, tentatively named IGFBP-5 and IGFBP-6, only N-terminal sequence data are available; these proteins have been purified from porcine ovarian follicular fluid and rat serum (Shimasaki *et al.*, 1991).

1.4.1. Methods for Identification of Binding Protein Complexes

1.4.1a. Size Exclusion Chromatography.   Early and present work on the characterization of IGFBPs employed size exclusion chromatography. The most commonly used approach involves the addition of radioiodinated IGF to serum followed by size fractionation at neutral pH (Zapf *et al.*, 1975;

Hintz and Liu, 1977). One limitation of this technique is that the radioiodinated IGF predominantly binds unsaturated binding sites. This can lead to misleading results because in serum, the most abundant binding protein is also the most fully saturated. For this reason, relative abundance of the IGFBPs cannot be determined by this method.

Another chromatographic method involves fractionation of the IGFBPs and subsequent addition of IGF to the fractions of interest. This method can also be limited by binding site saturation; however, dissociation of the endogenous IGF can be accomplished at low pH (e.g., 1 M acetic acid), then each fraction can be neutralized before assay and a measure of total binding sites obtained (Zapf et al., 1980).

1.4.1b. Ligand Blotting.   Another method for the analysis and characterization of IGFBPs from serum or cell supernatants involves separation of the sample by SDS–PAGE followed by transfer of the proteins onto nitrocellulose. The nitrocellulose is then incubated with radioiodinated IGF and autoradiography is used to detect IGFBPs (Hossenlopp et al., 1986). This procedure, termed ligand blotting or Western ligand blotting, allows for determination of the molecular weights of the IGFBPs and for the identification of glycosylation variants. The issue of unsaturated versus saturated binding sites is not a complicating factor with this method due to the separation of endogenous IGFs from the binding proteins under denaturing conditions. For this reason, semiquantitation of the IGFBPs can be accomplished with ligand blotting.

Covalent cross-linking of the IGFBP–IGF complex with disuccinimidyl suberate followed by separation of the products by SDS–PAGE or size exclusion chromatography provides another sensitive technique which allows further characterization of IGFBP–IGF complexes (Wilkins and D'Ercole, 1985; Ooi and Herington, 1986). One limitation of this method can be the formation of nonspecific bands resulting from random protein–protein interactions during the cross-linking process.

1.4.1c. Correlation of Chromatography Peak Components and Ligand Blot Bands.   When radiolabeled IGF is added to human serum and analyzed by neutral size exclusion chromatography, four peaks are observed which correspond to radiolabeled IGF bound in 150-kDa and 50-kDa IGFBP complexes, unbound radiolabeled IGF (7.5 kDa), and free isotope. In a "normal" well-fed mammal, the majority of IGFs have been shown to circulate bound in the 150-kDa complex (Hintz et al., 1981). This complex has been shown to be GH dependent (White et al., 1981; Hintz et al., 1981) with relatively few unsaturated binding sites and restricted capillary permeability. The 150-

kDa complex consists of IGF, an acid-labile subunit (ALS) of 85 kDa, and an acid-stable 53-kDa binding protein (Baxter and Martin, 1989). Using the ligand blot method, radiolabeled IGF binds to five binding proteins in human serum (Hardouin *et al.,* 1987). The proteins comprising the major bands at 41.5 and 38.5 kDa (IGFBP-3) occur predominantly in the 150-kDa region of the size exclusion column. These proteins, the largest of the group, are thought to represent glycosylation variants of the same protein. The 50-kDa peak seen by neutral size exclusion chromatography has been shown to be inversely related to GH concentrations (Drop *et al.,* 1984) and has a high turnover of its IGF pool. The protein(s) in the 50-kDa region, unlike the GH-dependent complex, are predominantly unsaturated with IGF-I or II; size exclusion chromatography of plasma which has been incubated with radioiodinated IGF-I shows that the majority of binding occurs in this 50-kDa region (White *et al.,* 1981). The binding proteins in the 50-kDa peak are thought to have greater capillary permeability than IGFBP-3 because of their smaller size. Minor ligand blot bands representing proteins of 34 kDa (IGFBP-2), 30 kDa (IGFBP-1), and 24 kDa (IGFBP-4) are present in the 50-kDa region of the column. From competitive binding studies, the 41.5- and 24-kDa proteins were found to preferentially bind to IGF-I while the 38.5-, 34-, and 30-kDa proteins bound preferentially to IGF-II (Hardouin *et al.,* 1987).

1.4.2. Characterization of the IGFBPs

1.4.2a. IGFBP-3.   The 53-kDa binding protein, referred to as IGFBP-3, has been cloned; it is a glycoprotein 291 amino acids long (Wood *et al.,* 1988) with three potential N-linked glycosylation sites. A number of size estimates have been reported for IGFBP-3, and these discrepancies may be accounted for by differences in the extent of glycosylation (Martin and Baxter, 1986; Wilkins and D'Ercole, 1985). IGFBP-3 has 18 cysteines and is thought to lack free sulfhydryl groups (Wilkins and D'Ercole, 1985); this results in complicated and extensive disulfide bonding which would make the protein very compact and could lead to the anomalous migration patterns observed on SDS gels under reducing conditions (Baxter *et al.,* 1987). It appears that there is only one species of mRNA of approximately 2.51 kb for IGFBP-3 in human liver, and there appears to be only one gene as well.

1.4.2b. IGFBP-2.   IGFBP-2, a 34-kDa protein, is present in high concentrations in cerebrospinal fluid (Drop *et al.,* 1984) and has also been isolated from fetal liver and rat and bovine cell lines (Binkert *et al.,* 1989). There is an abundance of IGFBP-2 in fetal tissues which suggests a role in developmental regulation. The mature human protein is 31,325 Da and 289 amino acids in length with a 39-amino-acid signal peptide (Binkert *et al.,* 1989). As reported for IGFBP-3 and IGFBP-1, IGFBP-2 has 18 cysteine residues whose positions are conserved. Also like IGFBP-1, IGFBP-2 has an Arg–Gly–Asp (RGD) sequence near the C-terminus (Binkert *et al.,* 1989; Rutanen and Pekonen, 1990) and is unglycosylated. IGFBP-2 is encoded by a single-copy gene, and a 1.5-kb mRNA for IGFBP-2 has been found in adult liver, brain, Jurkat, and kidney 293 cells, tissue types distinct from those where IGFBP-1 message is found (Binkert *et al.,* 1989).

1.4.2c. IGFBP-1. IGFBP-1 was first isolated from amniotic fluid and placenta tissue extract (Chochinov *et al.,* 1977). It has been purified to homogeneity, and the N-terminal amino acid sequence has been determined (Povoa *et al.,* 1984). It is a 25,274-Da protein containing 259 amino acids and there are a number of studies which show that IGFBP-1 contains no carbohydrate residues, and there are no N-linked glycosylation sites. As in IGFBP-3, 18 cysteine residues are present, and the positions they occupy are the same as in IGFBP-3. This would probably lead to the same compact structure and anomalous migration on SDS–PAGE mentioned above. In the human genome, there is a single gene encoding IGFBP-1 and a single species of mRNA which has been detected in fetal liver explants (Lewitt and Baxter, 1989), endometrium (Rutanen *et al.,* 1986), or Hep G2 cells (Lee *et al.,* 1988). Near the C-terminus of IGFBP-1, there is an Arg–Gly–Asp (RGD) sequence (Brewer *et al.,* 1988) which is a component of extracellular matrix proteins that can provide cell surface attachment sites (Ruoslahti and Pierschbacher, 1988).

1.4.2d. IGFBP-4. Rat IGFBP-4 consists of 233 amino acids while the human protein contains an additional 4-amino-acid fragment in the midregion of the molecule (Shimasaki *et al.,* 1991). Similar to IGFBP-1, -2, and -3, all of the cysteines are conserved except for two additional cysteines in IGFBP-4. Northern analysis of IGFBP-4 mRNAs in rat tissue demonstrated that transcription of the gene is highly active in the liver, although mRNA was also detected in ovary, testis, adrenal, spleen, heart, lung, stomach, and brain.

1.4.3. Regulation of IGFBPs

IGFBP-3 has been shown to be GH-dependent (Moses *et al.,* 1976). *In vitro* studies performed with serum from GH-deficient human subjects showed an increase in binding in the 150-kDa peak after GH treatment (White *et al.*, 1981). In addition, GH treatment in GH-deficient hypophysec-tomized rats resulted in a significant increase in the half-life of IGF-I activity in the circulation; this prolonged half-life corresponded to an increase in binding in the 150-kDa peak as well (Moses *et al.*, 1976). By ligand blotting, it has been shown that IGFBP-3 is inversely related to advancing age, and that the decrease in IGFBP-3 observed in elderly subjects can be accounted for by GH deficiency (Donahue *et al.*, 1990). The acid-labile subunit (ALS) of the 150-kDa complex also shows evidence of GH dependence. In acrome-galic patients, there is a 2.2-fold increase of the ALS over normal values, and there is a 70% decrease in GH-deficient subjects (Baxter, 1990). Baxter has also shown with serum from 170 children that there is a strong correlation between IGFBP-3 and ALS levels, despite the fact that most of the IGFBP-3 is in the 150-kDa complex and the ALS is in a free form.

IGFBP-2 is regulated by GH and insulin levels: GH and insulin sup-press IGFBP-2, probably by inhibiting its expression; for example, the ex-pression of mRNA encoding IGFBP-2 is greatly increased in diabetic rat liver (Zapf *et al.*, 1990). Although infusion of IGF-I appears to induce IGFBP-2, it is likely that this is not directly the effect of the IGF-I; instead, IGF-I infusion suppresses insulin and GH secretion, allowing an increase in IGFBP-2 expression. It may be that insulin deficiency and not hyperglyce-mia results in elevated IGFBP-2: in a study with insulin-dependent diabetics, it was found that during insulin infusion, IGFBP-2 levels decreased continu-ously, even after the establishment of steady-state blood glucose (Brismar *et al.*, 1988). In addition, IGFBP-2 is overexpressed in subjects with low serum GH levels and underexpressed in subjects with acromegaly (Zapf *et al.*, 1990). Ligand blots show a 2.5-fold increase in IGFBP-2 in elderly women, further supporting the inverse relationship of IGFBP-2 and GH levels (Don-ahue *et al.*, 1990).

IGFBP-1 is regulated by both insulin and glucose *in vivo* (Snyder and Clemmons, 1990). The concentration of this binding protein increases 3.5-to 12-fold after an overnight fast (Baxter and Cowell, 1987; Busby *et al.*, 1988) and falls to normal levels after a meal. It has also been shown that the infusion of glucose and/or insulin results in significant suppression of IGFBP-1 (Suikkari *et al.*, 1980). Since insulin increases the tissue uptake of circulating IGFBP-1 but not IGFBP-2, it has been suggested that insulin-facilitated changes could selectively mediate nutrient-dependent transport of IGF-I to peripheral tissues (Bar *et al.*, 1990).

Recent studies on the regulation of IGFBP-4 indicate that agents that increase bone cell proliferation (IGF-I, IGF-II, GH) decrease production of IGFBP-4 (Mohan *et al.*, 1991).

1.4.4. Effect of Binding Proteins on the Pharmacokinetics of IGF-I

Many studies have addressed the pharmacokinetics of IGF-I and have demonstrated altered clearance of IGF-I when it is bound to its specific binding proteins. When iodinated IGF-I was injected into rats, the majority of the protein-associated radioactivity in serum appeared initially (i.e., after 1–5 min) in the 50-kDa complex, but after 20 min to 4 hr it was found mainly in fractions corresponding to the 150-kDa complex (Kaufman *et al.,* 1977). Furthermore, it was found that the half-life of injected IGF-I in the 150-kDa complex was 3–4 hr, compared to only 10 min for free IGF (Cohen and Nissley, 1976). In a separate study, iodinated recombinant human IGF-I ($[^{125}I]$-rhuIGF-I) was injected into rats and four peaks of radioactivity were detected which corresponded to $[^{125}I]$-rhuIGF-I bound in 150- and 50-kDa binding protein complexes, unbound $[^{125}I]$-rhuIGF-I, and free ^{125}I (Mohler *et al.,* 1989). Clearance of unbound $[^{125}I]$-rhuIGF-I was very rapid (168–204 ml/min/kg); clearance of radioactivity associated with the low-molecular-weight complex was intermediate (15.5–48 ml/min/kg); and clearance of radioactivity associated with the high-molecular-weight complex was slowest (0.5–10 ml/min/kg). In hypophysectomized rats, which are deficient in the 150-kDa binding protein complex, IGF-I has a half-life of 20–30 min compared to 4 hr in normal rats (Zapf *et al.,* 1986).

1.4.5. Role of Binding Proteins in Regulation of Endogenous IGF-I

In an endocrine view of IGF-I actions, the IGFBPs and the IGFs were thought to be synthesized in the liver and then transported to target tissues. Implicit in this endocrine view is the assumption that the free form of the peptide is the form with biological activity, and that the binding proteins passively transport and inactivate the circulating pool of IGF-I (Ooi and Herington, 1990). The supporting evidence for this assumption is that although IGF-I serum concentrations are 1000-fold higher than those of insulin, and IGF-I has 7.5% of the potency of insulin, hypoglycemia does not result (Rutanen and Pekonen, 1990).

However, recent discoveries indicate that multiple tissues are involved in the synthesis of both IGFBPs and IGFs. IGFBPs may play an active role, including direct interaction with target cells and facilitation of delivery of the IGF-I to its receptor (Ooi and Herington, 1990). The role of the IGFBPs now appears to be more one of local regulation of IGF actions, defining a more paracrine or autocrine mechanism of action for IGF-I (Rutanen and Pekonen, 1990).

There are a number of studies which support an inhibitory role for IGFBP-1 and IGFBP-2; however, *in vitro* studies with IGFBP-3 suggest that this molecule might enhance the activity of IGF-I due not to direct interaction of the binding protein at the cell surface, but to a slower release of IGF to the cells in question (Rutanen and Pekonen, 1990). In general, many factors influence whether the IGFBPs function as promoters or inhibitors of IGF-I's actions, including cell type, IGFBP concentration at the site, and local physiological conditions (Donahue *et al.*, 1990).

The diversity observed among the IGFBPs may indicate that each serves a distinct function in regulating IGF-I activity, and a clear understanding of the structure and function of each binding protein will be necessary in defining their interaction with each other and with IGF-I (Zapf *et al.*, 1990). While the physiological role of the IGFBPs is not fully understood, it has been shown that the IGFBPs alter the clearance of IGFs from the circulation (Zapf *et al.*, 1986, 1989). From a metabolic and pharmacokinetic perspective, the IGFBPs may play a crucial role in protecting the organism against the acute insulinlike effects of the IGFs. In the interstitial fluid they may modulate the bioavailability of locally secreted IGFs to IGF target cells (Schmid *et al.*, 1989a,b; Elgin *et al.*, 1987; De Mellow and Baxter, 1988; Knauer and Smith, 1980), and they have also been shown to play a role in specific IGF targeting (Zapf *et al.*, 1989; Guler *et al.*, 1988). Elucidation of the mechanisms by which these proteins effect changes will be important for our further understanding of how IGFs potentiate cellular growth.

2. HUMAN GROWTH HORMONE

GH is a pituitary hormone responsible for postnatal growth. It is a 20- to 22-kDa polypeptide that exists in several molecular forms (isohormones). GH acts by binding to GH receptors in multiple tissues; it initiates a cascade of biochemical events, among which the generation of IGF-I is important for the growth process at the local tissue level. Other important actions of GH are those involved in anabolism (e.g., intracellular transport of amino acids and other cellular constituents, nitrogen retention, protein synthesis). The biochemical events immediately following GH binding to receptors (intracellular signaling) are only incompletely understood. The hepatic GH receptor has recently been cloned and structurally characterized; it is a single-chain glycoprotein with a 620-amino-acid backbone, of which 246 are extracellular, 24 form a hydrophobic transmembrane domain, and 350 are intracellular (Leung *et al.*, 1987).

Like all polypeptide hormones, GH was long thought to be circulating exclusively in the free form rather than being bound to plasma proteins (Berson and Yalow, 1966, 1968). This was in contrast to some other hormones of hydrophobic nature (e.g., steroids), which required protein binding for solubilization in the aqueous plasma environment. The first evidence that the free polypeptide hormone hypothesis was not universal was provided by the discovery of IGF-binding proteins (Zapf *et al.*, 1975; Megyesi *et al.*, 1975; Hintz and Liu, 1977) as well as binding proteins for other peptide growth factors. More recently, it was found that GH also circulates bound to specific GH-binding proteins in plasma (Baumann *et al.*, 1986; Herington *et al.*, 1986a; Baumann and Shaw, 1990a). At least two GH-binding proteins (GH-BP) have been identified in human plasma; one binds GH with high affinity, the other with lower affinity. This review will principally focus on the human GH-BPs.

2.1. High-Affinity Growth Hormone-Binding Protein

2.1.1. Discovery and Identification

The high-affinity GH-BP was independently described in two laboratories (Baumann *et al.*, 1986; Herington *et al.*, 1986a). One laboratory found it in the course of studies on the molecular nature of GH forms in plasma, the other as a result of investigations concerning the GH receptor. It should be noted that circumstantial evidence for binding of GH to plasma proteins existed previously, but was generally discounted as an artifact (see Baumann, 1989, for review). Two reasons account for this delayed recognition: (1) the above-mentioned "free hormone dogma" and (2) the fact that dissociation of the GH-BP complex during standard analytical procedures (such as gel filtration or PAGE) minimizes the residual bound fraction recognizable after analysis.

2.1.2. Chemical and Physical Properties

A relationship between the GH-BP and the GH receptor was suspected and speculated upon from the time of discovery. Such a relationship was suggested by the similar hormone specificities and binding kinetics of the BP and the receptor, by the previous demonstration of soluble forms of GH receptors in certain cell or tissue preparations (McGuffin *et al.*, 1976; Ymer and Herington, 1984), and by the relative solubility of both the BP and the receptor in polyethylene glycol (Ymer and Herington, 1984; Baumann *et al.*, unpublished). Further evidence showed that the BP was a glycoprotein that bound to wheat germ lectin, but much less well to concanavalin A—properties also observed with the GH receptor (Baumann and Shaw, unpublished). Strong circumstantial evidence of a connection between BP and receptor was provided by the observation that Laron dwarfs, which were known to be deficient in GH receptors (Eshet *et al.*, 1984), are also deficient in functional GH-BP (Baumann *et al.*, 1987b; Daughaday and Trivedi, 1987). Molecular cloning of the GH receptor and partial sequencing of the GH-BP in the rabbit showed that the BP corresponds to the extracellular portion of the GH receptor (Leung *et al.*, 1987). Although the structure of the human GH-BP has not been reported, immunochemical evidence indicates that the situation in man is analogous to that in the rabbit (Baumann and Shaw, 1988; Barnard *et al.*, 1989). Thus, the primary structure of the GH-BP is a polypeptide of $\sim$246 amino acids, encompassing the extracellular portion of the GH receptor (it should be noted that the BP has not been fully sequenced, and that the precise location of the carboxy terminus has not been reported). Little is presently known about the carbohydrate moieties of the receptor or BP, and how they compare. These moieties are responsible for a substantial part of the molecular mass. The GH-BP has a size of $\sim$ 61 kDa, as assessed by Western blots and gel filtration (Baumann, 1989), yet its amino acid sequence predicts a molecular mass of 28.5 kDa. Thus, at least half of the molecular mass is due to carbohydrate (and possibly other unrecognized moieties). The BP has an approximate isoelectric point of 5.0 (Baumann and Shaw, 1990a), similar to that predicted from its amino acid composition. This suggests that the carbohydrate portion is of a relatively neutral nature.

2.1.3. Biological Properties

The GH-BP binds human GH (22-kDa form) with high affinity ($K_a \sim 3 \times 10^8$ M^{-1}), but does not bind animal GHs or close relatives of human GH, such as human placental lactogen or prolactin (Baumann *et al.*, 1986; Herington *et al.*, 1986a). This is characteristic for the human GH receptor and consistent with the fact that animal GHs are biologically inactive in man. In contrast to 22-kDa GH, the 20-kDa variant of human GH binds very poorly to the GH-BP (Baumann *et al.*, 1986; Baumann and Shaw, 1990b), which is again consistent with the behavior of that variant in human liver GH receptor systems (McCarter *et al.*, 1990). Natural GH-BP binds GH with an affinity somewhat lower than the receptor (Baumann *et al.*, 1986; Herington *et al.*, 1986a,b; Barnard *et al.*, 1989), although recombinant BP preparations (see below) have an affinity closer to the receptor (Leung *et al.*, 1987; Fuh *et al.*, 1990). These discrepancies are minor and have been attributed to technical problems, such as analysis under nonequilibrium conditions or underestimation due to the presence of endogenous GH (Barnard *et al.*, 1989; Herington *et al.*, 1986b). Nevertheless, the affinity difference between BP and receptor is probably real (Spencer *et al.*, 1988; Leung *et al.*, 1987) and warrants further investigation. The BP binds 1 mole of GH per mole, forming a complex of 80–85 kDa. Its binding capacity in human serum averages 20 ng GH/ml serum, which corresponds to a calculated concentration of about 1 nM (Baumann *et al.*, 1986; Herington *et al.*, 1986a; Baumann and Shaw, 1990a).

2.1.4. Purification and Artificial Expression

The GH-BP has been highly purified from rabbit serum (Leung *et al.*, 1987; Spencer *et al.*, 1988). Rabbit serum is particularly rich in BP and thus a good source. Similar degrees of purification have not been reported in detail for the human BP, one reason being the lower concentration of BP in human serum. Partially purified preparations of human BP have been prepared (Baumann *et al.*, 1986, 1987a; Herington *et al.*, 1987) and used for a variety of biological studies (Baumann *et al.*, 1987a, 1988a,c; Mannor *et al.*, 1988; Jan *et al.*, 1990). The human BP has been expressed in cell systems bearing human GH receptor gene inserts. When expressed in mammalian cells, the resulting protein resembled the natural BP in size and function (Leung *et al.*, 1987). When a bacterial expression system was used, a much smaller ($\sim$31 kDa), nonglycosylated product was obtained (Fuh *et al.*, 1990). Interestingly, this product also resembled the natural BP in its function. Thus, it seems that the carbohydrate moiety is not essential for GH binding. The carbohydrate portion may have other important functions *in vivo*, such as affecting clearance.

2.1.5. Physiological Effects

The GH-BP binds about 40–50% of circulating 22-kDa GH under nonstimulated conditions (i.e., at serum GH concentrations below about 5 ng/ml) (Baumann *et al.*, 1988b). At higher GH concentrations, the fraction of bound GH gradually declines because of saturation of the BP. These relationships have been demonstrated both *in vitro* (Baumann *et al.*, 1988b) and *in vivo* (Baumann *et al.*, 1990). [A small proportion of the total GH bound is complexed with the low-affinity BP (see below), but the high-affinity BP is responsible for 85–90% of the complexed fraction (Baumann *et al.,* 1989a).] As indicated above, the complexes dissociate during analysis, and special techniques (such as frontal analysis) are required to measure or derive the true bound fraction.

The complexed fraction of GH exhibits different pharmacokinetics than the free hormone. In a rat model used to study *in vivo* kinetics, the distribution volume of bound GH is about twice the plasma volume, whereas that of free GH corresponds to the extracellular space (Baumann *et al.,* 1987a, 1988a). The metabolic clearance of bound GH is ten-fold slower than that of free GH (Baumann *et al.,* 1988a), and the chemical degradation of the bound fraction is also substantially lower than that of free GH (Baumann *et al.,* 1987a, 1988a; Moore *et al.,* 1988). This may be in part attributed to restricted access of GH to its main degradation site, the proximal nephron, because the GH-BP complex is too large to be filtered at the glomerulus (Baumann *et al.,* 1987a; Moore *et al.,* 1988). Another reason is the interference of the BP with binding of GH to receptors (see below) and hence receptor-mediated degradation.

As may be predicted, the BP competes with receptors for GH binding. *In vitro,* the BP inhibits GH binding to a variety of receptor preparations in a dose-dependent manner (Mannor *et al.,* 1988; Lim *et al.,* 1990). Human GH receptors are particularly sensitive to this effect. Substantial inhibition is seen at physiological BP concentrations. This phenomenon is responsible for the long-recognized but poorly understood "serum effect" in radioreceptor assays for GH. It is not entirely clear to what degree similar inhibition occurs *in vivo,* where the situation is considerably more complex, and where variables such as local receptor concentration, rate of diffusion of free GH out of the vascular space, and plasma transit times likely influence the interaction between BP, GH, and receptors. Nevertheless, it is probably safe to assume that the BP has a modulatory effect on receptor binding of GH even *in vivo.*

In view of the marked effect of the BP in radioreceptor assays, the question must also be asked whether the BP affects GH measurements in immunoassays. The affinity of antibodies used in immunoassays for GH is typically two to three orders of magnitude higher than that of the BP. Thus, the effect may be expected to be less than in receptor assays. This is indeed the case, as recently demonstrated for a variety of immunoassays (Jan *et al.,* 1990). In radioimmunoassays, the BP usually (but not always) inhibits GH binding to the antibodies, whereas in two-site immunoradiometric assays, it nonspecifically promotes association of the two antibodies. Thus, in both types of assays, it mimics the effect of GH and tends to be read as such. However, at physiological BP concentrations, these effects are very minor. In addition, they are usually corrected for by assay designs that subtract "nonspecific" or "zero-dose" binding (Jan *et al.,* 1990). For most immunoassays, therefore, the presence of the BP does not substantially alter GH measurements.

Little information is presently available about possible effects of the BP on growth. Its net effect on GH action *in vivo* is difficult to predict on the basis of available *in vitro* data. On the one hand, the BP prolongs the half-life of GH, thereby perhaps enhancing its effectiveness. On the other, it interferes with GH binding to receptors, which would decrease its effectiveness. One preliminary study in a hypophysectomized rat model showed that BP—in the limited doses given—had no effect on body weight gain induced by GH treatment (Mannor *et al.,* 1988). This issue requires further study.

2.1.6. Regulation and Tissues of Origin

The regulation of GH-BP levels in serum has been studied both in normal man and in various disease states. In all such studies, BP levels/activity show wide individual variation among subjects. There appears to be no sex difference in humans (Daughaday and Trivedi, 1987; Daughaday *et al.,* 1987; Baumann *et al.,* 1988c, 1989a; Amit *et al.,* 1990). BP levels are low in fetal and neonatal blood, rise rapidly during early childhood and more gradually in late childhood and adolescence to reach adult levels during the latter part of the second decade of life (Daughaday *et al.,* 1987; Baumann *et al.,* 1989a; Silbergeld *et al.,* 1989; Amit *et al.,* 1990; Merimee *et al.,* 1990). One study showed a positive correlation between age-adjusted height and BP activity in serum (Silbergeld *et al.,* 1989). The rise of BP after birth may be linked to a similar rise in GH receptors in tissues, consistent with the postnatal inception of GH responsiveness. BP levels/activity remain stable throughout adult life and may slightly decline in old age (Daughaday *et al.,* 1987; Baumann *et al.,* 1989a). There appears to be no diurnal variation of BP activity in serum (Snow *et al.,* 1990), excluding both fluctuations in BP levels and the presence in plasma of binding inhibitors or enhancers in healthy subjects.

Several physiological and pathological conditions have been examined with regard to serum BP levels. The results must be viewed with some caution because the wide interindividual variability makes it difficult to discern true differences. Pregnancy does not seem to affect BP activity in serum (Baumann *et al.,* 1988c, 1989a; Amit *et al.,* 1990). Chronic GH deficiency (hypopituitarism) or excess (acromegaly) also does not appear to change serum BP (Baumann *et al.,* 1986, 1988c, 1989a), suggesting that in the long term, BP levels are not regulated by GH. In contrast, acute GH exposure has been reported to increase BP levels in GH-deficient children in one preliminary report (Postel-Vinay *et al.,* 1990). Liver cirrhosis can result in either abnormally low or abnormally high BP levels, implicating the liver as a likely tissue source of the BP (Baumann *et al.,* 1989a; Amit *et al.,* 1990). Uremia also results in abnormally low serum binding activity (Baumann *et al.,* 1989a), possibly because of the presence of circulating binding inhibitors.

The tissue(s) of origin and mechanism of generation of the BP are not known with certainty. Liver is an obvious candidate because the BP is structurally related to the hepatic GH receptor. However, other tissues bearing GH receptors must also be considered. It is also not clear whether the BP arises by proteolytic cleavage of the receptor or by a separate pathway. There is evidence for both mechanisms. Cultured human lymphocytes bearing GH receptors shed BP into the medium (McGuffin *et al.,* 1976), and this process can be accelerated by sulfhydryl-inactivating agents (Trivedi and Daughaday, 1988). Because GH receptors on the cell surface are depleted in this process, proteolysis of the receptor seems to be involved. On the other hand, in the pregnant rodent (a model analogous to man) GH-BP appears to arise from a truncated GH receptor mRNA lacking an appropriately hydrophobic transmembrane domain as well as the intracellular domain (Smith *et al.,* 1989; Baumbach *et al.,* 1989). It is not known which mechanism predominates in man *in vivo.*

In a recent preliminary report, GH-BP was shown to be associated with chromatin in rabbit liver (Lobie *et al.,* 1990). The physiological significance of this intriguing finding remains to be elucidated.

2.1.7. Deficiency of GH-BP and Dwarfism

To date, two types of short stature have been shown to be associated with BP abnormalities. Laron dwarfism, a severely GH-resistant condition caused by an absent or abnormal GH receptor (Eshet *et al.,* 1984; Amselem *et al.,* 1989; Godowski *et al.,* 1989), is also associated with absent or nonfunctional BP (Baumann *et al.,* 1987b; Daughaday and Trivedi, 1987; Amselem *et al.,* 1989; Laron *et al.,* 1989). Whether part of the GH resistance is due to deficiency of the BP itself is unknown. African pygmies, who show a milder form of GH resistance, have decreased but detectable BP levels (Baumann *et al.,* 1989b), perhaps implying partial GH receptor deficiency or dysfunction. Also relevant in this context is the observation that even normal children show a correlation between height and BP level (Silbergeld *et al.,* 1989). Thus, the BP appears to be linked—directly or indirectly as an index of GH receptor complement—to the growth process.

2.1.8. Binding Proteins in Animals

Animals vary in their similarity to humans with respect to GH-BP (Shaw and Baumann, 1988). Rabbits, pigs, and pregnant murine species have high serum levels of a BP that corresponds in molecular size and binding affinity to the human BP (Peeters and Friesen, 1977; Ymer and Herington, 1985; Barnard and Waters, 1986; Shaw and Baumann, 1988; Lauterio *et al.,* 1988; Smith and Talamantes, 1988). A similar GH-BP has also been reported in rabbit milk (Djiane *et al.,* 1990). Nonpregnant mice and rats have undetectable or very low levels of this BP (Peeters and Friesen, 1977; Shaw and Baumann, 1988; Smith and Talamantes, 1988). However, they exhibit BPs of larger molecular mass forming complexes with GH of 110, 150, and 220 kDa (Emtner and Roos, 1990; Massa *et al.,* 1990; Bick *et al.,* 1990), whose connection to the GH receptor is presently not clear. In addition, murine species show a heterogeneous pattern of low-affinity binding in serum which remains poorly characterized (Shaw and Baumann, 1988; Baumann *et al.,* unpublished). Cow and sheep serum contain very little GH binding activity (Shaw and Baumann, 1988).

2.2. Low-Affinity Growth Hormone-Binding Protein

Human plasma also contains a low-affinity GH-BP. Much less is known about this BP. It has only recently been described in some detail and generally accepted as a real entity.

2.2.1. Discovery and Identification

During the early studies of the GH binding phenomenon in plasma, we consistently observed heterogeneity within the bound GH fraction (Baumann *et al.,* 1986). The bound GH fraction consisted primarily of the 80- to 85-kDa component described above, but a larger molecular size peak (>100 kDa) was also present when chromatography columns of sufficient resolving power were used. The latter was designated "Peak I" to differentiate it from the high-affinity GH-BP complex, named "Peak II." Since Peak I eluted in the void volume of the column and was not saturable within the physiological range of GH concentrations, it was initially considered an artifact of aggregation or "nonspecific binding" (Baumann *et al.,* 1986). However, Peak I was such a consistent phenomenon under a variety of circumstances that further scrutiny was warranted. Isolation and characterization of this plasma component showed it to be a second specific BP for human GH with relatively low-affinity specificity but high binding capacity (Baumann and Shaw, 1990a).

2.2.2. Chemical and Physical Properties

The low-affinity BP is a protein that can be separated from the high-affinity BP by differential precipitation with ammonium sulfate or polyethylene glycol (Baumann and Shaw, 1990a). It is a single-chain protein which binds 1 mole of human GH per mole to form a complex of ~125 kDa (Baumann and Shaw, 1990a). From that, a molecular mass of ~100 kDa for the BP itself can be derived. This BP is more basic than the high-affinity BP, with an isoelectric point of 7.1. Several lines of evidence indicate that it is not structurally related to the GH receptor or the high-affinity BP (see Baumann and Shaw, 1990a, for review).

2.2.3. Biological Properties

Like the high-affinity BP, this BP is highly specific for human GH. It does not bind animal GHs, placental lactogen, or prolactin. Its binding affinity for the 22-kDa form of human GH is about 10^5 M^{-1}; its binding capacity in plasma is 15 μg/ml, which corresponds to a calculated concentration of 0.7 μM (Baumann and Shaw, 1990a). The low-affinity BP binds the 20-kDa variant of human GH (20k) with equal or slightly higher affinity ($K_a \sim 2 \times 10^5$ M^{-1}) than the 22-kDa form (22k) (Baumann and Shaw, 1990b). This is in contrast to the high-affinity BP, which primarily binds 22k. In addition, the low-affinity BP appears to contain a 20k-specific binding site—or may be a heterogeneous mixture of two proteins binding 22k or 20k, respectively (Baumann and Shaw, 1990b).

2.2.4. Physiological Effects

The low-affinity BP participates in the complexing of GH in the circulation. For 22k GH, it contributes only a small amount to the total binding activity in plasma [10–15% of 22k is complexed with this BP (Baumann *et al.*, 1988b, 1989a)]. However, for 20k GH, the low-affinity BP is the major carrier in plasma (Baumann and Shaw, 1990b) because of the poor avidity of 20k for the high-affinity BP. The relatively abundant low-affinity BP thereby accounts for most of the complexed 20k.

Unlike the high-affinity BP, the low-affinity BP has little effect on binding of GH to receptors (Mannor *et al.*, 1988). This would be expected from the fact that its affinity is too low to effectively compete with receptors for ligand.

2.2.5. Regulation and Possible Origins

The low-affinity GH-BP appears to be regulated independently of the high-affinity BP. Levels of the low-affinity BP are higher in women than in men, and further rise during pregnancy (Baumann *et al.*, 1989a)—a pattern different from the high-affinity BP. This has led to the suggestion that the low-affinity BP may be estrogen dependent. Other evidence for separate regulation comes from studies of Laron dwarfs and pygmies, where the low-affinity BP level is normal despite the absence or decrease of the high-affinity BP (Baumann *et al.*, 1987b, 1989b; Amselem *et al.*, 1989).

The tissue source of the low-affinity BP is completely unknown; the liver has been suggested based on the observation that plasma levels of this BP are highly variable in cirrhosis (Baumann *et al.*, 1989a).

The new field of GH-BPs has seen much progress in a relatively short time. Despite that, the true physiological role of these BPs remains unknown. The concept of a "circulating receptor" in such high concentrations in plasma is particularly difficult to integrate into conventional endocrine thinking. It appears likely that circulating BPs for polypeptides have a biological role beyond simply serving as an intravascular storage pool.

3. HUMAN TISSUE PLASMINOGEN ACTIVATOR

A thrombus is a blood-derived mass composed of fibrin which may occur anywhere in the circulation and which results from the action of thrombin on fibrinogen. Local thrombus formation requires interaction between the processes that promote and inhibit coagulation or thrombosis. An increased understanding of the mechanisms of clot formation and dissolution has led to the development of several different fibrinolytic agents for the treatment of myocardial infarction, one of which is recombinant human tissue plasminogen activator (rt-PA). One advantage of rt-PA over other fibrinolytics is that it is thought to act preferentially on plasminogen that is associated with the blood clot (Hoylaerts *et al.*, 1982), which may result in less systemic consumption of plasminogen, α_2-antiplasmin, and fibrinogen.

3.1. Structure and Function

rt-PA consists of a serine protease domain at the carboxy-terminal end and several other domains typical of plasma proteases at the amino-terminal end (Patthy, 1985). It is composed of a fibronectin-type finger region of about 44 amino acid residues, a growth factor domain of 40 residues, two kringle domains of 81 residues each, and a protease region of about 260 residues (Higgins and Bennett, 1990). The rt-PA molecule contains 35 cysteines. The protease domain of rt-PA is about 35–40% homologous with prototypic serine proteases such as bovine trypsin and chymotrypsin. The amino-terminal "finger" region is about 25–30% homologous with corresponding regions in fibronectin, and the growth factor region is about 25% homologous with a region of human epidermal growth factor. The kringles are 30–50% homologous with kringle structures on other plasma proteins (Higgins and Bennett, 1990).

Tissue plasminogen activator has four potential N-linked glycosylation sites, and three (at positions 117, 184, and 448) are known to be glycosylated under some circumstances (Bennett, 1983; Pohl *et al.*, 1984). Recombinant CHO cell-derived t-PA exists in two subtypes, which are referred to as types I and II. Type I is glycosylated at positions 117, 184, and 448, whereas type II is glycosylated only at positions 117 and 448.

The accepted activity of rt-PA is to cleave the R561–V562 peptide bond of plasminogen, converting it to plasmin.

3.2. Metabolism and Pharmacokinetics of rt-PA

When [125I]-rt-PA was administered IV to rats and rabbits, the vast majority of the radioactivity was found sequestered by the liver (Bakhit *et al.*, 1988). Fractionation of supernatants obtained from liver homogenates by trichloroacetic acid (TCA) precipitation showed an increased conversion of intact [125I]-rt-PA to TCA-soluble material with time. In addition, there was a marked decrease of total radioactivity with time in the liver, and a concomitant increase in total radioactivity in the blood, kidney, and urine. This suggests a sequential process of degradation by the liver, secretion of degradation products into the blood, and transport to the kidney for excretion in the urine. In a separate study, the uptake, internalization, and intracellular degradation of [125I]-rt-PA were examined in rat hepatocytes (Bakhit *et al.*, 1987). The internalization of [125I]-rt-PA was followed by the appearance of TCA-soluble material in the surrounding media. This degradation was inhibited by lysosomal tropic agents, suggesting that $[^{125}I]$-rt-PA degradation occurs intracellularly, perhaps within the lysosome.

The half-life of fibrinolytic activity, as measured on fibrin plates, after a bolus t-PA injection in intact, partially, and totally hepatectomized rabbits was 2, 8, and 40 min, respectively (Nilsson *et al.*, 1985). In these studies, the disposition of radioactivity following $[^{125}I]$-t-PA dosing in normal rabbits was described by a three-compartment model. The initial $t_{1/2}$ coincided with the disappearance of the fibrinolytic activity in intact animals and probably reflects t-PA uptake by the eliminating organ(s) (e.g., liver). The half-life of the second compartment was 14 min, which was thought to be related to $[^{125}I]$-t-PA elimination as well as formation and elimination of specific and nonspecific inhibitor complexes. The third phase ($t_{1/2} \sim 2$ hr) reflected a complex function incorporating elimination, the formation of inhibitor complexes, and the generation of free ^{125}I.

3.3. Protease Inhibitors

Protease inhibitors represent nearly 10% of the total protein in blood plasma (Travis and Salvesen, 1983). They not only play a critical role in coagulation and fibrinolysis but are also associated with connective tissue turnover, complement activation, and inflammation. In most cases, protease inhibition is characterized by the formation of an irreversible complex between the active site on the protease and the protease inhibitor. Several high-molecular-weight plasma-derived inhibitors react with an inhibit t-PA. Relatively slow inhibitors, not considered physiologically important, are α_2-antiplasmin, C1-esterase inhibitor (Ranby *et al.*, 1982), and α_2-macroglobulin (Korninger and Collen, 1981). Moderate rates of inactivation are displayed by the protease nexin and the placental inhibitor, plasminogen activator inhibitor-2 (PAI-2) (Kruithof, 1988). A relatively fast-acting inhibitor of t-PA, usually referred to as plasminogen activator inhibitor-1 (PAI-1), is present in plasma (Chmielewska *et al.*, 1983), endothelial cells (Loskutoff and Edgington, 1981), and platelets (Murray *et al.*, 1974; Booth *et al.*, 1985).

3.3.1. Characterization of PA Inhibitors

3.3.1a. PAI-1.   Endothelial cell-type PA-inhibitor (PAI-1) is synthesized by endothelial cells (Dosne *et al.*, 1978) and has been identified in many cell lines (Sprengers *et al.*, 1985; Ny *et al.*, 1985; Schleef *et al.*, 1985; Wagner *et al.*, 1985), α-granules of blood platelets (Erickson *et al.*, 1984), and plasma

(Kruithof *et al.*, 1983; Chmielewska *et al.*, 1983). PAI-1 is a serpin-related protein (Ny *et al.*, 1986; Ginsberg *et al.*, 1986; Pannekoek *et al.*, 1986) and, as such, is cleaved by t-PA at the R346–M347 peptide bond (Andreasen *et al.*, 1986). PAI-1 reacts with and inhibits t-PA with a second-order rate constant of 10^6 to 10^7 moles/liter-sec (Kruithof *et al.*, 1983; Chmielewska *et al.*, 1983). PAI-1 has a molecular mass of 40–50 kDa, and the complex which is formed by PAI-1 and t-PA is usually SDS stable (Levine, 1983; Sprengers *et al.*, 1985). However, the activity of PAI-1 *in vitro* is unstable, possibly owing in part to its high susceptibility to oxidation (Lawrence and Loskutoff, 1986). The concentration of active PAI-1 in blood plasma in healthy subjects ranges from 0.0 to 1.3 nmoles/liter (Schleef *et al.*, 1985; Chmielewska *et al.*, 1983). Considering the second-order rate constant and the concentration of PAI-1 in blood, the half-life of t-PA in plasma due to inactivation by the inhibitor is about 100 sec (Sprengers and Kluft, 1987). While the normal concentration of PAI-1 would not be high enough to interfere significantly with t-PA therapy, its concentration can be markedly increased during myocardial infarction (Almer and Ohlin, 1987), as well as during certain disease-related or physiological states (Hersch *et al.*, 1987; Neerstrand *et al.*, 1987; Kruithof *et al.*, 1988). PAI-1 behaves as an acute phase reactant, and increases in this inhibitor have been shown to occur in association with deep-vein thrombosis, age, hypertriglyceridemia, pregnancy, septicemia, and surgery. This inhibitor is probably an important component of the intrinsic fibrinolytic system (Kruithof, 1988); as such, local concentrations of PAI-1 may be more important than its concentration in plasma.

3.3.1b. PAI-2.   PAI-2 was first purified from placental tissue (Astedt *et al.*, 1985) but was described as early as 1968 (Kawano *et al.*, 1968). The inhibitor has a molecular mass of 48 kDa and also forms SDS-stable complexes with t-PA (Kruithof *et al.*, 1986). It has been proposed that the placental inhibitor is of macrophage origin (Chapman and Stone, 1985). While PAI-2 is not found in pooled normal human blood plasma, concentrations of 2 μmoles/liter have been reported for women in their third trimester of pregnancy (Lecander and Astedt, 1986). With a second-order rate constant of about 10^6 moles/liter-sec, the half-life of t-PA in plasma due to PAI-2 inactivation is about 0.5 sec (Sprengers and Kluft, 1987).

3.3.1c. Protease Nexin 1.   Purified to homogeneity from human foreskin fibroblast-conditioned medium, protease nexin 1 (PN1) has a molecular mass of 43 kDa and contains approximately 6% carbohydrate (Scott and Baker, 1983; Scott *et al.*, 1985; Howard and Knauer, 1986). PN1 is synthesized by fibroblasts, heart muscle cells, and kidney epithelial cells (Eaton and Baker, 1983). It reacts with t-PA (k_{on} = 1.5–3.0 × 10³ moles/liter-sec) by forming a 1:1 complex. PN1 has not been detected in plasma. Since second-order rate constants of PN1 with thrombin and trypsin are much higher than with t-PA, PN1 cannot be considered a true t-PA inhibitor (Sprengers and Kluft, 1987).

3.3.1d. α_2-Macroglobulin. α_2-Macroglobulin (α_2M) plays an important role by binding proteases secondarily to their specific inhibitors (Barrett, 1980). It inhibits members of the four protease catalytic classes—serine, cysteine, aspartic, and metallo proteases; when t-PA is bound to α_2M, it loses its ability to cleave plasminogen. α_2M is a glycoprotein of molecular mass 725 kDa which contains 8–11% carbohydrate (Roberts *et al.*, 1974; Dunn and Spiro, 1967) and which is present in plasma at a concentration of 250 mg/100 ml. The purified protein is a tetramer of identical 185-kDa subunits which are linked in pairs by disulfide bonds. Two pairs associate by noncovalent binding to form the native, tetrameric molecule. The characteristics of the reaction of proteases with α_2M are unique. At least three types of binding reactions may occur during inhibitor–enzyme interactions, including (1) a steric trapping reaction, specific for proteases, (2) a covalent linking of proteases and other molecules containing nucleophilic groups, and (3) a noncovalent, nonsteric adherence reaction with a number of other proteins and other molecules, unrelated to proteolytic activity (Barrett, 1981; Salvesen *et al.*, 1981).

3.3.1e. α_2-Antiplasmin. α_2-Antiplasmin (α_2AP) is a single-chain glycoprotein with a molecular mass of 65–70 kDa which contains approximately 11% carbohydrate. The primary function of α_2AP is the inhibition of plasmin formed from circulating plasminogen. The normal plasma concentration of α_2AP is 6 mg/100 ml (Matsuda *et al.*, 1979). The half-life of t-PA in blood due to inactivation by α_2AP is 90 to 180 min (Korninger and Collen, 1981).

3.3.1f. Cl-esterase. Cl-esterase is a single-chain glycoprotein with a molecular mass of 96–104 kDa which contains approximately 35% carbohydrate (Pensky and Schwick, 1969; Haupt *et al.*, 1970). The normal concentration in plasma is 17 mg/100 ml. Cl-esterase is capable of neutralizing the factor XII-dependent plasminogen activator; the physiological importance of its role is poorly understood.

3.3.2. Role of t-PA Inhibitors in the Regulation of Endogenous t-PA

In vivo, the concentration of active t-PA is regulated by endothelial secretion, hepatic clearance, and PAI-1, which result in a relatively stable steady-state level of active t-PA. Plasma protease inhibitors tightly regulate systemic plasminogen activator activity, preventing its unwanted actions. PAI-1 is responsible for approximately 60% of the inactivation of t-PA in plasma, while the other 40% is attributed to other protease inhibitors. In addition to its physiological role in the control of plasminogen activators, α_2AP also plays a central role in the molecular mechanism of fibrinolysis. *In vivo,* the lysis of a thrombus requires a continuous replacement of plasminogen at the surface of the clot because of the inactivation of plasmin by α_2AP (Travis and Salvesen, 1983).

3.3.3. Role of t-PA Inhibitors in the Inactivation of Exogenous t-PA

To evaluate the impact of protease inhibitors on rt-PA in plasma, serial plasma samples from patients with acute myocardial infarction were taken after infusions of rt-PA (Lucore and Sobel, 1988). rt-PA was found to circulate not only as free rt-PA, but also in complexes with PAI-1 (100 kDa), α_2AP (110 kDa), and C_1-esterase (170 kDa); endogenous levels of t-PA in the same patients formed complexes only with PAI-1. After termination of infusions, levels of free rt-PA and rt-PA complexed with α_2AP and C_1-esterase declined, while concentrations of rt-PA associated with PAI-1 remained constant or increased. After infusion of rt-PA in normal rabbits, rt-PA was associated predominantly with α_2AP and C_1-esterase rather than PAI-1 (Lucore and Sobel, 1988). In a separate study, rt-PA/α_2AP complexes were absent after 2 h of thrombolytic therapy in 53% of patients; this was due to the depletion of α_2AP associated with the generation of plasmin in the circulation after administration of rt-PA (Bermann *et al.,* 1983). In summary, during infusions of rt-PA, fibrinolytic activity is maintained despite the fact that a small portion of the dose forms complexes with plasma protease inhibitors. Since PAI-I is an acute phase protein and quickly inactivates t-PA, it is possible that PAI-1 may be released in response to t-PA infusion or the formation of t-PA/PAI-1 complex. After infusions, when PAI-1 activity is increased due to myocardial infarction, PAI-1 attenuates the fibrinolytic activity of residual immunoreactive rt-PA.

The role of protease inhibitors in thrombolytic therapy is not fully understood. However, to maintain a steady-state concentration of plasminogen activator inhibitors of 1.0 nmole/liter, with a hypothetical half-life of 10 min in man, a rate of synthesis of 0.45 mg/hr is required. Since the amount of rt-PA (up to 60 mg/hr) usually administered during thrombolytic therapy is substantially greater, it is likely that protease inhibitors do not significantly modulate the efficacy of the pharmacologically administered dose of rt-PA (Sprengers and Kluft, 1987).

ACKNOWLEDGMENTS

Studies presented in the growth hormone section were supported in part by NIH grants DK 38128 and RR 05370, and by a grant from the Northwestern Memorial Foundation.

REFERENCES

Adashi, E. Y., Resnick, C. E., D'Ercole, A. J., Swoboda, M. E., and Van Wyck, J. J., 1985, Insulin-like growth factors as intraovarian regulators of granulosa cell growth and function, *Endocr. Rev.* **6**:400–420.

Almer, L., and Ohlin, H., 1987, Elevated levels of the rapid inhibitor of plasminogen activator (t-PAI) in acute myocardial infarction, *Thromb. Res.* **47**:335–339.

Amit, T., Barkey, R. J., Youdim, M. B. H., and Hochberg, Z., 1990, A new and convenient assay for growth hormone-binding protein activity in human serum, *J. Clin. Endocrinol. Metab.* **71**:474–479.

Amselem, S., Duquesnoy, P., Attree, O., Novelli, G., Bousnina, S., Postel-Vinay, M.-C., and Goossens, M., 1989, Laron dwarfism and mutations of the growth hormone-receptor gene, *N. Engl. J. Med.* **321**:989–995.

Andreasen, P. A., Nielson, L. S., Kristensen, P., Grondahl-Hansen, J., and Skriver, L., 1986, Plasminogen activator inhibitor from human fibrosarcoma cells binds urokinase-type plasminogen activator but not its proenzyme, *J. Biol. Chem.* **261**:7644–7651.

Astedt, B., Lecander, I., Brodin, T., Lundblad, A., and Low, K., 1985, Purification of a specific placental plasminogen activator inhibitor by monoclonal antibody and its complex formation with plasminogen activator, *Thromb. Haemost.* **53**:122–125.

Bakhit, C., Lewis, D., Billings, R., and Malfroy, B., 1987, Cellular catabolism of recombinant tissue-type plasminogen activator, *J. Biol. Chem.* **262**:8716–8720.

Bakhit, C., Lewis, D., Busch, U., Tanswell, P., and Mohler, M., 1988, Biodisposition and catabolism of tissue-type plasminogen activator in rats and rabbits, *Fibrinolysis* **2**:31–36.

Ballard, F. J., Baxter, R. C., Binoux, M., Clemmons, D. R., Drop, S. L., Hall, K., Hintz, R. L., Rechler, M. M., Rutanen, E. M., and Schwander, J. C., 1990, Report on the nomenclature of the IGF binding proteins. *J. Clin. Endocrinol. Metab.* **70**:817–818.

Bar, R. S., Boes, M., Clemmons, D., Busby, W. H., Sandra, A., Dake, B. L., and Booth, B. B., 1990, Insulin differentially alters transcapillary movement of intravascular IGFBP-1, IGFBP-2 and endothelial cell IGF-binding proteins in the rat heart, *Endocrinology* **127**:497–499.

Barnard, R., and Waters, M. J., 1986, Serum and liver cytosolic growth-hormone-binding proteins are antigenically identical with liver membrane 'receptor' types 1 and 2, *Biochem. J.* **237**:885–892.

Barnard, R., Quirk, P., and Waters, M. J., 1989, Characterization of the growth hormone-binding protein of human serum using a panel of monoclonal antibodies, *J. Endocrinol.* **123**:327–332.

Barrett, A. J., 1980, Protein degradation in health and disease, *Ciba Found. Symp.* **75**:1–13.

Barrett, A. J., 1981, α_2-Macroglobulin, *Methods Enzymol.* **80**:737–754.

Baumann, G., 1989, Circulating binding proteins for human growth hormone, in: *Advances in Growth Hormone and Growth Factor Research* (E. E. Müller, D. Cocchi, and V. Locatelli, eds.), Pythagora Press/Springer-Verlag, Rome/Berlin, pp. 69–83.

Baumann, G., and Shaw, M. A., 1988, Immunochemical similarity of the human plasma growth hormone-binding protein and the rabbit liver growth hormone receptor, *Biochem. Biophys. Res. Commun.* **152**:573–578.

Baumann, G., and Shaw, M. A., 1990a, A second, lower affinity growth hormone-binding protein in human plasma. *J. Clin. Endocrinol. Metab.* **70**:680–686.

Baumann, G., and Shaw, M. A., 1990b, Plasma transport of the 20,000 dalton variant of human growth hormone (20K): Evidence for a 20K-specific binding site, *J. Clin. Endocrinol. Metab.* **71**:1339–1343.

Baumann, G., Stolar, M. W., Amburn, K., Barsano, C. P., and DeVries, B. C., 1986, A specific growth hormone-binding protein in human plasma: Initial characterization, *J. Clin. Endocrinol. Metab.* **62**:134–141.

Baumann, G., Amburn, K. D., and Buchanan, T. A., 1987a, The effect of circulating growth hormone-binding protein on metabolic clearance, distribution, and degradation of human growth hormone, *J. Clin. Endocrinol. Metab.* **64**:657–660.

Baumann, G., Shaw, M. A., and Winter, R. J., 1987b, Absence of the plasma growth hormone-binding protein in Laron-type dwarfism, *J. Clin. Endocrinol. Metab.* **65**:814–816.

Baumann, G., Shaw, M. A., and Buchanan, T. A., 1988a, *In vivo* kinetics of a covalent growth hormone-binding protein complex, *Metabolism* **38**:330–333.

Baumann, G., Amburn, K., and Shaw, M. A., 1988b, The circulating growth hormone (GH)-binding protein complex: A major constituent of plasma GH in man, *Endocrinology* **122**:976–984.

Baumann, G., Shaw, M. A., and Amburn, K., 1988c, A rapid and simple assay for growth hormone-binding protein activity in human plasma, *Acta Endocrinol. (Copenhagen)* **119**:529–534.

Baumann, G., Shaw, M. A., and Amburn, K., 1989a, Regulation of plasma growth hormone-binding proteins in health and disease, *Metabolism* **38**:683–689.

Baumann, G., Shaw, M. A., and Merimee, T. J., 1989b, Low levels of high-affinity growth hormone-binding protein in African pygmies. *N. Engl. J. Med.* **320**:1705–1709.

Baumann, G., Vance, M. L., Shaw, M. A., and Thorner, M. O., 1990, Plasma transport of human growth hormone *in vivo, J. Clin. Endocrinol. Metab.* **71**:470–473.

Baumbach, W. R., Horner, D. L., and Logan, J. S., 1989, The growth hormone-binding protein in rat serum is an alternatively spliced form of the rat growth hormone receptor, *Genes Dev.* **3**:1199–1205.

Baxter, R. C., 1988, The insulin-like growth factors and their binding proteins, *Comp. Biochem. Physiol.* **91**:229–235.

Baxter, R. C., 1990, Circulating levels and molecular distribution of the acid-labile (alpha) subunit of the high molecular weight insulin-like growth factor binding protein complex, *J. Clin. Endocrinol. Metab.* **70**:1347–1353.

Baxter, R. C., and Cowell, C. T., 1987, Diurnal rhythm of growth hormone-independent binding protein for insulin-like growth factors in human plasma, *J. Clin. Endocrinol. Metab.* **65**:432–440.

Baxter, R. C., and Martin, J. L., 1989, Binding proteins for the insulin-like growth factors: Structure, regulation, and function, *Prog. Growth Factor Res.* **1**:49–68.

Baxter, R. C., Martin, J. L., and Wood, M. H., 1987, Two immunoreactive binding proteins for insulin-like growth factors in human amniotic fluid: Relationship to fetal maturity, *J. Clin. Endocrinol. Metab.* **65**:423–431.

Bennett, W., 1983, Two forms of tissue-type plasminogen activator differ at a single, specific glycosylation site, *Thromb. Haemost.* **50**:106 (abstract).

Bermann, S. R., Fox, K. A., Ter-Pogossian, M. M., Sobel, B. E., and Collen, D., 1983, Clot-selective coronary thrombolysis with tissue-type plasminogen activator, *Science* **220**:1181–1183.

Berson, S. A., and Yalow, R. S., 1966, State of human growth hormone in plasma and changes in stored solutions of pituitary growth hormone, *J. Biol. Chem.* **241**:5745–5749.

Berson, S. A., and Yalow, R. S., 1968, Peptide hormones in plasma, *Harvey Lect.* **62**:107–163.

Bick, T., Amit, T., Barkey, R. J., Hertz, P., Youdim, M. B. H., and Hochberg, Z., 1990, The interrelationship of growth hormone (GH), liver membrane GH receptor, serum GH-binding protein activity, and insulin-like growth factor I in the male rat, *Endocrinology* **126**:1914–1920.

Binkert, C., Landwehr, J., Mary, J. L., Schwander, J., and Heinrich, G., 1989, Cloning, sequence analysis and expression of a cDNA encoding a novel insulin-like growth factor binding protein, *EMBO J.* **8**:2497–2502.

Booth, N. A., Anderson, J. A., and Bennett, B., 1985, Platelet release protein which inhibits plasminogen activators, *J. Clin. Pathol.* **38**:825–830.

Brewer, M. T., Stetler, G. L., Squires, C. H., Thompson, R. C., Busby, W. H., and Clemmons, D. R., 1988, Cloning, characterization and expression of a human insulin-like growth factor binding protein, *Biochem. Biophys. Res. Commun.* **152**:1289–1297.

Brismar, K., Gutniak, M., Povoa, G., Werner, S., and Hall, K., 1988, Insulin regulates the 35 KDa IGF binding protein in patients with diabetes mellitus, *J. Endocrinol. Invest.* **11**:599–602.

Busby, W. H., Snyder, D. K., and Clemmons, D. R., 1988, Radioimmunoassay of a 26,000 dalton plasma insulin-like growth factor binding protein: Control by nutritional variables, *J. Clin. Endocrinol. Metab.* **67**:1231–1236.

Chapman, H. A., and Stone, O. L., 1985, Characterization of a macrophage derived plasminogen activator inhibitor: Similarities with placental urokinase inhibitor, *Biochem. J.* **230**:109–116.

Chmielewska, J., Ranby, M., and Wiman, B., 1983, Evidence for a rapid inhibitor to tissue plasminogen activator in plasma, *Thromb. Res.* **31**:427–436.

Chochinov, R. H., Mariz, I. K., Hajek, A. S., and Daughaday, W. H., 1977, Characterization of a protein in mid-term human amniotic fluid which reacts in the somatomedin-C radioreceptor assay, *J. Clin. Endocrinol. Metab.* **44**:902–908.

Cohen, K. L., and Nissley, S. P., 1976, The serum half-life of somatomedin activity: Evidence for growth hormone dependence, *Acta Endocrinol. (Copenhagen)* **83**:243–258.

Copeland, K. C., Underwood, L. E., and Van Wyk, J. J., 1980, Induction of immunoreactive somatomedin-C in human serum by growth hormone: Dose response relationships and effect on chromatographic profiles, *J. Clin. Endocrinol. Metab.* **50**:690–698.

Daughaday, W. H., and Rotwein, P., 1989, Insulin-like growth factors I and II: Peptide, messenger ribonucleic acid and gene structures, serum, and tissue concentrations, *Endocr. Rev.* **10**:68–91.

Daughaday, W. H., and Trivedi, B., 1987, Absence of serum growth hormone binding protein in patients with growth hormone receptor deficiency (Laron dwarfism), *Proc. Natl. Acad. Sci. USA* **84**:4636–4640.

Daughaday, W. H., Trivedi, B., and Andrews, B. A., 1987, The ontogeny of serum GH binding in man: A possible indicator of hepatic GH receptor development, *J. Clin. Endocrinol. Metab.* **65**:1072–1074.

De Mellow, J. S., and Baxter, R. C., 1988, Growth hormone dependent insulin-like growth factor (IGF) binding protein both inhibits and potentiates IGF-I stimulated DNA synthesis in human skin fibroblasts, *Biochem. Biophys. Res. Commun.* **156**:199–204.

Djiane, A., Tar, A., Belair, L., Postel-Vinay, M. C., and Kelly, P. A., 1990, Prolactin and growth hormone-binding proteins in milk, *Endocrinology* **126**(Suppl.):92 (abstract).

Donahue, L. R., Hunter, S. J., Sherblom, A. P., and Rosen, C., 1990, Age-related changes in serum insulin-like growth factor binding proteins in women, *J. Clin. Endocrinol. Metab.* **71**:575–579.

Dosne, A. M., Dupuy, E., and Bodevin, E., 1978, Production of a fibrinolytic inhibitor by cultured endothelial cells derived from human umbilical vein, *Thromb. Res.* **12**:377–387.

Drop, S. L. S., Kortleve, D. J., and Guyda, H. J., 1984, Isolation of a somatomedin binding protein from preterm human amniotic fluid: Development of a radioimmunoassay, *J. Clin. Endocrinol. Metab.* **59**:899–905.

Dunn, J. T., and Spiro, R. G., 1967, The α_2-macroglobulin of human plasma: Studies on the carbohydrate units, *J. Biol. Chem.* **242**:5556–5563.

Eaton, D. L., and Baker, J. B., 1983, Evidence that a variety of cultured cells secrete protease nexin and produce a distinct cytoplasmic serine-protease-binding factor, *J. Cell. Physiol.* **117**:175–182.

Elgin, R. G., Busby, W. H., and Clemmons, D. R., 1987, An insulin-like growth factor (IGF) binding protein enhances the biologic response to IGF-I, *Proc. Natl. Acad. Sci. USA* **84**:3254–3258.

Emtner, M., and Roos, P., 1990, Identification and partial characterization of a growth hormone-binding protein in rat serum, *Acta Endocrinol. (Copenhagen)* **122**:296–302.

Erickson, L. A., Ginsberg, M. H., and Loskutoff, D. J., 1984, Detection and partial characterization of an inhibitor of plasminogen activator in human platelets, *J. Clin. Invest.* **74**:1465–1472.

Eshet, R., Laron, Z., Pertzelan, A., Arnon, R., and Dintzman, M., 1984, Defect of human growth hormone receptors in the liver of two patients with Laron-type dwarfism, *Isr. J. Med. Sci.* **20**:8–11.

Fuh, G., Mulkerrin, M. G., Bass, S., McFarland, N., Brochier, M., Bourell, J. H., Light, D. R., and Wells, J. A., 1990, The human growth hormone receptor. Secretion from Escherichia coli and disulfide bonding pattern of the extracellular binding domain, *J. Biol. Chem.* **265**:3111–3115.

Ginsberg, D., Zeheb, R., Yang, A. Y., Rafferty, A. M., and Andreasen, P. A., 1986, cDNA cloning of human plasminogen activator inhibitor from endothelial cells, *J. Clin. Invest.* **78**:1673–1680.

Godowski, P. J., Leung, D. W., Meacham, L. R., Galgani, J. P., Hellmiss, R., Keret, R., Rotwein, P. S., Parks, J. S., Laron, Z., and Wood, W. I., 1989, Characterization of the human growth hormone receptor gene and demonstration of a partial gene deletion in two patients with Laron-type dwarfism, *Proc. Natl. Acad. Sci. USA* **86**:8083–8087.

Guler, H. P., Zapf, J., and Froesch, E. R., 1987, Short-term metabolic effects of recombinant insulin-like growth factor-I in healthy adults, *N. Engl. J. Med.* **317**:137–140.

Guler, H. P., Zapf, J., Scheiwiller, E., and Froesch, E. R., 1988, Recombinant insulin-like growth factor-1 stimulates growth and has distinct effects on organ size in hypophysectomized rats, *Proc. Natl. Acad. Sci. USA* **85**:4889–4893.

Hardouin, S., Hossenlopp, P., Segovia, B., Seurin, D., Portolan, G., Lassarre, C., and Binoux, M., 1987, Heterogeneity of insulin-like growth factor binding proteins and relationship between structure and affinity: 1. Circulating forms in man, *Eur. J. Biochem.* **170**:121–132.

Haupt, H., Heimburger, N., Kranz, T., and Schwick, H. G., 1970, A contribution to the characterization of the activated first component of complement inactivator in human plasma, *Eur. J. Biochem.* **17**:254–261.

Herington, A. C., Ymer, S., and Stevenson, J., 1986a, Identification and characterization of specific binding proteins for growth hormone in normal human sera, *J. Clin. Invest.* **77**:1817–1823.

Herington, A. C., Ymer, S. I., and Stevenson, J. L., 1986b, Affinity purification and structural characterization of a specific binding protein for human growth hormone in human serum, *Biochem. Biophys. Res. Commun.* **139**:150–155.

Herington, A. C., Smith, A. I., Wallace, C., and Stevenson, J. L., 1987, Partial purification from human serum of a specific binding protein for human growth hormone, *Mol. Cell. Endocrinol.* **53**:203–209.

Hersch, S. L., Kunelis, T., and Francis, R., 1987, The pathogenesis of accelerated fibrinolysis in liver cirrhosis: A critical role for tissue plasminogen activator inhibitor, *Blood* **69**:1315–19.

Higgins, D. L., and Bennett, W. F., 1990, Tissue plasminogen activator: The biochemistry and pharmacology of variants produced by mutagenesis, *Annu. Rev. Pharmacol. Toxicol.* **30**:91–121.

Hintz, R. L., and Liu, F., 1977, Demonstration of specific plasma protein binding sites for somatomedin, *J. Clin. Endocrinol. Metab.* **45**:988–995.

Hintz, R. L., Liu, F., Rosenfeld, R. G., and Kemp, S. F., 1981, Plasma somatomedin-binding proteins in hypopituitarism: Changes during growth hormone therapy, *J. Clin. Endocrinol. Metab.* **53**:100–104.

Hossenlopp, P., Seurin, D., Segovia-Quinson, B., Hardouin, S., and Binoux, M., 1986, Analysis of serum insulin-like growth factor binding proteins using Western blotting: Use of the method for titration of the binding proteins and competitive binding studies, *Anal. Biochem.* **154**:138–143.

Howard, E. W., and Knauer, D. J., 1986, Human protease nexin 1: Further characterization using a highly specific polyclonal antibody, *J. Biol. Chem.* **261**:684–685.

Hoylaerts, M., Rijken, D. C., Lijnen, H. R., and Collen, D., 1982, Kinetics of the activation of plasminogen by human tissue plasminogen activator, *J. Biol. Chem.* **257**:2912–2919.

Jan, T., Shaw, M. A., and Baumann, G., 1990, Effects of growth hormone binding proteins on serum growth hormone measurements, *J. Clin. Endocrinol. Metab.* **72**:387–391.

Kaufman, U., Zapf, J., Torretti, B., and Froesch, E. R., 1977, Demonstration of a specific carrier protein of nonsuppressible insulin-like activity *in vivo, J. Clin. Endocrinol. Metab.* **44**:160–166.

Kawano, T., Morimoto, K., and Uemura, Y., 1968, Urokinase inhibitor in human placenta, *Nature* **217**:253–254.

Knauer, D. J., and Smith, G. L., 1980, Inhibition of biologic activity of multiplication-stimulating activity by binding to its carrier protein, *Proc. Natl. Acad. Sci. USA* **77**:7252–7256.

Korninger, C., and Collen, D., 1981, Neutralization of human extrinsic tissue-type plasminogen activator in human plasma: No evidence for a specific inhibitor, *Thromb. Haemost.* **46**:662–665.

Kruithof, E. K. O., 1988, Plasminogen activator inhibitors: A review, *Enzyme* **40**:113–121.

Kruithof, E. K. O., Ransijn, A., and Bachmann, F., 1983, Inhibition of tissue plasminogen activator by human plasma, in: *Progress in Fibrinolysis,* Vol. VI (J. F. Davidson, F. Bachmann, C. A. Bouiver, and E. K. Kruithof, eds.), Churchill Livingstone, Edinburgh, pp. 365–378.

Kruithof, E. K. O., Vassalli, J. D., Schleuning, W. D., Mattaliano, R. J., and Bachmann, F., 1986, Purification and characterization of a plasminogen activator inhibitor from the histiocytic lymphoma cell line U-937, *J. Biol. Chem.* **261**:11207–11213.

Kruithof, E. K. O., Gudinchet, A., and Bachmann, F., 1988, Plasminogen activator inhibitor 1 and plasminogen activator inhibitor 2 in various disease states, *Thromb. Haemost.* **59**:7–12.

Laron, Z., Klinger, B., Erster, B., and Silbergeld, A., 1989, Serum GH binding protein activities identifies the heterozygous carriers for Laron type dwarfism, *Acta Endocrinol. (Copenhagen)* **121**:603–608.

Lauterio, T. J., Trivedi, B., Kapadia, M., and Daughaday, W. H., 1988, Reduced ^{125}I-hGH binding by serum of dwarf pigs but not by serum of dwarfed poodles, *Comp. Biochem. Physiol.* **91A**:15–19.

Lawrence, D. A., and Loskutoff, D. J., 1986, Inactivation of plasminogen activator inhibitors by oxidants, *Biochemistry* **25**:6351–6355.

Lecander, I., and Astedt, B., 1986, Isolation of a new specific plasminogen activator inhibitor from pregnancy plasma, *Br. J. Haematol.* **62**:221–228.

Lee, Y. L., Hintz, R. L., James, P. M., Lee, P. D. K., Shively, J. E., and Powell, D. R., 1988, Insulin-like growth factor (IGF) binding protein complimentary deoxyribonucleic acid from human HEP G2 hepatoma cells: Predicted protein sequence suggests an IGF binding domain different from those of the IGF-I and IGF-II receptors, *Mol. Endocrinol.* **2**:404–411.

Leof, E. B., Wharton, W., Van Wyck, J. J., and Pledger, W. J., 1982, Epidermal growth factor (EGF) and somatomedin-C regulate G-1 progression in competent Balb-c-3T3 cells, *Exp. Cell Res.* **141**:107–116.

Leung, D. W., Spencer, S. A., Cachianes, G., Hammonds, R. G., Collins, C., Henzel, W. J., Barnard, R., Waters, M. J., and Wood, W. I., 1987, Growth hormone receptor and serum binding protein: Purification, cloning and expression, *Nature* **330**:537–543.

Levine, G., 1983, Latent tissue plasminogen activator produced by bovine endothelial cells in culture: Evidence for an enzyme–inhibitor complex, *Proc. Natl. Acad. Sci. USA* **80**:6804–6808.

Lewitt, M. S., and Baxter, R. C., 1989, Regulation of growth hormone-independent insulin-like growth-factor binding protein-1 production in human granulosa-luteal cells, *J. Clin. Endocrinol. Metab.* **69**:1174–1179.

Lim, L., Spencer, S. A., McKay, P., and Waters, M. J., 1990, Regulation of growth hormone (GH) bioactivity by a recombinant human GH-binding protein, *Endocrinology* **127**:1287–1291.

Lin, T., Haskell, J., Vinson, N., and Terracio, L., 1986, Characterization of insulin-like growth factor-I receptors of purified Leydig cells and their role in steroidogenesis in primary culture: A comparative study, *Endocrinology* **119**:1641–1647.

Lobie, P. E., Barnard, R., and Waters, M. J., 1990, The nuclear growth hormone receptor/binding protein, *Endocrinology* **126**(Suppl.):247 (abstract).

Loskutoff, D. J., and Edgington, T. S., 1981, An inhibitor of plasminogen activator in rabbit endothelial cells, *J. Biol. Chem.* **256**:4142–4145.

Lucore, C. L., and Sobel, B. E., 1988, Interactions of tissue-type plasminogen activator with plasma inhibitors and their pharmacologic implications, *Circulation* **3**:660–669.

McCarter, J., Shaw, M. A., Winer, L., and Baumann, G., 1990, The 20,000 dalton variant of human growth hormone does not bind to growth hormone receptors in human liver, *Mol. Cell. Endocrinol.* **73**:11–14.

McGuffin, W. L., Jr., Gavin, J. R., III, Lesniak, M. A., Gorden, P., and Roth, J., 1976, Water-soluble specific growth hormone binding sites from cultured human lymphocytes: Preparation and partial characterization, *Endocrinology* **98**:1401–1407.

Maiter, D., Maes, M., Underwood, L., Flieson, T., Gerard, G., and Ketelslegers, J. M., 1988, Early changes in serum concentrations of somatomedin-C induced by dietary protein deprivation in rats: Contributions of growth hormone receptor and post-receptor defects, *J. Endocrinol.* **118**:113–120.

Mannor, D. A., Shaw, M. A., Winer, L. M., and Baumann, G., 1988, Circulating growth hormone-binding proteins inhibit growth hormone (GH) binding to GH receptors but not *in vivo* GH action, *Clin. Res.* **36**:870A (abstract).

Martin, J. L., and Baxter, R. C., 1986, Insulin-like growth factor-binding protein from human plasma: Purification and characterization, *J. Biol. Chem.* **261**:8754–8760.

Massa, G., Mulumba, N., Ketelslegers, J. M., and Maes, M., 1990, Initial characterization and sexual dimorphism of serum growth hormone-binding protein in adult rats, *Endocrinology* **126**:1976–1980.

Matsuda, M., Wakabayashi, K., Aoki, N., and Morioka, Y., 1979, Alpha 2-plasmin inhibitor is among acute phase reactants, *Thromb. Res.* **17**:527–532.

Megyesi, K., Kahn, C. R., Roth, J., and Gorden, P., 1975, Circulating NSILA-s in man: Preliminary studies of stimuli *in vivo* and of binding to plasma components, *J. Clin. Endocrinol. Metab.* **41**:475–484.

Merimee, T. J., Baumann, G., and Daughaday, W., 1990, Growth hormone-binding protein: II. Studies in pygmies and normal statured subjects, *J. Clin. Endocrinol. Metab.* **71**:1183–1188.

Mohan, S., Tremollieres, F., Campbell, M., Wergedal, J., and Baylink, D., 1991, IGF-I, IGF-II and human growth hormone markedly reduce production of inhibitory IGF binding protein (hIGFBP-4) in human bone cells, *2nd International Symposium on Insulin-Like Growth Factors/Somatomedins,* San Francisco, p. 252 (abstract).

Mohler, M. A., Cohen, S., Cook, J., Roth, M., and Ferraiolo, B., 1989, Metabolic clearance of Insulin-like growth factor-1 in rats, *American Association of Pharmaceutical Scientists. Western Regional Meeting,* Reno, NV., pp. 18. (abstract).

Moore, J., Vandlen, R., MacKay, P., and Spencer, S. A., 1988, Serum clearance of human growth hormone bound to growth hormone binding protein, *Endocrinology* **122**(Suppl.):121 (abstract).

Morera, A., Benahmed, M., and Chauvin, M. A., 1986, Somatomedin C: A factor in the differentiation of adrenal cortex cells, *C. R. Acad. Sci. Ser. C* **303**:581–584.

Moses, A. C., Nissley, S. P., and Cohen, K. L., 1976, Specific binding of a somatomedin like polypeptide in rat serum depends on growth hormone, *Nature* **263**:137.

Murphy, L. J., and Friessen, H. G., 1988, Differential effects of estrogen and growth hormone on uterine and hepatic insulin-like growth factor 1 gene expression in the ovariectomized hypophysectomized rat, *Endocrinology* **122**:325–332.

Murray, J., Crawford, G. M., Ogston, D., and Douglas, A. S., 1974, Studies on an inhibitor of plasminogen activators in human platelets, *Br. J. Haematol.* **26**:661–668.

Neerstrand, H., Hedner, U., Lutzen, O., Hauch, O., and Jorgensen, L., 1987, The influence of surgery on plasminogen activator inhibitor, *Thromb. Haemost.* **58**:557 (abstract).

Nilsson, C., Emarsson, M., Ekvarn, S., Haggroth, L., and Mattsson, C., 1985, Turnover of tissue plasminogen activator in normal and hepatectomized rabbits, *Thromb. Res.* **39**:511–521.

Nissley, S. P., and Rechler, M. M., 1984, Insulin-like growth factors: Biosynthesis, receptors, and carrier proteins, in: *Hormonal Proteins and Peptides* (H. C. Li, ed.), Academic Press, New York, pp. 127–203.

Ny, T., Bjersing, L., Hsuek, A. J., and Loskutoff, D. J., 1985, Cultured granulosa cells produce two plasminogen activators and an antiactivator, each regulated differently by gonadotropins, *Endocrinology* **116**:1666–1668.

Ny, T., Sawdey, M., Lawrence, D., Millan, J. L., and Loskutoff, D. J., 1986, Cloning and sequence of a cDNA coding for the human b-migrating endothelial cell-type plasminogen activator inhibitor, *Proc. Natl. Acad. Sci. USA* **83**:6776–6780.

Ooi, G. T., 1990, Insulin-like growth factor-binding proteins (IGFBPs): More than just 1, 2, 3, *Mol. Cell. Endocrinol.* **71**:C39–C43.

Ooi, G. T., and Herington, A. C., 1986, Covalent cross-linking of insulin-like growth factor-1 to a specific inhibitor from human serum, *Biochem. Biophys. Res. Commun.* **137**:411–417.

Ooi, G. T., and Herington, A. C., 1990, Recognition of insulin-like-growth-factor-binding proteins in serum and amniotic fluid by an antiserum against a low-molecular-mass insulin-like-growth-factor-inhibitor/binding protein, *Biochem. J.* **267**:615–620.

Pannekoek, H., Veerman, H., Lambers, H., Diergaarde, P., and Verweij, C. L., 1986, Endothelial plasminogen activator inhibitor (PAI): A new member of the serpin gene family, *EMBO J.* **5**:2539–2544.

Patthy, L., 1985, Evolution of the proteases of blood coagulation and fibrinolysis by assembly from modules, *Cell* **41**:657–663.

Peeters, S., and Friesen, H. G., 1977, A growth hormone binding factor in the serum of pregnant mice, *Endocrinology* **101**:1164–1179.

Pensky, J., and Schwick, H. G., 1969, Human serum inhibitor of C'1 esterase: Identity with alpha-2-neuraminoglycoprotein, *Science* **163**:698–699.

Philips, L. S., and Vassilopoulou-Selin, R., 1979, Nutritional regulation of somatomedin, *Am. J. Clin. Nutr.* **32**:1082–1096.

Pohl, G., Kallstrom, M., Bergsdorf, N., Wallen, P., and Jornvall, H., 1984, Tissue plasminogen activator: Peptide analyses confirm an indirectly derived amino acid sequence, identify the active site serine residue, establish glycosylation sites, and localize variant differences, *Biochemistry* **23**:3701–3707.

Postel-Vinay, M. C., Tar, A., Hocquette, J. F., Brauner, R., and Rappaport, R., 1990, Growth hormone increases and testosterone decreases the GH-binding activity in human plasma, *Endocrinology* **126**(Suppl.):248 (abstract).

Povoa, G., Enberg, G., Jornvall, H., and Hall, K., 1984, Isolation and characterization of a somatomedin binding protein from mid-term human amniotic fluid, *Eur. J. Biochem.* **144**:199–204.

Prewitt, T., D'Ercole, J., Switzer, B., and van Wyk, J., 1982, Relationship of serum somatomedin C to dietary protein and energy in growing rats, *J. Nutr.* **112**:144–150.

Ranby, M., Bergdorf, N., and Wallen, P., 1982, Isolation of two variants of native one-chain tissue plasminogen activator, *FEBS Lett.* **146**:289–292.

Rechler, M. M., and Nissley, S. P., 1985, The nature and regulation of the receptors for insulin-like growth factors, *Annu. Rev. Physiol.* **47**:425–442.

Rechler, M., Zapf, J., Nissley, S., Froesch, E., Moses, A., Podskalny, J., Schilling, E., and Humbel, R., 1980, Interactions of insulin-like growth factors I and II and multiplication-stimulating activity with receptors and serum carrier proteins, *Endocrinology* **107**:1451–1459.

Rinderknecht, E., and Humbel, R. E., 1978a, The amino acid sequence of human insulin-like growth factor I and its structural homology with proinsulin, *J. Biol. Chem.* **253**:2769–2776.

Rinderknecht, E., and Humbel, R. E., 1978b, Primary structure of human insulin-like growth factor II, *FEBS Lett.* **89**:283–286.

Riss, T. L., Karey, K. P., Burleigh, D. B., Parker, D., and Sirbasku, D. A., 1988, Human recombinant insulin-like growth factor-I. Development of a serum-free medium for clonal density assay of growth factors using BALB/c3T3 mouse embryo fibroblasts, *In Vitro Cell Dev. Biol.* **24**:1099–1106.

Roberts, R. C., Riesen, W. A., and Hall, P. K., 1974, Proteinase inhibitors, in: *Proc. 5th Bayer Symp.* (H. Fritz, H. Tscheche, L. J. Greene, and E. Truscheit, eds.), Springer, Berlin, pp. 63–71.

Ruoslahti, E., and Pierschbacher, M., 1988, A versatile recognition signal, *Cell* **44**:517–518.

Rutanen, E. M., and Pekonen, F., 1990, Insulin-like growth factors and their binding proteins, *Acta Endocrinol. (Copenhagen)* **123**:7–13.

Rutanen, E. M., Koistinen, R., Sjoberg, J., Julkunen, M., Wahlstrom, T., Bohn, H., and Seppala, M., 1986, Synthesis of placental protein 12 by human endometrium, *Endocrinology* **118**:1067–1071.

Salvesen, G. S., Sayers, C. A., and Barrett, A. J., 1981, Further characterization of the covalent linking reaction of alpha-2-macroglobulin, *Biochem. J.* **195**:453–461.

Schleef, R. R., Sinha, M., and Loskutoff, D. J., 1985, Immunoradiometric assay to measure the binding of a specific inhibitor to tissue-type plasminogen activator, *J. Lab. Clin. Med.* **106**:408–415.

Schmid, C. J., Zapf, J., and Froesch, E. R., 1989a, Production of carrier proteins for insulin-like growth factors (IGFs) by rat osteoblastic cells: Regulation by IGF-I and cortisol, *FEBS Lett.* **244**:328–332.

Schmid, C. J., Ernst, M., Zapf, J., and Froesch, E. R., 1989b, Release of insulin-like growth factor carrier proteins by osteoclasts: Stimulation by estradiol and growth hormone, *Biochem. Biophys. Res. Commun.* **160**:788–794.

Scott, R. W., and Baker, J. B., 1983, Purification of human protease nexin, *J. Biol. Chem.* **258**:10439–10444.

Scott, R. W., Bergman, B. L., Bajpai, A., Hersh, R. T., Rodriguez, H., Jones, B. N., Barreda, C., Watts, S., and Baker, J. B., 1985, Protease nexin: Properties and a modified purification procedure, *J. Biol. Chem.* **260**:7029–7034.

Shaw, M. A., and Baumann, G., 1988, Growth hormone-binding proteins in animal plasma: A survey, *Endocrinology* **122**(Suppl.):80 (abstract).

Shimasaki, S., Shimonaka, M., Bicsak, T., and Ling, N., 1991, Isolation and molecular characterization of three novel IGFBPs, *2nd International Symposium on Insulin-Like Growth Factors/Somatomedins,* San Francisco, p. 255 (abstract).

Silbergeld, A., Lazar, L., Erster, B., Keret, R., Tepper, R., and Laron, Z., 1989, Serum growth hormone binding protein activity in healthy neonates, children and young adults: Correlation with age, height and weight, *Clin. Endocrinol.* **31**:295–303.

Smith, W. C., and Talamantes, F., 1988, Gestational profile and affinity cross-linking of the mouse serum growth hormone-binding protein, *Endocrinology* **123**:1489–1494.

Smith, W. C., Kuniyoshi, J., and Talamantes, F., 1989, Mouse serum growth hormone (GH) binding protein has GH receptor extracellular and substituted transmembrane domains, *Mol. Endocrinol.* **3**:984–990.

Snow, K. J., Shaw, M. A., Winer, L. M., and Baumann, G., 1990, Diurnal pattern of plasma growth hormone binding protein in man, *J. Clin. Endocrinol. Metab.* **70**:417–420.

Snyder, D. K., and Clemmons, D. R., 1990, Insulin-dependent regulation of insulin-like growth factor binding protein-1, *J. Clin. Endocrinol. Metab.* **71**:1632–1636.

Spencer, S. A., Hammonds, R. G., Henzel, W. J., Rodriguez, H., Waters, M. J., and Wood, W. I., 1988, Rabbit liver growth hormone receptor and serum binding protein, *J. Biol. Chem.* **263**:7862–7867.

Sprengers, E. D., and Kluft, C., 1987, Plasminogen activator inhibitors, *Blood* **69**:381–387.

Sprengers, E. D., Princen, H. M., Kooistra, T., and Van Hinsbergh, V. W., 1985, Inhibition of plasminogen activators by conditioned medium of human hepatocytes and hepatoma cell line Hep G2, *J. Lab. Clin. Med.* **105**:751–758.

Suikkari, A. M., Koistinen, V. A., Rutanen, E. M., Jarvinen, H., Karonen, S. L., and Seppala, M., 1980, Insulin regulates serum levels of low molecular weight insulin-like growth factor binding proteins, *J. Clin. Endocrinol. Metab.* **66**:266–273.

Tramontano, D., Cushing, G. W., Moses, A. C., and Ingbar, S. H., 1986, Insulin-like growth factor-I stimulates the growth of rat thyroid cells in culture and synergizes the stimulation of DNA synthesis induced by TSH and Graves IgG, *Endocrinology* **119**:940–942.

Travis, J., and Salvesen, G. S., 1983, Human plasma proteinase inhibitors, *Annu. Rev. Biochem.* **52**:655–709.

Trivedi, B., and Daughaday, W. H., 1988, Release of growth hormone binding protein from IM-9 lymphocytes by endopeptidase is dependent on sulfhydryl group inactivation, *Endocrinology* **123**:2201–2206.

Underwood, L. E., Clemmons, D. R., Maes, M., D'Ercole, J., and Ketelslegers, J. M., 1986, Regulation of somatomedin C/insulin-like growth factor 1 by nutrients, *Horm. Res.* **24**:166–177.

Wagner, O., Korninger, C., Speiser, W., Schwaiger, N., Beckmann, R., and Binder, B. R., 1985, Isolation and characterization of a fast acting PA-inhibitor from a human melanoma cell line, *Thromb. Haemost.* **54**:48 (abstract).

White, R. M., Nissley, S. P., Moses, A. C., Rechler, M. M., and Johnsonbaugh, R. E., 1981, The growth hormone dependence of a somatomedin binding protein in human serum, *J. Clin. Endocrinol. Metab.* **53**:49.

Wilkins, J. R., and D'Ercole, A. J., 1985, Affinity-labeled plasma somatomedin-C/insulin-like growth factor-I binding proteins: Evidence for growth hormone dependence and subunit structure, *J. Clin. Invest.* **75**:1350–1358.

Wood, W. I., Cathianes, G., Henzel, W. J., Winslow, G. A., Spencer, S. A., Hellmiss, R., Martin, J. L., and Baxter, R. C., 1988, Cloning and expression of the GH dependent insulin-like growth factor binding protein, *Mol. Endocrinol.* **2**:1176–1185.

Ymer, S. I., and Herington, A. C., 1984, Water-soluble hepatic growth hormone receptors: Structural studies using gel chromatography and chemical cross-linking, *Endocrinology* **114**:1732–1739.

Ymer, S. I., and Herington, A. C., 1985, Evidence for the specific binding of growth hormone to a receptor-like protein in rabbit serum, *Mol. Cell. Endocrinol.* **41**:153–161.

Zapf, J., Waldvogel, M., and Froesch, E. R., 1975, Binding of nonsuppressible insulinlike activity to human serum: Evidence for a carrier protein, *Arch. Biochem. Biophys.* **168**:638–645.

Zapf, J., Schoenle, E., and Froesch, E., 1978, Insulin-like growth factors I and II: Some biological actions and receptor binding characteristics of two purified constituents of non-suppressible insulin-like activity of human serum, *Eur. J. Biochem.* **87**:285–296.

Zapf, J., Morell, B., Walter, H., Laron, Z., and Froesch, E. R., 1980, Serum levels of insulin-like growth factor (IGF) and its carrier protein in various metabolic disorders, *Acta Endocrinol. (Copenhagen)* **95**:505–517.

Zapf, J., Hauri, C., Waldvogel, M., and Froesch, E. R., 1986, Acute metabolic effects and half-lives of intravenous insulin-like growth factor I and II in normal and hypophysectomized rats, *J. Clin. Invest.* **77**:1768–1775.

Zapf, J., Hauri, C., Waldvogel, M., Futo, E., Hasler, H., Binz, K., Guler, H. P., Schmid, C., and Froesch, E. R., 1989, Recombinant insulin-like growth factor-I induces its own specific carrier protein in hypophysectomized and diabetic rats, *Proc. Natl. Acad. Sci. USA* **86**:3813–3817.

Zapf, J., Schmid, C., Guler, H. P., Waldvogel, M., Hauri, C., Futo, E., Hossenlopp, P., Binoux, M., and Froesch, E. R., 1990, Regulation of binding proteins for insulin-like growth factors (IGF) in humans: Increased expression of IGF binding protein 2 during IGF-I treatment of healthy adults and in patients with extrapancreatic tumor hypoglycemia, *J. Clin. Invest.* **86**:952–961.

Chapter 3

Potential Effects of Antibody Induction by Protein Drugs

Peter K. Working

1. INTRODUCTION

The immunogenicity of proteins has long been exploited medically, both to inactivate plant and animal toxins after accidental exposure (by passive immunization with antibodies that function as antitoxins to neutralize the biological effects of the toxin) and to provide protection from diseases (by vaccination with living attenuated or killed infectious agents to stimulate adaptive resistance). More recently, considerable research has been devoted to using synthetic or recombinant peptides and proteins, which do not pose the disease risk of attenuated living agents, in an attempt to stimulate the production of neutralizing antibody to a variety of disease vectors (Mackett and Arrand, 1985; Wroblewska *et al.*, 1985; Chanh *et al.*, 1986; Putney *et al.*, 1986; Francis *et al.*, 1987; Krohn *et al.*, 1987; Tomley *et al.*, 1987; Britt *et al.*, 1988; Goudsmit *et al.*, 1988; Michel *et al.*, 1988).

Antibody formation can be almost routinely expected in preclinical studies of human proteins in animals, but is also observed in clinical studies with protein drugs in humans. Although a desirable property of disease-causing proteins, immunogenicity can be a problem in the study of protein drugs, since they, like all proteins, can induce antibodies. The presence of

Peter K. Working • Department of Pharmacology and Toxicology, Liposome Technology, Inc., Menlo Park, California 94025.

Protein Pharmacokinetics and Metabolism, edited by Bobbe L. Ferraiolo *et al.* Plenum Press, New York, 1992.

antibodies can confound the interpretation of the results of preclinical and clinical studies by inactivating (neutralizing) the biological activity of the protein drug, by affecting its distribution, metabolism, and excretion, or, potentially, even by causing secondary adverse effects and disease. In this chapter, we will discuss some of the basics of protein immunogenicity, as well as the potential and actual effects of antibodies, both neutralizing and nonneutralizing, on preclinical and clinical studies.

2. ANTIBODIES TO RECOMBINANT PROTEINS

It was originally expected that the use of recombinant human proteins in humans would not lead to the development of antibodies because of the species specificity of the molecules. However, as early as 1981, Vallbracht *et al.* (1981) reported the presence of neutralizing antibodies in a patient who had been treated with natural interferon-β (nIFN-β). Since that study, there have been numerous other reports of antibody formation after treatment of humans with a wide variety of human proteins, including nIFN-α (Trown *et al.*, 1983), recombinant (r) IFN-α_{2a} (Quesada *et al.*, 1985; Jones and Itri, 1986), rIFN-α_{2b} (Spiegel *et al.*, 1986), rIFN-β (Hawkins *et al.*, 1985; Sarna *et al.*, 1986), recombinant insulin (Fineberg *et al.*, 1983), recombinant growth hormone (Kaplan *et al.*, 1986), and various recombinant plasma proteins (Laurian *et al.*, 1984; Nilsson *et al.*, 1990).

2.1. What Are Neutralizing Antibodies?

The distinction of an antibody as "neutralizing" is a purely empirical one, usually based on its inactivation of some biological activity of the protein as measured in a bioassay. Neutralizing antibodies may inactivate a protein by blocking its active site or by binding at a more distant site on the molecule and inactivating because of alteration of the tertiary structure of the protein, i.e., by acting through steric effects (Sell, 1987); the net result is the same: loss of biological activity.

2.2. Detection of Antibodies

Whatever the mechanism of action of neutralizing antibodies, their apparent presence or absence in preclinical and clinical studies is dependent, first and foremost, upon the sensitivity, accuracy, and type of bioassay used to detect them. In most studies of recombinant protein drugs, the bioassays employed are conducted *in vitro,* e.g., the expression of antiviral activity *in vitro,* as is used to assess the activity of the interferons (Murasko and Blankenhorn, 1984) or the stimulation of cAMP production in cultured human uterine cells *in vitro* by relaxin (Lofgren and Fei, unpublished). In the cAMP assay, antibodies are classified as neutralizing if they decrease the amount of cAMP generated by more than two standard deviations from that produced in the presence of relaxin alone, demonstrating the empirical nature of the definition.

For practical purposes, an antibody is defined as neutralizing if it interferes with the activity of the protein in the bioassay. Thus, the accuracy of the definition will depend, in part, upon whether the antibody behaves the same *in vivo* as it does *in vitro.* In other words, an antibody that neutralizes a particular activity of a protein *in vitro* may not do so in the whole animal, and, therefore, the neutralizing activity of an antibody measured *in vitro* may not be an accurate reflection of its ability to nullify the action of the drug at its effector site *in vivo.* For this reason, *in vivo* bioassays may give a more accurate measure of the true neutralizing activity of the antibody.

The sensitivity of the assay employed is paramount. Of the various types of assays utilized to measure the titer of neutralizing antibodies to rIFN-α, for example, the enzyme immunoassay (EIA) is believed to be more sensitive than the antiviral neutralization bioassay (ANB), which is believed to be more sensitive than the immunoradiometric assay (IRMA) (Koeller, 1989). Itri and co-workers (1989) reported a false-negative rate of nearly 50% when samples were screened for activity using an IRMA rather than an EIA and ANB, i.e., half of those positive in both of the two more sensitive assays were negative in the IRMA. Different versions of rIFN-α produce markedly different incidences of antibody formation (Clark and Longo, 1987; Zoon, 1987; Figlin and Itri, 1988), and differences in assay sensitivity are thought by some investigators to be responsible for most of the variations reported (Itri *et al.,* 1989). Other investigators, however, report a negligible false-negative rate with the IRMA and suggest that the differences in antibody induction may be due to the different amino acid sequences of the rIFN-α forms (see Section 3.1) or to manufacturing and formulation differences (see Section 4.5.2) (Spiegel *et al.,* 1989). Nonetheless, the differences in sensitivity of the assays suggest that direct comparison of data from studies in which the antibodies were measured by different types of assays may be risky.

2.3. Real-Life Effects of Neutralizing Antibodies

The association of naturally occurring neutralizing antibodies with human disease is well documented. Antibodies may interfere directly with biologically active molecules, but may also act indirectly by blocking or inducing changes in cell surface receptors. In human disease, examples are seen of each type of interference. Diseases such as diabetes, aplastic and pernicious anemia, and myasthenia gravis all result from immune neutralization or inactivation of active proteins, specific receptors, or specific cell types. Specific antigen-associated human diseases are summarized in Table I.

3. FACTORS CONTRIBUTING TO IMMUNOGENICITY

A complete antigen is one that can both induce an immune response and react with the products of that response. Drugs that are complete antigens have the highest potential of producing antibodies, but even small molecules that are incomplete antigens (haptens), if combined with larger molecules (carriers), can be immunogenic. Complete antigens usually have a high molecular weight, although naturally occurring proteins of fairly low molecular weight (e.g., insulin, ribonuclease, and angiotensin) can also function as complete antigens.

The properties that cause a protein drug to be immunogenic are those that cause any protein to be immunogenic (e.g., size, primary and tertiary structure, and structural complexity). Antibodies are most likely to be induced when the protein is foreign to the host, as when porcine or ovine insulin is used in human diabetics, when mouse-derived monoclonal antibodies are administered to humans, or when human proteins are tested for safety in nonhuman species.

Table I
Antigens in Immune Neutralization Diseases[a]

Disease	Antigen type
Diabetes mellitus	Intracellular (islet cell)
	Receptors (insulin)
	Cell membrane components (islet cell)
	Hormones (insulin)
Hemolytic anemia	Cell membrane components (RBC)
Aplastic anemia	Hormone (erythropoietin)
Pernicious anemia	Intracellular (parietal cells)
Thyroid disease	Receptors (TSH)
	Hormone (triiodothyronine)
Myasthenia gravis	Receptors (acetylcholine)
Rheumatoid arthritis	Cell membrane components
Systemic lupus	Intracellular
erythematosus	Cell membrane components
	Plasma proteins

[a] Adapted from Sell (1987) and Golub (1987).

3.1. Primary Structure

Of particular importance in studies of human proteins in animals is the amino acid sequence (primary structure) of the protein, in large part because the degree of sequence homology of the human protein to the native protein of the test species relates directly to the immunogenicity of the human protein in the test species. In other words, the more alike the sequences of human and animal protein, the less intense is the immune response, and the less alike (the more evolutionary distance between human and test species), the greater the response. This was evident in studies that compared the relative immunogenicity of natural human and bovine growth hormone (hGH and bGH) in hypophysectomized rats (Groesbeck and Parlow, 1987). Based on an assessment of the binding of radioiodinated hGH or bGH by sera from rats exposed by intraperitoneal injection, hGH induced antibody titer levels nearly fivefold higher than did bGH. Similar species-specific immune responses are seen in humans. Diabetic humans treated with bovine or ovine insulin may develop resistance to insulin derived from one species, but not to the other, because of species-specific antibodies developed toward that insulin.

Even single amino acid differences in the primary sequence may affect the relative immunogenicity of the protein drugs in humans. rIFN-α_{2a} and rIFN-α_{2b} are distinct gene products, differing from one another only at amino acid residue 23, where rIFN-α_{2a} has an arginine rather than a lysine. In clinical studies in which humans received rIFN-α_{2a}, an approximately 25% overall incidence of neutralizing antibody formation was reported (Itri et al., 1987). In contrast, a much lower incidence of antibody formation (<3%) has been reported in patients treated with rIFN-α_{2b} (Spiegel et al., 1986). Recently, however, higher frequencies of rIFN-α_{2b} neutralizing antibodies (ca. 15–20%) have been measured in patients with chronic myelogenous leukemia (Von Wussow et al., 1987; Freund et al., 1989) and carcinoid tumors (Oberg and Alm, 1989), clouding the issue somewhat. One must also consider that differences in study design may have contributed to some of the observed differences in apparent immunogenicity of the IFNs (see Section 4).

3.2. Tertiary Structure

The tertiary structure, or spatial folding, of the protein can also affect its relative immunogenicity. Antibodies produced to the α chain of insulin do not recognize intact insulin, which is a dimer of α and β chains, and some, but not all, antisera to native proteins may not cross-react with the same protein if it is denatured and subsequently refolded, differing only in tertiary structure (Sell, 1987). Other antisera may still be reactive with the refolded protein, and, perhaps more importantly from the standpoint of commercially produced proteins, antibodies that are generated to misfolded proteins produced in the manufacturing or formulation process, often cross-react with the native version (see Section 4.5.2).

3.3. Structural Complexity

More structurally complex proteins are generally better antigens. Proteins may be purposely made more complex to enhance their immunogenicity, as when bovine serum albumin is complexed with p-azobenzoate, stimulating antibodies to both p-azobenzoate and albumin, or when poly-L-lysine is conjugated with dinitrophenol. In real world terms for recombinant proteins, this means, for example, that the degree and type of posttranslational modifications of the protein (e.g., glycosylation) will directly affect its relative immunogenicity.

The source of the recombinant protein can affect the degree and type of posttranslational modifications that it undergoes. Plasma from mice, rabbits, newborn calves, and normal humans contains high titers of IgG-mediated endogenous neutralizing activity directed against high-mannose glycosylated recombinant murine rIFN-β derived from yeast (Sedmak and Grossberg, 1989). No neutralizing activity was detected against nonglycosylated murine rIFN-β made in *E. coli,* suggesting that the high-mannose carbohydrate moieties of the yeast-derived rIFN were responsible for the immune response and the neutralization observed. Thus, a protein produced by recombinant methods in *E. coli,* which will not be glycosylated, may induce a qualitatively and quantitatively different immune response than the same protein produced in a yeast or mammalian cell line, which likely will be glycosylated. Further, proteins produced in different mammalian cell lines, which may be glycosylated differently, could have different relative immunogenicities in test animals.

4. EFFECTS OF STUDY DESIGN

An assortment of study design issues can affect the apparent immunogenicity of a protein drug, including the route of administration, the dose administered, the duration of dosing, the strain of animal used, and, in human studies, the underlying disease state.

4.1. Route of Administration

Potential routes of administration include intradermal (ID), subcutaneous (SC), intramuscular (IM), intraperitoneal (IP), intravenous (IV), intratracheal (IT), and others, but for practical purposes protein drugs are given most often by the IV, SC, or IM routes, with the latter two routes more likely to produce an immune response.

Route effects were evident in a Phase I clinical trial of human interleukin 2 (IL-2), in which the cytokine was administered to patients by either the SC or IV route (Krigel *et al.*, 1988). Antibody titers to IL-2 were detected in patients who were treated by both routes, but the highest titers were measured in patients who received SC injections. In this study, due both to the occurrence of neutralizing antibodies in the patients administered IL-2 SC and to the modest immunogenicity of IL-2 administered IV, the route of administration of the drug was changed to IV-only for the remainder of the Phase I clinical trial. IV treatment of patients with rIFN-β led to antibody formation in only 2 of 36. In contrast, 20 of 25 who received IM administrations at the same dose became antibody-positive (Konrad *et al.*, 1987). Similarly, only 3 of 335 patients (0.9%) receiving IV rIFN-β_{ser} became antibody-positive, as compared to 13 of 136 (9.6%) who were treated subcutaneously (Larocca *et al.*, 1989).

The greater antibody response observed when proteins are administered by SC or IM routes in these studies is likely attributable to the aggregation of the drug after injection. Precipitation of antigens is known to occur after SC and IM administration, and aggregated proteins generally induce a more vigorous antibody response than do soluble ones (Sell, 1987).

4.2. Dosing Considerations

The dose administered, the frequency of dosing, and the duration of administration are also important. Dosing at higher levels, more frequently, or for a longer duration increases the cumulative dose, increases the probability of an immunogenic response, and raises the titer of antibody produced.

4.2.1. Duration of Dosing

Groups of pregnant female rhesus monkeys received daily IV infusions of synthetic human relaxin (hRlx-2) for 2 to 14 days prior to parturition (Working *et al.*, 1991). Antibody titers were measured at 3, 6, and 12 weeks after the final treatment. Three of six females that received eight or more doses of 0.1 mg/kg hRlx-2 produced significant titers of antibodies, but none were neutralizing. In the high-dose group, which received multiple doses of 2.0 mg/kg hRlx-2, all females that received eight or more treatments produced antibodies. Antibody titers were higher, and, moreover, three of the four antibody-positive animals produced neutralizing antibodies. Qualitatively, then, in this study it appeared that the higher the dose and the longer the treatment, the greater the likelihood of generating neutralizing antibodies.

In the patients receiving IL-2, there was a correlation between the number of days of dosing and the antibody titer in SC-dosed, but not IV-dosed, patients (Krigel *et al.*, 1988). In a study of rIFN-α_{2a}, the duration of treatment was significantly longer for antibody-positive patients than for antibody-negative patients (Itri *et al.*, 1987). The mean duration of treatment was 301 days for patients receiving rIFN-β_{ser} that produced neutralizing antibodies, as compared to a mean of 121 days for the negative patients (Larocca *et al.*, 1989). Clearly, prolonged treatment raises the probability of antibody development.

4.2.2. Cumulative Dose

Cumulative dose also plays a role, even when the duration of dosing is similar. Patients with hairy cell leukemia who were administered rIFN-α_{2a} at a daily dose of 3×10^6 U for 6 months received over twice the dose of those who received 3.4×10^6 U three times weekly. The cumulative dose was considerably higher for the patients treated daily, and they exhibited a higher incidence of antibody formation as a result (Steis *et al.*, 1988; Golomb *et al.*, 1988). As a consequence of this phenomenon, it is sometimes observed that the length of patient survival in clinical studies is actually longer in neutralizing antibody-positive patients (Quesada *et al.*, 1985; Itri *et al.*, 1987) than in antibody-negative patients. Rather than suggesting that the antibodies have a positive therapeutic effect, these observations simply demonstrate that the likelihood of antibody formation is increased with longer treatment with the protein and the higher cumulative dose that results from the longer survival time.

4.3. Genetic Considerations

In preclinical studies, even the strain of animal used may affect the immunological outcome, presumably due to the differing expression of major histocompatibility complexes among strains. To demonstrate this, strains of rats bearing one of five different haplotypes of the major histocompatibility locus RT-1 were tested for their ability to produce neutralizing antibodies to mouse interferon (Murasko and Blankenhorn, 1984). Rats representing three of the haplotypes were responders (i.e., produced antibodies), but the other two were nonresponders. The authors were unable to explain the genetic mechanism for this difference, but the study demonstrates the highly specific nature of the immune response and suggests that preclinical studies in different strains may not be directly comparable.

4.4. The Effect of Disease on Antibody Induction

Comparison of the frequency of antibody occurrence in different patient populations must be made with care. It has been noted by several clinical investigators, particularly those working with the IFNs, that the underlying disease can affect the immunologic outcome of treatment with recombinant human proteins. Patients with different diseases often exhibit different incidences of antibody formation after treatment, and even undergo different durations of treatment before antibodies are formed. Patients with renal cell carcinoma, for example, have a median time to appearance of antibodies of about 2 months (range 1–15 months), whereas patients with hairy cell leukemia have a median time of nearly 1 year (range 4–30 months) after treatment with rIFN-α_{2a} (Itri *et al.,* 1989).

The treatment of different diseases, of course, requires different durations of treatment and different dosages, thereby resulting in different cumulative doses, which will affect the incidence of antibody formation (see Section 4.2). However, some diseases, such as B-cell myelomas, may actually disrupt immune function and thereby directly hinder the production of antibodies (Itri *et al.,* 1989).

4.5. Miscellaneous Effects

4.5.1. Sample Collection Effects

The timing of collection of samples, relative both to administration of the drug and to the cessation of treatment, is critical to the accurate detection of antibodies. Measurable antibody levels fall precipitously 1 to 6 hr after injection of IFN because of interference of IFN with the assays. Preinjection levels are seen again only after 48 hr without further treatment (Von Wussow *et al.*, 1987, 1989), so it is important to choose a standardized time after treatment for collection of samples to ensure accurate measurement of antibody titers.

Similar care should be taken when serum is collected for antibody analysis after cessation of treatment. After treatment with rIFN-α, antibody titers continued to rise after treatment ended, reached a maximum at 3 weeks, and declined steadily thereafter over the subsequent months (Von Wussow *et al.*, 1989). The continuing presence of antigen is necessary to maintain antibody titers, and comparisons of different studies must take this into account.

4.5.2. Formulation Effects

Formulation and manufacturing procedures can also affect the induction of antibodies by protein drugs. Just as routes of administration that lead to aggregation can increase the relative antibody response (see Section 4.1), preparation procedures that lead to aggregation may also increase the immunogenic response. Spiegel *et al.* (1989) attribute the differences in relative immunogenicity of rIFN-α_{2a} and rIFN-α_{2b} seen in clinical studies to differences in preparation procedures. rIFN-α_{2b} is purified from a soluble nonaggregated protein fraction using no denaturing processes (Nagabhushan *et al.*, 1984), whereas the manufacture of rIFN-α_{2a} utilizes more rigorous conditions that could result in some aggregation (Sidney *et al.*, 1984). Significant antibody responses were reported in recipients of antithymocyte globulin (ATG) made in horses or rabbits, which is typically aggregated (Martin *et al.*, 1976; Satoh *et al.*, 1979), but more modest responses were seen in recipients of more recent preparations of ATG, which is deaggregated and administered IV (Hoitsma *et al.*, 1982).

5. EFFECTS OF ANTIBODIES ON STUDIES
WITH PROTEIN DRUGS

Antibodies to therapeutically administered proteins, if neutralizing, can directly nullify the pharmacologic action of the drug. Since many toxicities associated with recombinant proteins are expressions of the action of the drug at suprapharmacologic concentrations, the production of neutralizing antibodies during a safety study could lessen this type of test article-related toxicity and lead to an erroneous conclusion of "no effect". The binding of drugs to antibodies can also affect their catabolism, alter their distribution in the body, and potentially affect their clearance from the circulation and, thus, their efficacy.

5.1. Direct Nullification of Drug Action

The induction of neutralizing antibodies does not necessarily lead to neutralization of the activity of a protein drug, but the possibility exists. Evidence for direct neutralization is limited in preclinical studies and somewhat controversial in clinical studies.

5.1.1. Effects in Preclinical Studies

Direct nullification of biological activity has been seen in studies of recombinant human growth hormone (hGH) in rats when neutralizing antibodies are induced (Moore and Fong, unpublished). Hypophysectomized rats received daily IP injections of recombinant hGH or saline only for 15 days. A nonhypophysectomized rat served as a control. Two rats given hGH maintained a normal rate of body weight gain as compared to the control until day 11, when the growth rate of one of them decreased dramatically. Titers of antibodies were measured in all rats on day 16 by indirect ELISA. Both hypophysectomized rats that received hGH developed antibodies to the protein, but titers in the rat that had stopped growing were markedly higher than in the rat that continued to grow. These data demonstrate that not only the presence, but also the activity of neutralizing antibodies induced by protein drugs can affect the biological activity of the protein.

5.1.2. Effects in Clinical Studies

Clinically, the appearance of neutralizing antibodies in patients receiving rIFN-α or rIFN-β has been associated with the cessation or lessening of drug-related side effects and, in some cases, the reoccurrence of disease in antibody-positive patients. Quesada *et al.* (1985) reported that 12 of 41 patients being treated for renal cell carcinoma with rIFN-α_{2a} achieved partial or complete remission of tumor. Seven of the twelve (58%) developed neutralizing antibodies, which coincided with an improvement in drug-induced side effects, such as fatigue and anorexia, as well as a normalization of blood counts and liver function tests. Median survival rate in the antibody-positive patients was 2 months, compared to 10 months in the antibody-negative patients, leading the authors to suggest that the antibodies may also have abrogated the biologic (antitumor) effects of rIFN-α_{2a} *in vivo*.

In a small study, patients with carcinoid tumors were treated with rIFN-α_{2b} for six months. Eleven of twenty showed objective tumor response to treatment and six more had stable disease with no progression (Oberg *et al.*,

1989). Three patients developed neutralizing antibody to rIFN-α_{2b}. One had progressive disease from the start of therapy, whereas the other two had initially exhibited an objective response, which lasted only until antibodies developed. In one of the latter, a change to natural human leukocyte IFN was followed by the disappearance of neutralizing rIFN-α_{2b} antibodies and a second objective response. Natural human leukocyte IFN contains at least 12 subtypes of IFN-α, and a specific antibody to rIFN-α_{2b} is not likely to neutralize it completely, accounting for the activity of the natural preparation. This has important clinical implications, since patients who have developed neutralizing antibodies to rIFN-α may be switched to natural IFN-α (nIFN-α) to circumvent the antibodies. The response of such patients to nIFN-α, as in the study described above, is also considered good proof that neutralizing antibodies can nullify the biological effectiveness of recombinant molecules (Von Wussow et al., 1989).

Other studies suggest that disease relapse may not be related to the appearance of neutralizing antibodies. Itri and co-workers (1987) reported on the treatment of over 1600 cancer patients with rIFN-α_{2a}. Partial or complete remission occurred in 28% of the antibody-positive patients and 24% of the antibody-negative patients, and the authors concluded that the presence of neutralizing antibodies did not affect the rate of remission. In a later study, it was reported that there were no significant differences between antibody-positive patients and antibody-negative patients in terms of time required for response to treatment to occur, i.e., the time to partial or complete remission, or in the duration of the response, i.e., the time to relapse (Itri et al., 1989). Both these studies included patients with a variety of different cancers and, therefore, who were subjected to a variety of different treatment regimens with rIFN-α_{2a}, which may complicate interpretation (see Section 4.4). However, in a study confined to assessing the antitumor activity of rIFN-α_{2a} in patients with renal cell carcinoma, there was also no difference in the incidence of neutralizing antibody formation in responders and nonresponders, nor did the duration of response differ between antibody-positive and antibody-negative patients (Figlin et al., 1988).

5.2. Interference with Pharmacokinetics

Antibodies may indirectly affect the efficacy of a protein drug by altering its pharmacokinetic profile. Clearance rates may be either increased or decreased by antibody formation and binding. Either outcome can have significant effects on drug efficacy and safety.

5.2.1. Increased Clearance of Antigen–Antibody Complexes

Protein drugs may be effectively inactivated by binding to *any* antibody, neutralizing or not, if the antigen–antibody complex that results is cleared or catabolized more rapidly than the unbound protein itself. Antigen-antibody complexes in excess may be cleared rapidly by the reticuloendothelial system because of the formation of aggregates, and a nonneutralizing antibody can effectively reduce the activity of a protein by this mechanism (Sell, 1987).

An increased rate of plasma clearance after the formation of antibodies was observed in a study in which murine monoclonal antibody 4D5 was administered IV or IP to female cynomolgous monkeys (Maneval and Baughman, unpublished). After a single IV injection, plasma concentrations of 4D5 declined slowly over the course of the 3-week study. Between 12 and 17 days after treatment, plasma concentrations in nine of ten animals in the mid- and high-dose groups fell precipitously within a 48-hr period. This precipitous drop coincided with the appearance in the circulation of monkey anti-mouse antibody specific for 4D5. In the low-dose group, only one of three animals developed antibodies to 4D5 (on day 18); concurrently, plasma levels of the monoclonal antibody also began to drop in this animal.

In a study of the immunogenicity of rIFN-α_{2a} in cancer patients, 3 of 37 patients developed antibodies while on treatment. The appearance of the antibodies coincided with striking decreases in peak serum concentrations of the IFN in 2 of the patients (Quesada and Gutterman, 1983). Patients given mouse monoclonal antibodies to tumor-associated antigens often produce anti-mouse antibodies (Van Kroonenburgh and Pauwels, 1988). The production of these antibodies may be a limitation to repeated administration of monoclonal antibodies, since the immune complexes formed are rapidly cleared to the liver and spleen (Larson *et al.,* 1983; Pimm *et al.,* 1985). Hepatic clearance was increased once the monoclonal antibody was bound to antibody and altered the distribution of the labeled monoclonal antibody. Interestingly, antibodys to the murine monoclonal antibody also prevented the localization of the monoclonal antibody in tumors in nude mice, suggesting that monoclonal antibodies that evoke antibodies may not be effective in tumor targeting (Pimm and Baldwin, 1990).

5.2.2. Decreased Clearance of Antigen–Antibody Complexes

Not all immune complexes are cleared more rapidly. Serum half-life of a protein may be prolonged if the clearance of the immune complex is decreased relative to that of the unbound drug. If binding to antibody does not negate the biological activity of the protein, the duration of drug action may in fact be extended, as demonstrated when human leukocyte IFN-α, either bound or unbound to an anti-IFN-α monoclonal antibody, is administered to rats. Plasma clearance of the IFN-α activity is 15 times slower for the bound drug than for the unbound drug (Rosenblum *et al.*, 1985). Moreover, the biological activity of the interferon was unaltered when bound to the antibody, suggesting that its therapeutic activity could be prolonged by such treatment.

Molecular biologists have taken advantage of such effects by purposely altering recombinant proteins with antibody fragments to increase their serum half-life. Soluble recombinant human CD4, a T-cell cell surface receptor for HIV-1 that has been tested as a means of blocking HIV infectivity, was combined with a portion of a human IgG molecule, with the aim of both increasing its half-life *in vivo* and incorporating certain other functions of antibody molecules (e.g., Fc receptor binding, placental transfer) that might increase the value of the drug as a potential barrier to HIV infection (Capon *et al.*, 1989). By lessening the rate of rCD4 clearance, and thereby increasing its effective serum concentration, the efficacy of the molecule should be increased, if its HIV-1 binding activity is unaltered. After such molecular modifications, these investigators found that the plasma half-life of one CD4-antibody hybrid in rabbits was nearly 200 times longer than that of soluble rCD4 and was, in fact, comparable to that of human IgG in rabbits. The hybrid rCD4 molecule retained the ability to block HIV-1 infection of T cells and monocytes *in vitro*.

Similar effects are seen when antibodies are accidentally induced during the course of a study. Recombinant human tissue plasminogen activator (t-PA) was administered to dogs as a continuous daily IV infusion for 14 days. Plasma concentrations of t-PA were determined daily, as were levels of antibody to t-PA. Steady-state plasma t-PA concentrations measured by ELISA remained approximately constant for the first 7 days of the study, but began to rise by day 10 of treatment. Concomitantly, the titer of anti-t-PA antibodies increased, suggesting that antibody binding of the protein had markedly decreased its clearance. Whether these were neutralizing antibodies is not clear, since a t-PA activity assay was not performed on the samples. It is also possible that the antibodies generated interfered with the ELISA used to analyze t-PA concentrations, i.e., that the increased plasma concentration was artifactual.

It is important to note that even when the biological activity of a drug is neutralized, its serum concentration will still be increased if binding to an antibody slows its rate of clearance. Thus, biological function is lost and serum concentration is raised, but effectiveness is not. The unwary pharmacologist may be misled by relying upon assays that do not measure biological activity to quantitate drug concentrations in the serum.

5.3. Immune Complex Reactions

Immune complex reactions are another possible complication when antibodies to an administered protein drug are produced in excess in test animals or humans. These reactions are caused by antibody reacting directly with tissue antigens or, of potentially more significance to our subject, by the reaction of soluble antigen with antibody in the blood to form antigen–antibody complexes that subsequently are deposited in various tissues, particularly arterial walls, small vessels, and the kidney glomerulus (see review in Sell, 1987). Subsequent complement-mediated inflammation may cause significant damage at these sites, leading to immune complex diseases such as serum sickness, glomerulonephritis, and Arthus reaction.

It is unclear whether immune complex-related diseases are a likely consequence of treatment with protein drugs. Oberg *et al.* (1989) reported that there was no relationship between development of IFN antibodies and development of autoimmune phenomena in patients being treated with rIFN-α_{2b}. Similarly, there are no reports of adverse sequelae associated with the formation of antigen–antibody complexes after treatment of humans with rIFN-α_{2a} (Quesada *et al.,* 1985; Figlin *et al.,* 1988). There are also as yet no reports of immune-mediated diseases or damage in preclinical studies of recombinant proteins, and the subject will not be further discussed here.

6. OVERVIEW

Proteins used as drugs are, like all proteins, capable of inducing antibodies. The antibodies may nullify the biological activity of the drug and may or may not affect the outcome of preclinical and clinical studies. Whether an antibody is neutralizing or not is determined empirically, usually by its ability to interfere with the outcome of an *in vitro* bioassay. The sensitivity of the assay employed to measure antibodies is of foremost importance. False-negative rates vary among assay types, and, for this reason, the comparison of studies in which antibodies are detected with different types of assays must be done with caution.

Both the characteristics of the protein and of the study design affect the induction of antibodies. The molecular weight, structural complexity, and primary and tertiary structure of a protein will determine its relative immunogenicity in a test system, and recombinant proteins produced in different cell types may exhibit differential immunogenicity in animals and humans. The route and duration of administration, the cumulative dose of drug administered, and, in clinical studies, the underlying disease of the patient population being studied will also affect antibody induction.

Neutralizing antibodies may directly interfere with the action of protein drugs, as has been seen in clinical studies of the IFNs. In fact, the appearance of antibodies may be the limiting factor in the clinical usefulness of many proteins, particularly murine monoclonal antibodies. The appearance of neutralizing antibodies is associated with the cessation or lessening of drug-related side effects, and, in some cases, the reappearance of disease in antibody-positive patients.

In addition to direct neutralization of activity, the disposition of a protein drug may be dramatically altered once antibodies are formed. Clearance from the circulation may be increased, particularly if the antigen–antibody complexes form aggregates, or decreased, leading to a potential prolongation of activity if the antibodies are not neutralizing. In either case, the resultant effects on plasma drug concentrations will often seriously limit the correlation of plasma concentrations with toxicity or efficacy measurements.

In summary, the induction of antibodies may markedly affect the outcome of preclinical and clinical studies of recombinant proteins. The antibody status of the animals or humans in these studies must be carefully monitored to assess fully the significance of the study results. Careful attention to details of study design, including route and duration of exposure to the drug, the source and formulation of the test protein, the type of assay for antibodies utilized, and potential effects of the antibody on pharmacokinetic parameters, is also critical to accomplish this goal.

REFERENCES

Britt, W. J., Vugler, L., and Stephens, E. B., 1988, Induction of complement-dependent and -independent neutralizing antibodies by recombinant-derived human cytomegalovirus gp55-116 (gB), *J. Virol.* **62:**3309–3318.

Capon, D. J., Chamow, S. M., Mordenti, J., Marsters, S. A., Gregory, T., Mitsuya, H., Byrn, R. A., Lucas, C., Wurm, F. M., Groopman, J. E., Broder, S., and Smith, D. H., 1989, Designing CD4 immunadhesins for AIDS therapy, *Nature* **337:**525–531.

Chanh, T. C., Dreesman, G. R., Kanda, P., Linette, G. P., Sparrow, J. R., Ho, D. D., and Kennedy, R. C., 1986, Induction of anti-HIV neutralizing antibodies by synthetic peptides, *EMBO J.* **5:**3065–3071.

Clark, J. W., and Longo, D. L., 1987, Interferons in cancer therapy, *Cancer Princ. Pract. Oncol. Update* **4**:1–16.

Figlin, R. A., and Itri, L. M., 1988, Anti-interferon antibodies: A perspective, *Semin. Hematol.* **25**(Suppl. 3):9–15.

Figlin, R. A., deKernion, J. B., Mukamel, E., Palleroni, A. V., Itri, L. M., and Sarna, G. P., 1988, Recombinant interferon alfa-2a in metastatic renal cell carcinoma: Assessment of antitumor activity and anti-interferon antibody formation, *J. Clin. Oncol.* **6**:1604–1610.

Fineberg, S., Galloway, J., Fineberg, N., Rathbun, M., and Hufferd, S., 1983, Immunogenicity of recombinant DNA human insulin, *Diabetologia* **25**:465–469.

Francis, M. J., Hastings, G. Z., Sangar, D. V., Clark, R. P., Syred, A., Clarke, B. E., Rowlands, D. J., and Brown, F., 1987, A synthetic peptide which elicits neutralizing antibody against human rhinovirus type 2, *J. Gen. Virol.* **68**:2687–2691.

Freund, M., Von Wussow, P., Diedrich, H., Eisert, R., Link, H., Wilke, H., Buchholz, F., LeBlanc, S., Fonatsch, C., Deicher, H., and Poliwoda, H., 1989, Recombinant human interferon (IFN) alpha-2b in chronic myelogenous leukaemia: Dose dependency of response and frequency of neutralizing anti-interferon antibodies, *Br. J. Haematol.* **72**:350–356.

Golomb, H., Ratain, M., Fefer, A., Thomason, J., Golde, D. W., Ozer, H., Portlock, C., Silber, R., Rappeport, J., and Bonnem, E., 1988, Randomized study of the duration of treatment with interferon alpha-2b in patients with hairy cell leukemia, *J. Natl. Cancer Inst.* **80**:369–373.

Golub, E. S., 1987, *Immunology: A Synthesis,* Sinauer Associates, Sunderland, Mass.

Goudsmit, J., Debouck, C., Meloen, R. H., Smit, L., Bakker, M., Asher, D. M., Wolff, A. V., Gibbs, C. J., Jr., and Gajdusek, D. C., 1988, Human immunodeficiency virus type 1 neutralization epitope with conserved architecture elicits early type-specific antibodies in experimentally infected chimpanzees, *Proc. Natl. Acad. Sci. USA* **85**:4478–4482.

Groesbeck, M. D., and Parlow, A. F., 1987, Highly improved precision of the hypophysectomized female rat body weight gain bioassay for growth hormone by increased frequency of injections, avoidance of antibody formation, and other simple modifications, *Endocrinology* **120**:2582–2590.

Hawkins, M., Horning, S., Konrand, M., Anderson, S., Sielaff, K., Rosno, S., Schiesel, J., Davis, T., DeMets, D., and Merigan, T., 1985, Phase I evaluation of a synthetic mutant of β-interferon, *Cancer Res.* **45**:5914–5920.

Hoitsma, A. J., Reekers, P., Kreeftenberg, J. G., VanLier, H. J. J., Capel, P. J. A., and Koene, R. A. P., 1982, Treatment of acute rejection of cadaveric renal allografts with rabbit antilymphocyte globulin, *Transplantation* **33**:12–16.

Itri, L. M., Campion, M., Dennin, R. A., Palleroni, A. V., Gutterman, J. U., Groopman, J. E., and Trown, P. W., 1987, Incidence and clinical significance of neutralizing antibodies in patients receiving recombinant interferon-alfa-2a by intramuscular injection, *Cancer* **59**:668–674.

Itri, L. M., Sherman, M. I., Palleroni, A. V., Evans, L. M., Tran, L.-L., Campion, M., and Chizzonite, R., 1989, Incidence and clinical significance of neutralizing antibodies in patients receiving recombinant interferon-alpha-2. *J. Interferon Res.* **9**(Suppl. 1):S9–S15.

Jones, G. J., and Itri, L. M., 1986, Safety and tolerance of recombinant interferon-alfa-2A (Roferon-A) in cancer patients, *Cancer* **57**:1709–1715.

Kaplan, S. L., August, G. P., Blethen, S. L., Brown, D. R., Hintz, R. L., Johansen, A., Plotnick, L. P., Underwood, L. E., Bell, J. J., Blizzard, R. M., Foley, T. P., Hopwood, N. J., Kirkland, R. T., Rosenfeld, R. G., and Van Wyck, J. J., 1986, Clinical studies with recombinant-DNA-derived methionyl human growth hormone in growth hormone deficient children, *Lancet* **1**:697–700.

Koeller, J. M., 1989, Biologic response modifiers: The interferon alfa experience, *Am. J. Hosp. Pharm.* **46**(Suppl. 2):S11–S15.

Konrad, M., Childs, A., Merigan, T., and Bordon, E., 1987, Assessment of the antigenic response in humans to a recombinant mutant interferon beta, *J. Clin. Immunol.* **7**:365–375.

Krigel, R. L., Padavic-Shaller, K. A., Rudolph, A. R., Litwin, S., Konrad, M., Bradley, E. C., and Comis, R. L., 1988, A Phase I study of recombinant interleukin 2 plus recombinant β-interferon, *Cancer Res.* **48**:3875–3881.

Krohn, K., Robey, W. G., Putney, S., Arthur, L., Nara, P., Fischinger, P., Gallo, R. C., Wong-Staal, F., and Ranki, A., 1987, Specific cellular immune response and neutralizing antibodies in goats immunized with native or recombinant envelope proteins derived from human T-lymphocyte virus type III$_B$ and in human immunodeficiency virus-infected men, *Proc. Natl. Acad. Sci. USA* **84**:4994–4998.

Larocca, A. P., Leung, S. C., Marcus, S. G., Colby, C. B., and Borden, E. C., 1989, Evaluation of neutralizing antibodies in patients treated with recombinant interferon-β$_{ser}$, *J. Interferon Res.* **9**(Suppl. 1):S51–S60.

Larson, S. M., Brown, J. P., Wright, P. W., Carrasquillo, J. A., Hellstrom, I., and Hellstrom, K. E., 1983, Imaging of melanoma with I-131-labeled monoclonal antibodies, *J. Nucl. Med.* **24**:123–129.

Laurian, Y., Girma, J. P., Lambert, T., Meyer, D., and Larrieu, M. J., 1984, Incidence of immune responses following 102 infusions of autoplex in 18 hemophilic patients with antibody to factor VIII, *Blood* **63**:457–462.

Mackett, M., and Arrand, J. R., 1985, Recombinant vaccinia virus induces neutralising antibodies in rabbits against Epstein–Barr virus membrane antigen gp340, *EMBO J.* **4**:3229–3234.

Martin, W. J., II, Blumenthal, M., Roitman, M. S., Condie, R., Simmons, R., and Najarian, J., 1976, Antilymphoblast globulin in renal transplant patients: No allergic reactions, *J. Am. Med. Assoc.* **236**:1729–1735.

Michel, M.-L., Mancini, M., Sobczak, E., Favier, V., Guetard, D., Bahraqui, E. M., and Tiollais, P., 1988, Induction of anti-human immunodeficiency virus (HIV) neutralizing antibodies in rabbits immunized with recombinant HIV–hepatitis B surface antigen proteins, *Proc. Natl. Acad. Sci. USA* **85**:7957–7961.

Murasko, D. M., and Blankenhorn, E. P., 1984, Genetic control of neutralizing antibody response to mouse interferon in rats, *J. Interferon Res.* **4**:435–440.

Nagabhushan, T. L., Surprenant, H., Le, H. V., Kosecki, R., Levine, A., Reichert, P., Sharma, B., Tsai, H., Trotta, P., Bausch, J., Foster, J., Gruber, S., Hoogerheide, J., and Mecorelli, S., 1984, Characterization of genetically engineered α_2 interferon, in: *Interferons: Research. Clinical Application and Regulatory Consideration* (K. C. Zoon, P. D. I. Noguchi, and T.-Y. Liu, eds.), Elsevier, Amsterdam, pp. 79–88.

Nilsson, I. M., Berntorp, E., Zettervall, O., and Dahlback, B., 1990, Noncoagulation inhibitory factor VIII antibodies after induction of tolerance to factor VIII in hemophilia A patients, *Blood* **75**:378–383.

Oberg, K., and Alm, G. V., 1989, Development of neutralizing interferon antibodies after treatment with recombinant interferon-α_{2b} in patients with malignant carcinoid tumors, *J. Interferon Res.* **9**(Suppl. 1):S45–S49.

Oberg, K., Alm, G., Magnusson, A., Lundqvist, G., Theodorsson, E., Wide, L., and Wilander, E., 1989, Treatment of malignant carcinoid tumors with recombinant interferon alfa-2b: Development of neutralizing interferon antibodies and possible loss of tumor activity, *J. Natl. Cancer Inst.* **81**:531–535.

Pimm, M. V., and Baldwin, R. W., 1990, Syngeneic anti-idiotypic antibody prevents localization of a murine monoclonal antibody in human tumour xenografts, *Eur. J. Cancer* **26**:567–568.

Pimm, M. V., Perkins, A. C., Armitage, N. C., and Baldwin, R. W., 1985, The characteristics of blood-borne radiolabels and the effect of anti-mouse IgG antibodies on the localization of radiolabeled monoclonal antibody in cancer patients, *J. Nucl. Med.* **26**:1011–1023.

Putney, S. D., Matthews, T. J., Robey, W. G., Lynn, D. L., Robert-Guroff, M., Mueller, W. T., Langlois, A. J., Ghrayeb, J., Petteway, S. R., Jr., Weinhold, K. J., Fischinger, P. J., Wong-Staal, F., Gallo, R. C., and Bolognesi, D. P., 1986, HTLV-III/LAV-neutralizing antibodies to an E. coli-produced fragment of the virus envelope, *Science* **234**:1392–1395.

Quesada, J. R., and Gutterman, J. U., 1983, Clinical study of recombinant DNA-produced leukocyte interferon (clone A) in an intermittent schedule in cancer patients, *J. Natl. Cancer Inst.* **70**:1041–1046.

Quesada, J. R., Rios, A., Swanson, D., Trown, P., and Gutterman, J. U., 1985, Antitumor activity of recombinant-derived interferon alpha in metastatic renal carcinoma, *J. Clin. Oncol.* **3**:1522–1528.

Rosenblum, M. G., Unger, B. W., Gutterman, J. U., Hersh, E. M., David, G. S., and Fincke, J. M., 1985, Modification of human leucocyte interferon pharmacology with monoclonal antibody, *Cancer Res.* **45**:2421–2424.

Sarna, G., Pertcheck, M., Figlin, R., and Ardalan, B., 1986, Phase I study of recombinant β ser 17 interferon in the treatment of cancer, *Cancer Treat. Rep.* **70**:1365–1372.

Satoh, P., Elberg, A., Davis, J., Uittenbogaart, C., Fine, R., and Hardy, M. A., 1979, Effects of different ATG (ATGAM) regimens on the levels of rosette-forming cells (RFC), serum horse IgG (Hol), and anti-horse antibodies in kidney transplant recipients, *Transplant. Proc.* **11**:1427–1439.

Sedmark, J. J., and Grossberg, S. E., 1989, High levels of circulating neutralizing antibody in normal animals to recombinant interferon-β_{ser}, *J. Interferon Res.* **9**(Suppl. 1):S61–S65.

Sell, S., 1987, *Immunology, Immunopathology and Immunity,* 4th ed., Elsevier, Amsterdam.

Sidney, P., Burde, K., and Ortaldo, J., 1984, The human interferons and their biologic activities, in: *Interferons: Research. Clinical Application and Regulatory Consideration* (K. C. Zoon, P. D. I. Noguchi, and T.-Y. Liu, eds.), Elsevier, Amsterdam, pp. 59–77.

Spiegel, R. J., Spicehandler, J. R., Jacobs, S. L. and Oden, E. M., 1986, Low incidence of neutralizing factors in patients receiving recombinant-alfa-2b interferon (Intron-A), *Am. J. Med.* **80**:223–228.

Spiegel, R. J., Jacobs, S. L., and Treuhaft, M. W., 1989, Anti-interferon antibodies to interferon-α_{2b}: Results of comparative assays and clinical perspective, *J. Interferon Res.* **9**(Suppl. 1):S17–S24.

Steis, R. G., Smith, J. W., II, Urba, W. J., Clark, J. W., Itri, L. M., Evans, L. M., Schoenberger, C., and Longo, D. L., 1988, Resistance to recombinant interferon alfa-2a in hairy-cell leukemia associated with neutralizing anti-interferon antibodies, *N. Engl. J. Med.* **318**:1409–1413.

Tomley, F. M., Mockett, A. P. A., Boursnell, M. E. G., Binns, M. M., Cook, J. K. A., Brown, T. D. K., and Smith, G. L., 1987, Expression of the infectious bronchitis virus spike protein by recombinant vaccinia virus and induction of neutralizing antibodies in vaccinated mice, *J. Gen. Virol.* **68**:2291–2298.

Trown, P. W., Kramer, M. J., Dennin, R. A., Connel, E. V., Palleroni, A. V., Quesada, J., and Gutterman, J. U., 1983, Antibodies to human leukocyte interferon in cancer patients, *Lancet* **1**:81–84.

Vallbracht, A., Treuner, J., Flehmig, B., Joester, K. E., and Niethammer, D., 1981, Interferon neutralizing antibodies in a patient treated with human fibroblast interferon, *Nature* **289**:496–497.

Van Kroonenburgh, N. J. P. G., and Pauwels, E. K. J., 1988, Human immunological response to mouse monoclonal antibodies in the treatment or diagnosis of malignant diseases, *Nucl. Med. Commun.* **9**:919–930.

Von Wussow, P., Freund, M., Block, B., Diedrich, H., Poliwoda, H., and Deicher, H., 1987, Clinical significance of anti-IFN-antibody titers during interferon therapy, *Lancet* **2**:635–636.

Von Wussow, P., Jakschies, D., Freund, M., and Deicher, H., 1989, Humoral response to recombinant interferon-α_{2b} in patients receiving recombinant interferon-α_{2b} therapy, *J. Interferon Res.* **9**(Suppl. 1):S25–S31.

Working, P. K., Golub, M. S., and Green, J. D., 1991, Maternal and neonatal safety of synthetic human relaxin administered to near-term pregnant rhesus monkeys, *Toxicologist* **11**:343.

Wroblewska, Z., Gilden, D., Green, M., Devlin, M., and Vafai, A., 1985, Affinity-purified varicella-zoster glycoprotein gp1/gp3 stimulates the production of neutralising antibody, *J. Gen. Virol.* **66**:1795–1799.

Zoon, K. C., 1987, *Human Interferons: Structure and Function,* Academic Press, New York.

Chapter 4

Pharmacokinetics and Metabolism of Protein Hormones

Jerome A. Moore and Victor J. Wroblewski

1. INTRODUCTION

The therapeutic potential of biosynthetic protein hormones is often determined by the half-life of the administered protein in the circulation. Most of the proteins tested display rapid serum clearance following intravenous administration relative to more conventional therapeutics. Tissue localization studies with radiolabeled hormones indicate that the liver and kidney are the primary organs of elimination for exogenously administered proteins. This observation, as will be discussed below, must be cautiously interpreted, however, as it is often subject to experimental artifact. The pharmacokinetic profiles of exogenously administered protein hormones are influenced by interactions with specific binding proteins and clearance by receptor- and non-receptor-mediated uptake and degradation (Maack *et al.,* 1979; Posner *et al.,* 1982; Walton *et al.,* 1989; Baumann *et al.,* 1990). Peripheral metabolism of protein hormones involves, predominantly, the hydrolysis of the amino acid backbone by a combination of exo- and endoproteinases (Bond and Butler, 1987). Proteolysis is involved in terminating the action of the molecule and may be related to the pharmacology of the hormone by producing molecules having biological activity. These differences may be re-

Jerome A. Moore • Department of Safety Evaluation, Genentech, Inc., South San Francisco, California 94080. *Present address:* Celtrix Pharmaceuticals, Santa Clara, California 95052. *Victor J. Wroblewski* • Lilly Research Laboratories, Lilly Corporate Center, Indianapolis, Indiana 46285.

Protein Pharmacokinetics and Metabolism, edited by Bobbe L. Ferraiolo *et al.* Plenum Press, New York, 1992.

lated to processing at target versus nontarget tissues. Therefore, determining peripheral sites of degradation and the cellular mechanisms involved in regulating the catabolism or processing of an administered protein may contribute to the design of molecules possessing desirable kinetic or pharmacological properties.

The "metabolism" of proteins is not as clearly defined an area of research as is classical xenobiotic metabolism. Preclinical studies related to the absorption, distribution, metabolism and elimination of new drug entities are necessary to ensure safety in clinical trials. The interpretation of absorption and particularly metabolism studies with proteins is influenced by the approach taken in this evaluation. Information obtained with the use of immunoassay and immunoextraction techniques depends on the ability of an antibody to recognize proteolytic fragments. Similarly, information gleaned with the use of radiotracers is limited by both the type of label and the position(s) within the primary structure which are labeled. As a result, the complexity of protein degradation *in vivo* is not completely understood.

The enzyme systems involved in the metabolism of exogenously administered protein hormones are poorly understood, but are most likely distinct from characterized prohormone processing enzymes (Harris, 1989). Studies with purified enzymes, subcellular fractions, whole cells, or isolated organs have been helpful in defining proteolytic mechanisms and cleavage products which may be important *in vivo*. However, the physiological relevance of *in vitro* information to a protein administered to an animal or human is not easy to establish experimentally. The most complete assessment of the metabolism of a protein *in vivo* requires a combination of approaches. As mentioned above, important considerations include the selection of an appropriate radiolabeled tracer, characterization of multiple antisera which will specifically and sensitively detect different portions of the protein, and the ability to chemically separate the parent molecule from degradation products. This section will review how *in vitro* and *in vivo* studies have been integrated to shape our understanding of protein hormone metabolism. Furthermore, preclinical pharmacokinetic data of some protein hormones will be discussed and related to metabolism, efficacy, and predictability of performance in human studies.

2. HUMAN INSULIN

2.1. Pharmacokinetics

Insulin provides an excellent example of the effects of analytical methods on the interpretation of the results of pharmacokinetic studies. In the long history of research into the biology of insulin activity, a wide variety of methods have been employed. Early studies with iodinated insulin produced extremely variable estimates of the serum clearance. Highly labeled insulin appeared to be cleared more slowly than insulin with one or no iodines attached (Genuth, 1972; Navalesi et al., 1978; Orskov and Christensen, 1969). Studies conducted since the development of specific immunoassays indicate that the serum clearance of insulin in humans is in the range of 700 to 800 ml/min, and the volume of distribution approximates extracellular space.

Studies of the organs of elimination suggest that 40 to 50% of circulating insulin is removed by the first pass through the liver (Chamberlain and Stimmler, 1967; Duckworth and Kitabchi, 1981; Ferrannini et al., 1983). The majority of the balance (approximately 30% of the total) appears to be cleared by the kidneys (Chamberlain and Stimmler, 1967; Ferrannini et al., 1983). Several factors influence the absorption of insulin following subcutaneous administration. The site of injection (Berger et al., 1982; Binder, 1969; Hildebrandt et al., 1984; Koivisto and Felig, 1980), temperature, exercise (Berger et al., 1982; Kawamori and Vranic, 1977; Koivisto et al., 1981), and the physical state of the insulin injected can have an effect on the rate and extent of absorption from the subcutaneous site. Estimates of degradation at the site of injection vary from 10 to 70% (Kraeger and Chisholm, 1985; Kobayashi et al., 1983). Insulin absorption following subcutaneous injection may be increased by coadministration with aprotinin (Berger et al., 1982; Binder et al., 1984).

2.2. Metabolism

The mechanism(s) of degradation of human insulin has been the most extensively studied of any protein hormone and has been recently reviewed (Duckworth, 1988). Binding of insulin to the insulin receptor is thought to be the essential step in the clearance of insulin from the circulation with subsequent degradation of the molecule (Terris and Steiner, 1975). There has been extensive investigation of insulin metabolism in vitro, with biochemical characterization of enzymes involved and products formed by these enzymes. In contrast, there is a relative paucity of information regarding the metabolism of insulin in vivo, reflecting, as pointed out above, the technical difficulty of these studies. As a result, the cellular mechanisms and biological significance of insulin degradation remain poorly understood.

The major enzyme believed to be involved in the cellular degradation of insulin is insulin protease, also referred to as insulin-degrading enzyme. This protease has been purified from rat and pig skeletal muscle (Duckworth et *al.,* 1972; Yokono *et al.,* 1980). A similar enzyme has also been purified from human erythrocytes, cultured fibroblasts and lymphocytes (Roth *et al.,* 1984; Shii *et al.,* 1986; Stentz *et al.,* 1989). Human insulin labeled with ^{125}I at tyrosine 14 of the A chain and tyrosine 26 of the B chain has been used as a probe to characterize peptide bonds in insulin sensitive to cleavage by the purified protease (Fig. 1). Cleavage sites have been identified between Leu(13)–Tyr(14) and Tyr(14)–Gln(15) of insulin A chain (Hamel *et al.,* 1986; Duckworth *et al.,* 1987). B-chain cleavage sites have been described between Ser(9)–His(10), Glu(13)–Ala(14), Tyr(16)–Leu(17), Phe(25)–Tyr(26), His(10)–Leu(11), Ala(14)–Leu(15), and Phe(24)–Phe(25) (Duckworth *et al.,* 1988). Similar results have been obtained with insulin labeled with ^{3}H at the A1 and B1 positions (Muir *et al.,* 1986; Davies *et al.,* 1988). The nondescript nature of the susceptible peptide bonds suggests that the proteolytic specificity may be related to the structure of the insulin molecule rather than specific amino acid residues. In this regard, synthetic insulin analogs have been used to implicate the importance of the hydrophobic composition of the three-amino-acid sequence Phe–Phe–Tyr (amino acids 24–26) in the insulin B chain in its high-affinity interaction with insulin-degrading enzyme (Affholter *et al.,* 1990). In general, most kinetic and immunological evidence supports the role of insulin protease as the major enzyme involved in the cellular degradation of insulin. Studies in whole cells and isolated organs employing ^{125}I-labeled insulin indicate a similar pattern of insulin cleavages as observed with purified insulin protease (Duckworth *et al.,* 1988, 1989). Furthermore, after administration of [^{125}I]insulin to rats, ^{125}I-labeled insulin degradation products isolated from liver endosomes are chromatographically similar to products generated by insulin protease (Hamel *et al.,* 1988).

In addition to proteolytic cleavage in the A and B chains of insulin, another potential route of metabolism is reduction of the disulfide bonds of insulin. The enzyme, protein disulfide isomerase (also known as glutathione insulin transhydrogenase), catalyzes disulfide interchange between thiols and protein disulfides (Roth and Koshland, 1981; Bjelland *et al.,* 1983). This enzyme was initially thought to be involved in cleavage of insulin into its A and B chains as an obligatory step to subsequent proteolysis (Varandani *et al.,* 1972). More recent evidence would suggest that this is not the case, although the interaction between proteolytic and reductive enzyme systems in the degradation of insulin in different species has not been exhaustively examined. Stentz *et al.* (1989) have identified multiple two-chain insulin intermediates from the digestion of unlabeled insulin with purified insulin protease from human fibroblasts. The possibility that protein disulfide isomerase could be involved in the secondary metabolism of these two-chain species cannot be excluded (Poole *et al.,* 1982; Misbin and Almira, 1989).

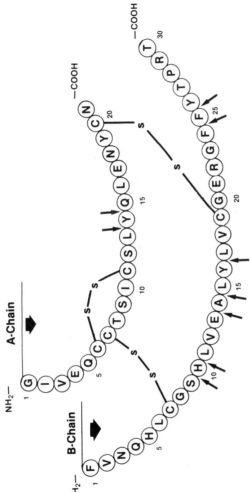

Figure 1. Structure of human insulin. Arrows indicate sites in the A and B chains that have been identified as sensitive to hydrolysis by insulin protease.

The significance of protein disulfide isomerase in the clearance of insulin was examined by Striffler (1987) in the perfused rat liver. The results suggested that prolonged fasting of rats decreased the activity of this enzyme and this correlated with a decreased removal of insulin from the perfusate. Therefore, the physiological role of this enzyme in metabolizing insulin or two-chain insulin protease degradation products needs to be more clearly established.

3. RELAXIN

3.1. Background

Another member of the structurally related insulinlike family of protein hormones is relaxin. This molecule has been shown to facilitate cervical ripening and parturition, and is absorbed systemically after parenteral administration with the expression of pharmacological activity (MacLennan *et al.*, 1986; Ferraiolo *et al.*, 1989). Like insulin, relaxin is synthesized from a single-chain prohormone to yield a two-chain molecule having two interchain and one interchain disulfide bridge.

3.2. Metabolism

The structural similarity between insulin and relaxin prompted Pilistine and Varandani (1986) to examine the degradation of relaxin by neutral thiol protease, protein disulfide isomerase (PDI), and neutral metalloendopeptidase. Purified neutral thiol protease and PDI both degraded [^{125}I]relaxin, although kinetic analysis revealed that insulin was the preferred substrate. PDI converted relaxin into individual A and B chains, but cleavage sites susceptible to metabolism by the protease were not identified. As with insulin, the significance of these enzyme systems in the pharmacological activity of relaxin awaits further investigation.

3.3. Pharmacokinetics

Pharmacokinetic studies employing a variety of relaxins, including human relaxin (hRlx), and a variety of methods and assays have been reported in the literature for mice, rats, gilts (Sherwood *et al.*, 1980), rabbits (Zarrow and Money, 1948), pregnant rhesus monkeys (Lucas *et al.,* 1989), baboons (O'Byrne and Steinetz, 1976), pregnant heifers (Musah *et al.,* 1987), and near-term pregnant humans (MacLennan *et al.,* 1986). As has been observed with other protein therapeutics (Tam *et al.,* 1982), the dosage regimen employed for hRlx delivery appears to be an important determinant of the expression of its pharmacodynamic effects. Ferraiolo *et al.* (1989) studied the effects of dose, route, regimen, and the presence or absence of a repository vehicle (benzopurpurine, BPP) for synthetic hRlx in estrogen-primed female CFW albino mice in the mouse pubic symphysis bioassay. Administration of 88 mg/kg hRlx subcutaneously (SC) in 1% BPP resulted in delayed, prolonged absorption. Although peak hRlx concentrations were lower, serum concentrations remained elevated longer in the presence of BPP as compared to a single SC administration of hRlx in saline at the same dose. The SC bioavailabilities with and without BPP were similar (109 and 96%, respectively). The observation of the effect of BPP to prolong absorption of relaxin in mice confirmed a report by O'Byrne and Steinetz (1976) who showed that purified porcine relaxin injected in a 20% gelatin solution had delayed absorption in a female baboon. While the pharmacodynamic effect (i.e., lengthening of the pubic ligament in estrogen-primed mice) was near maximal at 88 mg/kg hRlx SC with BPP, single SC administration of hRlx without BPP (up to 264 mg/kg) had no effect on pubic ligament length. This increase in apparent potency in the presence of the repository vehicle was previously observed by Steinetz *et al.* (1960). In the absence of the BPP vehicle, manipulation of the regimen (e.g., multiple SC doses) to emulate the serum concentration versus time profile observed for hRlx in the presence of BPP resulted in similar pharmacodynamic effects. It appears that BPP delays the absorption of hRlx after SC administration, resulting in prolonged, elevated hRlx serum concentrations. In addition, hRlx was shown to be efficacious in this model without BPP if it was administered by a multidose SC schedule.

In addition to the published reports discussed above, the absorption, distribution, metabolism, and elimination of recombinant human relaxin (rhRlx) and two forms of synthetic human relaxin (hRlx and hRlx-2) were examined in a series of comprehensive studies in female animals at Genentech, Inc., as part of the preclinical drug development program for rhRlx.

rhRlx is the product of the combination of A and B chains made by recombinant DNA methods; the A and B chains contain 24 and 29 amino acids, respectively. The A and B chains of the synthetic human relaxins hRlx and hRlx-2 are made by solid-phase methods. The amino acid sequences of rhRlx and hRlx are identical to that of the predominant form of relaxin in human corpora lutea and pregnancy sera (Winslow *et al.,* 1989). The synthetic human relaxin hRlx-2 was deduced to be composed of A and B chains containing 24 and 33 amino acids, respectively, based on sequence homology with porcine relaxin (Hudson *et al.,* 1984; Johnston *et al.,* 1985). rhRlx, hRlx, and hRlx-2 have indistinguishable bioactivities in the mouse pubic symphysis (MPS) bioassay. rhRlx and hRlx-2 have been evaluated in preclinical safety studies. rhRlx is the form of human relaxin proposed for use in clinical trials.

rhRlx was studied after single-dose intravenous or intravaginal administration in nonpregnant rhesus monkeys. Studies using hRlx employed single intravenous administration in mice and rats, single-dose intravenous, subcutaneous, or intravaginal administration in pregnant rabbits, and single-dose intravenous administration in pregnant and nonpregnant rhesus monkeys. Additional studies with hRlx-2 entailed single-dose intravenous administration in pregnant and nonpregnant rhesus monkeys.

Nonpregnant rhesus monkeys received single doses of 100 mg/kg rhRlx by the intravenous and intravaginal routes. The intravenous clearance of rhRlx was 4.1 ± 0.6 ml/min per kg. The initial and steady-state volumes of distribution were 78 ± 25 and 692 ± 215 ml/kg, respectively. The initial, intermediate, and terminal half-lives were 2.0 ± 0.5, 24 ± 7, and 249 ± 49 min. Intravaginal administration of rhRlx resulted in very low levels of systemic exposure. The maximum intravaginal bioavailability of rhRlx was 1.4%. hRlx was also administered as a single intravenous dose of 100 mg/kg to nonpregnant rhesus monkeys. The clearance of hRlx was 3.1 ± 0.5 ml/min per kg, and the initial and steady-state volumes of distribution were 64 ± 2 and 311 ± 32 ml/kg, respectively. Half-lives of 3.6 ± 1.6, 29 ± 4, and 134 ± 10 min were calculated. The differences in the clearance, initial, and steady-state volumes of distribution, and terminal half-life between hRlx and rhRlx were statistically significant. Despite these differences, the serum concentration data indicate that both forms have generally similar and predictable pharmacokinetic behavior.

The intravenous pharmacokinetics of the synthetic human relaxins hRlx and hRlx-2 were compared in pregnant and nonpregnant rhesus monkeys after single doses of 88–100 mg/kg. The clearance of hRlx was almost twofold slower than the clearance of hRlx-2 in both groups. Although these differences were statistically significant, their biological significance is considered to be minimal because clearance in normal human populations may vary as much as twofold. There were no significant differences in the pharmacokinetics of either hRlx or hRlx-2 in pregnant versus nonpregnant rhesus monkeys after intravenous administration.

hRlx was administered as single intravenous bolus, subcutaneous, and intravaginal doses of 72–88 mg/kg in pregnant rabbits. The intravenous clearance was 2.6 ± 0.8 ml/min per kg. The initial and steady-state volumes of distribution were 79 ± 19 and 184 ± 26 ml/kg, respectively. The half-lives were 8.4 ± 2.7, 59.1 ± 8.7, and 279 ± 128 min. The subcutaneous bioavailability was 76.0 ± 20.6%. Intravaginal administration of a single dose of hRlx in pregnant rabbits resulted in a bioavailability of 1.2 ± 1.0%. It should be noted that the low intravaginal bioavailability of hRlx in pregnant rabbits is similar to the intravaginal bioavailability observed for rhRlx in rhesus monkeys (maximum of 1.4%).

Intravenous pharmacokinetic studies with hRlx in rats following single doses of 88–100 mg/kg yielded serum clearances of 5.8–5.9 ml/min per kg. The initial and steady-state volumes of distribution were 66–72 and 294–622 ml/kg, respectively. Half-lives of 2.0–2.1, 20.7–39.4, and 76–293 min were calculated.

The clearance of hRlx in mice after intravenous administration of a single dose of 88 mg/kg was 15.7 ml/min per kg. The initial and steady-state volumes of distribution were 224 and 672 ml/kg, respectively. Half-lives were 5.3, 36, and 513 min.

3.4. Tissue Distribution

Tissue distribution studies using radiolabeled relaxins in mice, guinea pigs (O'Byrne *et al.*, 1982), and rats (Cheah and Sherwood, 1980; Weiss and Bryant-Greenwood, 1982) support observations that the kidney is the primary site of accumulation of radioactivity. Other important sites of accumulation were reported to be liver, lung, uterus, pituitary, and spleen.

3.5. Summary

In summary, the intravenous pharmacokinetics of all three forms of relaxin in all species tested were considered to be similar and were described by triexponential equations. Clearance (ml/min per kg) was species specific and decreased as species increased in size. Initial distribution volumes approximated the plasma volume in most species. There appeared to be no difference in the pharmacokinetics of relaxins in pregnant versus nonpregnant animals. There was little or no evidence of accumulation of relaxin following daily 60-min infusions and placental transfer was limited. The maximum bioavailability of relaxins after intravaginal administration in rabbits and rhesus monkeys was 1–1.4%. After administration of radiolabeled relaxins, radioactivity was distributed primarily to highly perfused organs and/or presumed organs of elimination.

4. HUMAN PROINSULIN

4.1. Background

The availability of biosynthetic human proinsulin has resulted in much interest in the use of this insulin precursor as a therapy for diabetes mellitus. Compared to insulin, human proinsulin has a longer half-life in the circulation, lower affinity for insulin receptors, and lower biological potency (Stoll *et al.*, 1971; Revers *et al.*, 1984; Peavy *et al.*, 1985; Tillil *et al.*, 1990). During the processing of proinsulin to insulin in the pancreas by the action of several endopeptidases, the two-chain intermediates des(64,65)-proinsulin and des(31,32)-proinsulin are formed (Fig. 2) (Steiner *et al.*, 1968; Tager *et al.*, 1980). These split intermediates have been demonstrated to have a higher affinity for insulin receptors and greater biological activity than proinsulin itself (Peavy *et al.*, 1985). Exogenously administered proinsulin may thus act as a prodrug requiring proteolytic activation for some of its pharmacological activity. Understanding the mechanism(s) of proinsulin proteolysis may provide insight into how metabolism can influence the application of peptides or proteins as therapeutic agents.

4.2. Metabolism

The prolonged half-life of proinsulin in the circulation compared to insulin appears to be related to its lower affinity for insulin receptors and poorer extraction by the liver (Peavy *et al.*, 1985). The metabolic clearance rates and liver extraction of split proinsulin metabolites were intermediate between those of insulin and proinsulin, again apparently related to relative affinity for insulin receptors. However, the intrahepatic half-lives of these molecules vary in an inverse order, with $T_{1/2}$ for proinsulin > des(31,32)-HPI > des(64,65)-HPI > insulin (Sodoyez-Goffaux *et al.*, 1988; Tillil *et al.*, 1990). This observation could be related to the reported greater local effects of proinsulin at the liver than other peripheral sites. Differences in the rate of proteolytic degradation of proinsulin compared to insulin and proinsulin metabolites in the liver could contribute to these effects. However, the pathways for the metabolism of proinsulin in the liver and other peripheral sites are poorly understood. Proinsulin appears to be a poor substrate for insulin protease (Duckworth *et al.*, 1972) and the conversion of proinsulin to insulin *in vivo* or by tissue preparations *in vitro* has not been demonstrated. Given *et al.* (1985) examined circulating forms of proinsulin in humans after subcutaneous and intravenous infusion by a combination of RP-HPLC and insulin-specific radioimmunoassays. After subcutaneous, but not intravenous infusion, 4–11% of proinsulin was processed to intermediates having properties of des(31,32)-HPI and des(64,65)-HPI. The results would suggest that peripheral organs of extraction, such as the liver and kidney, cannot form and/or release split intermediates of proinsulin at a significant rate. In contrast to insulin, proteolytic metabolism of proinsulin at the subcutaneous depot could also be involved in the release of more active forms of the molecule to the circulation. In addition, the action of protein disulfide isomerase in the metabolism of proinsulin or proinsulin split intermediates cannot be excluded. Both des(64,65)-HPI and des(31,32)-HPI act as substrates for the purified enzyme *in vitro* while proinsulin is a very poor substrate (Wroblewski, unpublished observation). Understanding the interaction of proteolytic as well as nonproteolytic enzyme systems in the bioactivation or inactivation of proinsulin is an area which deserves more extensive investigation.

5. GROWTH HORMONE AND PROLACTIN

5.1. Growth Hormone Background

Growth hormone (GH) is an endocrine hormone produced and stored in the anterior pituitary gland. It is secreted into the circulation in a pulsatile pattern. Deficiency of GH results in abnormally poor growth. This condition, hypopituitary dwarfism, has been recognized for several years. In 1958, Raben reported that a hypopituitary dwarf given injections of extracts of human pituitary tissue grew remarkably well. Unfortunately, since the only source of hGH was from cadaver pituitaries, the supply was extremely limited. The equivalent of approximately 50 pituitaries was required to supply enough hGH for one patient for a year. Specific criteria were established for identifying patients who were clearly GH deficient and who stood the best chance of benefiting from therapy. Once identified, a child would receive intramuscular injections of these pituitary extracts three times a week. The dosing levels were limited and the continuing therapy was always threatened by shortages of material. Nevertheless, this therapeutic regime proved to be beneficial for the vast majority of those treated.

In addition to being in short supply, the problem of purifying the material posed a challenge. As with any therapeutic agent which is derived from human tissue, contaminants of other protein hormones as well as those of viral origin had to be removed. The availability of hGH produced through recombinant DNA technology greatly reduced these concerns and provided

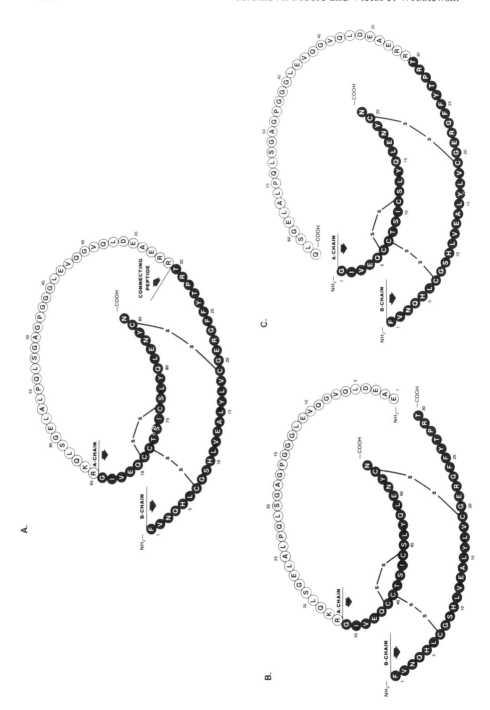

a plentiful supply of hormone. As part of the process of proving the safety and efficacy of the use of hGH in humans, pharmacokinetic studies have been conducted in a number of species including mice, rats, monkeys, and man. The pharmacokinetic profiles have been compared and the value of animal studies as a predictor of human pharmacokinetics has been evaluated.

5.2. Growth Hormone Pharmacokinetics

The disappearance rates of hGH and a 20-kDa variant over 10 min following IV injection of iodinated material were studied in male and female mice (Sigel *et al.,* 1982). The half-lives of hGH and the 20-kDa variant in males were 4.1 and 3.9 min, respectively; the half-lives in females were 4.7 and 4.0 min. The differences in the protein half-lives were not statistically significant. The authors concluded that there was no apparent relationship between the equal rates of disappearance, the dissimilar binding characteristics that have been shown for the hormones, and their equipotent growth-promoting activities.

The pharmacokinetics of biosynthetic human growth hormone (B-hGH) and pit-hGH were compared in rats (Jorgensen *et al.,* 1988). Normal and hypophysectomized male and female rats were injected subcutaneously and intramuscularly with 0.10 mg/kg B-hGH and pit-hGH and intravenously with 0.06 mg/kg B-hGH and pit-hGH. A statistically significantly smaller distribution volume and a slower metabolic clearance rate (MCR) were found for B-hGH compared to pit-hGH. A statistically significantly smaller distribution volume and MCR were found for hypophysectomized rats compared to normal rats for both proteins. The plasma half-lives were estimated to be about 3 to 7 min (initial phase) and 29 min (terminal phase). The plasma levels of hGH were higher after subcutaneous compared to intramuscular administration for both proteins. The authors interpreted these results to imply that an extensive local degradation took place at the subcutaneous and intramuscular injection sites. Comparative tissue distribution studies were performed with radioiodinated B-hGH and pit-hGH. No differences were observed between the growth hormones. Most of the labeled TCA-precipitable material accumulated in the liver. The females accumu-

←──

Figure 2. Structures of various human proinsulins. (A) Proinsulin; (B) des(31,32)-proinsulin, lacking Arg(31)–Arg(32) of the C-peptide; (C) (64,65)-proinsulin, lacking Lys(64)–Arg(65) of the C-peptide.

lated relatively more label in the liver than the males, while the males accumulated relatively more label in the kidneys.

The serum clearance (CL) of synthetic hGH was studied in eight adult rhesus monkeys (Dubey *et al.*, 1988). Four monkeys were lean (<20% body fat), and four were obese (>35% body fat). The monkeys were given a single IV bolus injection of GH (2.5 mg/kg), followed by a constant infusion of GH (250 mg/hr) for 2.5 hr. Venous blood samples were collected before the infusion and every 10 min during the infusion. In both groups, steady state was reached 70 min after the start of the infusion. The CL of GH was calculated from the ratio of the constant GH infusion rate and the steady-state plasma GH concentration in each monkey. The mean CL (±S.D.) of synthetic GH was 12.7 ± 1.7 liters/24 hr in the lean group and 19.5 ± 2.9 liters/24 hr in the obese group ($p < 0.007$); however, the CL/kg ratio was the same in both groups. The authors concluded the following: (1) the CL of GH is directly proportional to body weight and (2) the lower plasma GH levels in obesity may be due to an increase in its CL not compensated for by an appropriate increase in the rate of GH secretion.

5.3. Species Scaling of Growth Hormone Pharmacokinetics

Pharmacokinetic data collected in laboratory animals can often be used to predict pharmacokinetic parameters in humans by extrapolation based on body weight. Application of this method assumes that the pharmacokinetics are dose proportional over the dosage range under consideration. The allometric approach uses the following power function:

$$Y = aW^b$$

where Y is the pharmacokinetic parameter, W is body weight in kg, a is the allometric coefficient, and b is the allometric exponent. A log transformation simplifies this function to the linear equation:

$$\log Y = \log a + b(\log W)$$

where $\log a$ is the y intercept and b is the slope of the log–log plot of Y versus W. Values for $\log a$ and b are derived from linear regression analysis of unweighted $\log Y$ versus $\log W$ data.

In the early phases of product development, data from laboratory animals can sometimes be used to predict the pharmacokinetic profile in humans (Mordenti and Chappell, 1989). Dosing levels for initial clinical trials can thus be selected. In the later stages of development, when human data are available, allometric analysis is useful for evaluating the clinical relevance of pharmacological data collected in laboratory animals.

The CL, initial volume of distribution (V_c), and volume of distribution at steady state (V_{ss}) for rhGH in mice, rats, cynomolgous monkeys, and humans are listed in Table I. Linear regression analysis was performed for each parameter, testing the hypothesis that the slope of the regression line (b) = 0, i.e., the parameter did not correlate with weight. The results of these analyses and resulting allometric equations for each parameter are shown in Table II and Fig. 3. A significant t value ($p < 0.01$, two degrees of freedom) and the high r^2 indicate a strong association between the parameters and body weight. This good interspecies agreement supports the use of pharmacokinetic data collected in laboratory animals to predict the disposition of rhGH and other proteins in humans (Mordenti *et al.*, 1991).

5.4. Metabolism of Growth Hormone and Prolactin

Prolactin is a pituitary hormone which shows considerable structural and sequence homology to GH as well as overlap in its biological activity (Lewis, 1984; Mittra, 1984). Both prolactin and GH exist as a number of molecular forms which may be related to the wide variety of physiological actions of these molecules (Baumann *et al.*, 1983, 1985; Sinha *et al.*, 1985). Proteolytic cleavage of GH by plasmin, trypsin, and subtilisin *in vitro* results in the formation of two-chain molecules linked by a disulfide bridge. These molecules, which are cleaved in the large disulfide loop, have been demonstrated to possess increased or altered biological activities (Yadley and Chrambach, 1973; Lewis *et al.*, 1977; Maciag *et al.*, 1980). Similarly cleaved, two-chain forms of prolactin have been described in the pituitary of rodents and humans and in human plasma (Mittra, 1980; Sinha *et al.*, 1985). Proteolysis of prolactin in the pituitary may result from the action of glandular

Table I
Pharmacokinetic Parameters of rhGH in Various Species[a]

	Weight (kg)	CL (ml/min)	V_c (ml)	V_{ss} (ml)
Mouse	0.016	0.32	2.26	3.8
Rat	0.127	2.03	11.88	15.7
Monkey	3.8	14.7	199	314
Human	73.7	152.4	2432	4149

[a] CL, serum clearance; V_c, initial volume of distribution; V_{ss}, volume of distribution at steady state.

Table II
Linear Regression Analysis of Pharmacokinetic Parameters[a]

	Slope	p	r^2	Allometric equations
CL	0.7095	<0.01	0.9945	$6.8W^{0.71}$
V_c	0.8284	<0.01	0.9999	$68W^{0.83}$
V_{ss}	0.8389	<0.01	0.9981	$105W^{0.84}$

[a] CL, serum clearance; V_c, initial volume of distribution; V_{ss}, volume of distribution at steady state.

kallikrein or an acidic protease to yield 22-, 16-, and 8-kDa fragments upon reduction (Casabiell *et al.*, 1989; Powers and Hatala, 1990). The 16-kDa fragment of prolactin binds to prolactin receptors and has mitogenic and lactogenic activity (Clapp *et al.*, 1988, 1989). When administered therapeutically, the proteolytic processing of these molecules may be of physiological significance locally or after release of forms which have altered biological activities and slower clearance rates (Baumann, 1979). Understanding the diverse pharmacological activities of these molecules may be facilitated by studying their metabolism at target and nontarget tissues.

As with other protein hormones, the metabolic fate of GH and prolactin in peripheral tissues has not been clearly defined. The clearance of GH, and

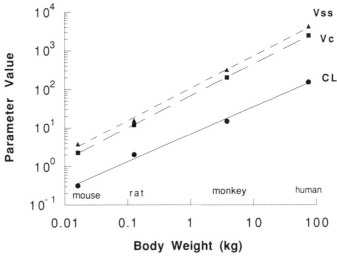

Figure 3. Allometric plot of human growth hormone pharmacokinetic parameters versus body weight for four species. Each line represents the least-squares linear regression line for unweighted log-transformed data.

presumably prolactin, from the circulation is thought to occur by receptor-mediated uptake at the liver and filtration at the kidney (Carone *et al.,* 1979; Josefsberg *et al.,* 1979; Postel-Vinay *et al.,* 1982). After intravenous injection of [^{125}I]hGH to female rats, hGH was specifically taken up by the liver with radioactivity ultimately becoming associated with the lysosomal subfraction (Postel-Vinay *et al.,* 1982). Groves *et al.* (1972) reported a similar localization for [^{125}I]bGH in liver and kidney after administration to rats. The specific uptake of [^{125}I]prolactin into rat liver was associated with accumulation in the Golgi apparatus (Josefsberg *et al.,* 1979). Interestingly, in each of these studies radioactivity in the liver was characterized by TCA solubility and membrane binding as intact hormone. These findings suggest that receptor-mediated uptake of GH and prolactin in the liver is not directly linked to hydrolytic inactivation of the hormone. We have observed a similar stability of hGH during *in vitro* incubations with rat liver and kidney subcellular fractions at neutral pH (Wroblewski, unpublished). In contrast, Scheppert *et al.* (1984) presented evidence for the limited proteolysis of GH by rabbit liver plasmalemma. These investigators suggested that a trypsinlike enzyme acted to specifically cleave GH in the large disulfide loop to generate a two-chain molecule which had a greater affinity for rabbit liver receptors. However, after intravenous administration of [^{125}I]hGH to rhesus monkeys, Baumann and Hodgen (1976) found no evidence for proteolytically modified forms of hGH in the circulation. The possibility that proteolytically cleaved products were formed locally but not released into the circulation could not be determined from the experimental protocol.

In contrast to GH, there have been a number of studies attempting to characterize the proteolytic metabolism of prolactin by target tissues *in vitro.* Shiu (1980) demonstrated the metabolism of prolactin by cultured human breast cells to small fragments that was inhibited by trypsin inhibitors and lysosomotropic agents. Other characterization of the metabolism and its significance were not evaluated. Compton and Witorsch (1984) indicated that *in vitro,* rat ventral prostate was the most active tissue in processing prolactin at the disulfide loop to relatively stable products. These products had approximate molecular masses of 16 and 8 kDa by reducing SDS–PAGE. Kidney, spleen, and lung were also active, but liver, among other tissues, possessed poor activity. The low activity in liver appears consistent with most studies described above for GH. Prolactin processing activity was present in all subcellular fractions of rat prostate but was most active in the 25,000*g* pellet fraction at acid pH. This activity was further localized to the lysosomes. In addition, the activity in the various fractions showed differential sensitivity to protease inhibitors. Wong *et al.* (1986) and Clapp (1987) also demonstrated the processing of rat prolactin to similar products by the 25,000*g* pellet from rat mammary gland. The activity was greater than observed in the 25-kDa pellet of male prostate, and in contrast to this activity was insensitive to serine and metalloenzyme proteinase inhibitors. Prolactins and growth hormones of other species were metabolized to a variety of low-molecular-mass forms, suggesting a species and hormone specificity of the proteolysis. The physiological significance of this processing is unclear since cleaved prolactin had the same binding activity to rat liver prolactin receptors as did intact prolactin, while the 16-kDa fragment actually had much reduced binding affinity (Clapp, 1987). If the proteolytic cleavage of GH and prolactin in peripheral tissues has physiological relevance, it seems obvious that the enzyme(s) involved would be rather specific. At this time the protease(s) involved in this peripheral processing have not been characterized. An important area of research will be to identify such proteases, if they exist, to better characterize the functional significance of their activity.

6. PARATHYROID HORMONE

6.1. Background

Studies with the 1–34 fragment of parathyroid hormone (PTH) in osteoporotic patients, demonstrating increased cancellous bone, have increased interest in the application of PTH as a therapeutic agent for the treatment of osteoporosis (Slovik *et al.,* 1981, 1986). The biological activity of PTH(1–84) has been attributed to the amino-terminal 34 residues, with COOH-terminal and midregion fragments of the molecule lacking biological activity (Tregear *et al.,* 1973). In the parathyroid gland, fragments of PTH are generated after biosynthesis of the hormone (Flueck *et al.,* 1977; MacGregor *et al.,* 1986) with subsequent secretion of intact hormone and fragments into the circulation. Therefore, PTH exists in the circulation as a heterogeneous group of molecules, comprised predominantly of intact hormone and COOH-terminal PTH fragments (Habener *et al.,* 1984). The contribution of peripheral metabolism of PTH to the circulating levels of biologically active or inactive fragments has been a topic of intensive investigation. Interesting approaches have been taken in *in vitro* and *in vivo* systems to address both the contribution of metabolism to the profile of circulating PTH forms, and the biological and pharmacological significance of this metabolism.

6.2. Pharmacokinetics

PTH in the circulation is rapidly cleared from the plasma by the liver and kidney which are the principal organs involved in the peripheral metabolism of the hormone (Habener *et al.*, 1984). The liver appears to be more selective for the extraction of intact (1–84) PTH over COOH-terminal or N-terminal (1–34) fragments of the hormone. In contrast, the kidney is effective in clearing intact PTH and its fragments (Martin *et al.*, 1976; Daugaard *et al.*, 1990). Other studies have shown that PTH(1–34) but not PTH(1–84) is extracted by bone (Martin *et al.*, 1978). These differences in the extraction of PTH fragments have been proposed to explain the presence of much higher levels of biologically inactive COOH-terminal compared to active N-terminal fragment(s) in the circulation.

6.3. Metabolism

Several investigations have directly addressed the metabolism of PTH in isolated cells, organs, or the whole animal (Fig. 4). One approach has been to analyze metabolites with the combined use of HPLC separation and radioimmunoassays specific for amino-terminal, midmolecule, or COOH-terminal fragments of PTH. Other studies used ^{3}H-, ^{125}I-, or ^{35}S-labeled PTH as a tracer with subsequent HPLC separation. Serge *et al.* (1977) examined the metabolism of ^{125}I-labeled bovine PTH in dogs and provided evidence for circulating COOH-terminal fragments formed as a result of cleavages between Asn(33)–Phe(34), Ala(36)–Leu(37), Ser(40)–Ile(41), and Ala(42)–Tyr(43). The fact that no cleavage sites closer to the N-terminal region were observed suggested that biologically active N-terminal fragments may indeed be formed and released into the circulation by peripheral tissues. Isolated rat liver Kupffer cells also metabolized [^{125}I]bPTH and [^{3}H]bPTH to products with the same N-terminal residues as observed *in vivo*, indicating a similar pattern of cleavage sites. In the same series of experiments, the use of [^{35}S]bPTH as substrate indicated that N-terminal fragments were present along with COOH-terminal fragments. However, the N-terminal fragments appeared to undergo a rapid degradation subsequent to their formation (Bringhurst *et al.*, 1982). Pillai and Zull (1986) suggested that in Kupffer cells PTH was cleaved between Phe(34)–Val(35) and Leu(37)–Gly(38) by cathepsin D (Zull and Chuang, 1985) to yield stable N-terminal peptides of the composition 1–34 and 1–37. The inconsistencies in the stability of N-terminal peptides in these studies may evolve from differences in the nature of the Kupffer cell preparations used in the two studies which may expose the protein to different sets of proteolytic activities. The exact nature of the PTH-degrading protease(s) in liver and kidney have yet to be established. Cathepsin D from bovine and rat kidney have been shown to cleave PTH into N-terminal and C-terminal fragments (Pillai *et al.*, 1983). In addition, a high-molecular-weight metalloproteinase which degrades PTH has been de-

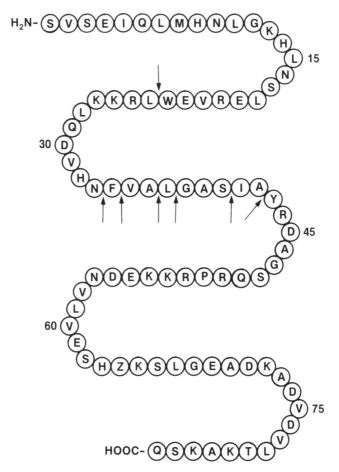

Figure 4. Primary structure of human parathyroid hormone. Arrows indicate sites of proteolytic cleavage that have been identified *in vivo* and *in vitro*.

scribed in rat kidney microsomes (Maruyama *et al.,* 1970; Fujita *et al.,* 1986). Yamaguchi *et al.* (1988) have also described a chymotrypsin-like endopeptidase in opossum kidney cells which cleaved human PTH between Trp(23)–Leu(24) and Phe(34)–Val(35). The physiological relevance and interaction between the various endopeptidases and lysosomal enzymes which can degrade PTH require further study.

Investigations in isolated organs and the whole animal are consistent with the proteolytic cleavage of PTH at peripheral sites and subsequent release of large COOH-terminal fragments into the circulation. Regardless of the radiolabel used in the studies, only fragments with amino-termini similar to those observed by Serge *et al.* (1977) were consistently detected. Most evidence only demonstrates the presence of amino-terminal fragments in plasma which are too small to be of pharmacological significance (Bringhurst *et al.*, 1988; Daugaard *et al.*, 1988, 1990). The data suggest that COOH-terminal, but not amino-terminal fragments of PTH are stable to subsequent proteolysis upon their formation. However, very low levels of amino-terminal fragments were observed in the liver and especially the kidney of rats after IV infusion of [^{35}S]-PTH (Bringhurst *et al.*, 1988). Thus, it remains possible that metabolism of PTH within specific target tissues may generate fragments of pharmacological importance at these sites. The significance of local metabolism of PTH in peripheral target tissues should be a subject of important future investigations.

7. ATRIAL NATRIURETIC FACTOR

7.1. Background

The protein hormone, atrial natriuretic factor [ANF(99–126)], atrial natriuretic peptide, may be a useful therapeutic agent for some cardiovascular disorders because of its physiological effects on the kidney and cardiovascular system. ANF is believed to play a role in the regulation of blood pressure, fluid volume, and electrolyte balance (Laragh, 1985; Ackermann, 1986). The predominant form of ANF in the circulation is a 28-amino-acid peptide [ANF(99–126)] which is the biologically active form of the molecule (Fig. 5) (Ackermann, 1986). ANF is synthesized in the heart as a 126-amino-

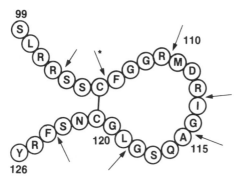

Figure 5. Sequence of α-human atrial natriuretic peptide. Arrows indicate points of hydrolysis by endopeptidase-24.11. Asterisk indicates proposed initial site of hydrolysis.

acid precursor and is thought to be processed in atrial tissue to the COOH-terminal and N-terminal peptides prior to their release into the circulation (Michener *et al.*, 1986).

7.2. Pharmacokinetics

The pharmacokinetics of ANF have been studied in a number of species using various methodologies and assays with very consistent results. In rats, ANF is cleared from the circulation extremely rapidly, approximately 100 ml/min per kg, following IV administration (Krieter and Trapani, 1989; Katsube *et al.*, 1986; Almeida *et al.*, 1989). Similar clearance was reported in rabbits (King *et al.*, 1989) and dogs (Cernacek *et al.*, 1988; Bie *et al.*, 1988; Verburg *et al.*, 1986; Woods, 1988). Likewise, in humans ANF has been shown to have rapid serum clearance and relatively large volume of distribution (Cusson *et al.*, 1988; Yandle *et al.*, 1986). Evidence that the fast clearance and large volume of distribution are related to receptor binding events has been reported (Almeida *et al.*, 1989). This rapid clearance of ANF(99–126) from the circulation is a major limitation to the potential use of this agent in cardiovascular disease.

7.3. Metabolism

Studies on the degradation *in vitro* have attempted to identify potentially important tissue sites of metabolism with subsequent characterization of the major proteolytic activity involved. Sites of cleavage have been identified to understand the role of metabolism as an activating or inactivating process. The biochemical characterization of ANF metabolism *in vitro* has provided a basis by which to test the physiological significance of these mechanisms to the pharmacology of ANF *in vivo*.

There is a significant amount of evidence implicating the involvement of endopeptidase 24.11 of kidney in the metabolism of ANF(99–126) (Stephenson and Kenny, 1987; Sonnenberg *et al.*, 1988; Vanneste *et al.*, 1990). This enzyme is distributed in brain, lung, kidney, thyroid and parts of the intestine (Ronco *et al.*, 1988), making it a potentially major mechanism by which ANF is inactivated. Vanneste *et al.* (1988) demonstrated purified endopeptidase 24.11 to cleave human ANF(99–126) between Arg(102)–Ser(103), Cys(105)–Phe(106), Arg(109)–Met(110), Arg(112)–Ile(113), Gly-(114)–Ala(115), Gly(118)–Leu(119), and Ser(123)–Phe(124). The initial site of metabolism by endopeptidase 24.11 occurred within the disulfide-linked loop between Cys(105)–Phe(106) to generate a two-chain product lacking biological activity (Currie *et al.,* 1984). The product cleaved between Cys(105)–Phe(106) was also identified as the predominant degradation product formed after incubation of rat ANF with kidney cortex membranes (Koehn *et al.,* 1987; Bertrand and Doble, 1988; Sonnenberg *et al.,* 1988). Metabolism of ANF in the kidney therefore appears to represent degradation resulting in inactivation of the peptide. Woods (1988), however, showed limited ($\sim$14%) contribution of the kidney to the overall metabolic clearance of synthetic human ANF(1–28) in dogs. ANF appears to be relatively stable to plasma proteinases. Murthy *et al.* (1986) reported a 10–15% conversion of ANF to ANF(Ser103–Tyr126), a molecule with lower biological activity, after a 30-min incubation with plasma *in vitro.* This result is consistent with the presence of this metabolite as a minor circulating form of ANF in the rat (Schwartz *et al.,* 1985). After infusion of [^{125}I]-ANF in rats, Vanneste *et al.* (1990) identified three radioactive metabolites in the plasma. The C-terminal tripeptide, Phe–Arg–Tyr, was present in significant quantity, with the major metabolite identified as the Cys–Phe cleaved form. Both of these metabolites could be predicted from the *in vitro* studies which indicated the involvement of endopeptidase 24.11 in the metabolism of ANF. Selective proteinase inhibitors have been useful in delineating the significance of specific proteolytic activities on the plasma half-life and pharmacological activity of ANF. Gros *et al.* (1989) demonstrated that the potent enkephalinase (EC 3.4.24.11) inhibitor, acetorphan, elicited a prolonged elevation of levels of intact ANF in the circulation of mice and humans. This elevation paralleled the time course for enzyme inhibition and was accompanied by significant increases in urinary volume and sodium excretion. Increases in ANF plasma levels, $t_{1/2}$, and biological responses were also observed upon coadministration of the endopeptidase-24.11 inhibitors, phosphoramidon and SCH 39370 with ANF (Sybertz *et al.,* 1989; Vanneste *et al.,* 1990). Inhibitor studies have also provided evidence for the secondary metabolism of endopeptidase-24.11-generated intermediates by a kallikrein-like proteinase (Vanneste *et al.,* 1990). Evidence also exists for the local processing of ANF-related peptides at the amino- and carboxy-termini to molecules which differ in biological potency (Harris and Wilson, 1985; Baxter *et al.,* 1986). This processing has been studied in atrial tissue and has been suggested to be involved in the maturation of the natural hormone. Whether this type of proteolysis has some role in the activation or inactivation of the peptide after interaction with specific receptors has not been established.

8. OVERVIEW

At present our understanding of the significance of protein degradation is at its infancy. As more biosynthetic protein hormones are applied as therapeutic agents, it is likely that modifications will be made in the natural sequence of the protein to enhance pharmacological or pharmacokinetic parameters. This will make it imperative to characterize the mechanisms by which protein hormones are degraded, so differences between natural sequence and modified molecules can be evaluated. In turn, a better understanding of protein degradation will likely lead to the design of modified molecules or appropriate adjuncts for protein hormone therapy.

It would appear that, if toxicological findings are insignificant, metabolism studies will not be critical to the development of protein hormones as drugs. A particularly difficult question to address in the preclinical stage is the significance of the metabolism of a recombinant human protein in animals. However, as noted in the examples above, the findings of metabolism studies can be relevant to the pharmacological activity of the hormone. Therefore, the application of protein metabolism to the discovery process should be an exciting and interesting approach in the future.

The difficulties from a regulatory standpoint are also in study design and interpretation. As mentioned previously, the sensitivity and specificity of the analytical approach will influence the interpretation of study results. Certainly, the emphasis should be on evaluating the production of relatively stable metabolites which may have pharmacological and, potentially, toxicological significance. In this regard, the evaluation of initial proteolytic events and processing at parenteral sites of administration warrant investigation. Although difficult to translate to the *in vivo* situation, *in vitro* approaches can provide information quickly with simpler analytical development. As illustrated in the examples above, *in vitro* studies allowed the identification of cleavage sites and major proteolytic fragments of several hormones by sequencing techniques. The importance of the metabolites was then studied with subsequent analysis by receptor binding or other bioassays. Hormone metabolites identified *in vitro* have also been used to characterize antibodies used for radioimmunoassay and to develop antibodies to critical portions of the molecule. Based on this information, immunological techniques can then be established for the sensitive and specific analysis of chromatographic metabolite profiles or for the immunopurification of *in vivo* metabolic products. The use of specifically radiolabeled proteins in combination with information from immunological assays can provide a rather comprehensive analysis of protein metabolism *in vivo. In vitro* studies can also provide a characterization of the protease(s) involved in the metabolism, and their tissue and subcellular distribution. The application of specific protease inhibitors may then help to evaluate the relevance of this activity *in vivo.* It is clear that much needs to be learned about the metabolism of therapeutic protein hormones. This process will be facilitated with advances in analytical techniques and our understanding of the function of proteolytic enzymes.

REFERENCES

Ackermann, U., 1986, Structure and function of atrial natriuretic peptides, *Clin. Chem.* **32**:241–247.

Affholter, J. A., Cascieri, M. A., Bayne, M. L., Brange, J., Casaretto, M., and Roth, R. A., 1990, Identification of residues in the insulin molecule important for binding to insulin-degrading enzyme, *Biochemistry* **29**:7727–7733.

Almeida, F. A., Suzuki, M., Scarborough, R. M., Lewicki, J. A., and Maack, T., 1989, Clearance function of type C receptors of atrial natriuretic factor in rats, *Am. J. Physiol.* **256**:R469–R475.

Baumann, G., 1979, Metabolic clearance rates of isohormones of human growth hormone in man, *J. Clin. Endocrinol. Metab.* **49**:495–499.

Baumann, G., and Hodgen, G., 1976, Lack of in vivo transformation of human growth hormone to its "activated" isohormones in peripheral tissues of the rhesus monkey, *J. Clin. Endocrinol. Metab.* **43**:1009–1014.

Baumann, G., MacCart, J. G., and Amburn, K., 1983, The molecular nature of circulating growth hormone in normal and acromegalic man: Evidence for a principal and minor monomeric forms, *J. Clin. Endocrinol. Metab.* **56**:946–952.

Baumann, G., Stolar, M. W., and Amburn, K., 1985, Molecular forms of circulating growth hormone during spontaneous secretory episodes and in the basal state, *J. Clin. Endocrinol. Metab.* **60**:1216–1220.

Baumann, G., Vance, M. L., Shaw, M. A., and Thorner, M. O., 1990, Plasma transport of human growth hormone *in vivo, Endocrinology* **71**:470–473.

Baxter, J. H., Wilson, I. B., and Harris, R. A., 1986, Identification of an endogenous protease that processes atrial natriuretic peptide at its amino terminus, *Peptides* **7**:407–411.

Berger, M., Cuppers, H. J., Hegner, H., Jorgens, U., and Berchtold, P., 1982, Absorption kinetics and biological effects of subcutaneously injected insulin preparations, *Diabetes Care* **5**:77–91.

Bertrand, P., and Doble, A., 1988, Degradation of atrial natriuretic peptides by an enzyme in rat kidney resembling neutral endopeptidase 24.11, *Biochem. Pharmacol.* **37**:3817–3821.

Bie, P., Wang, B. C., Leadley, R. J., Jr., and Goetz, K. L., 1988, Hemodynamic and renal effects of low-dose infusions of atrial peptide in awake dogs, *Am. J. Physiol.* **254**:R161–R169.

Binder, C., 1969, Absorption of injected insulin, *Acta Pharmacol. Toxicol.* **27**(Suppl.):1–84.

Binder, C., Lauritzen, T., Faber, O., and Pramming, S., 1984, Insulin pharmacokinetics, *Diabetes Care* **5**:188–199.

Bjelland, S., Wallevik, K., Kroll, J., Dixon, J. E., Morin, J. E., Freedman, R. B., Lambert, N., Varandani, P. T., and Nafz, M. A., 1983, Immunological identity between bovine preparations of thiol:protein-disulfide oxidoreductase, glutathione-insulin transhydrogenase and protein disulfide isomerase, *Biochim. Biophys. Acta* **747**:197–199.

Bond, J. S., and Butler, P. E., 1987, Intracellular proteases, *Annu. Rev. Biochem.* **56**:333–364.

Bringhurst, F. R., Serge, G. V., Lampman, G. W., and Potts, J. T., Jr., 1982, Metabolism of parathyroid hormone by Kupffer cells: Analysis by reverse-phase high-performance liquid chromatography, *Biochemistry* **21**:4252–4258.

Bringhurst, F. R., Stern, A. M., Yotts, M., Mizrahi, N., Serge, G. V., and Potts, J. T., Jr., 1988, Peripheral metabolism of PTH: Fate of biologically active amino terminus *in vivo, Am. J. Physiol.* **255**:E886–E893.

Carone, F. A., Peterson, D. R., Oparil, S., and Pullman, T. N., 1979, Renal tubular transport and catabolism of proteins and peptides, *Kid. Int.* **16**:271–278.

Casabiell, C., Robertson, M. C., Friesen, H. G., and Casanueva, F. F., 1989, Cleaved prolactin and its 16K fragment are generated by an acid protease, *Endocrinology* **125**:1967–1972.

Cernacek, P., Maher, E., Crawhall, J. C., and Levy, M., 1988, Renal dose response and pharmacokinetics of atrial natriuretic factor in dogs, *Am. J. Physiol.* **255**:R929–R935.

Chamberlain, M. J., and Stimmler, L., 1967, The renal handling of insulin, *J. Clin. Invest.* **46**:911–919.

Cheah, S. H., and Sherwood, O. D., 1980, Target tissues for relaxin in the rat: Tissue distribution of injected ^{125}I-labeled relaxin and tissue changes in adenosine 3′,5′-monophosphate levels after in vitro relaxin incubation, *Endocrinology* **106**:1203–1209.

Clapp, C., 1987, Analysis of the proteolytic cleavage of prolactin by the mammary gland and liver of the rat: Characterization of the cleaved and 16K forms, *Endocrinology* **121**:2055–2064.

Clapp, C., Sears, P. S., Russell, D. H., Richards, J., Levey-Young, B. K., and Nicoll, C. S., 1988, Biological and immunological characterization of cleaved and 16K forms of rat prolactin, *Endocrinology* **122**:2892–2898.

Clapp, C., Sears, P. S., and Nicoll, C. S., 1989, Binding studies with intact rat prolactin and a 16K fragment of the hormone, *Endocrinology* **125**:1054–1059.

Compton, M. M., and Witorsch, R. P., 1984, Proteolytic degradation and modification of rat prolactin by subcellular fractions of the rat ventral prostate gland, *Endocrinology* **115**:476–484.

Currie, M. G., Geller, D. M., Cole, B. R., Siegel, N. R., Folk, K. F., and Adams, S. P., 1984, Purification and sequence analysis of bioactive peptides (atriopeptins), *Science* **233**:67–69.

Cusson, J. R., Du Souich, P., Hamet, P., Schiffrin, E. L., Kuchel, O., Tremblay, J., Cantin, M., Genest, J., and Larochelle, P., 1988, Effects and pharmacokinetics of bolus injections of atrial natriuretic factor in normal volunteers, *J. Cardiovasc. Pharmacol.* **11**:635–642.

Daugaard, H., Egfjord, M., and Olgaard, K., 1988, Metabolism of intact parathyroid hormone in isolated perfused rat liver and kidney, *Am. J. Physiol.* **254**:E740–E748.

Daugaard, H., Egfjord, M., and Olgaard, K., 1990, Metabolism of parathyroid hormone in isolated perfused rat kidney and liver combined, *Kidney Int.* **38**:55–62.

Davies, G., Muir, A. V., Rose, K., and Offord, R. E., 1988, Identification of radioactive insulin fragments liberated by insulin proteinase during the degradation of semisynthetic [[³H]Gly^{A1}]insulin and [[³H]Phe^{B1}]insulin, *Biochem. J.* **249**:209–214.

Dubey, A. K., Hanukoglu, A., Hansen, B. C., and Kowarski, A. A., 1988, Metabolic clearance rates of synthetic human growth hormone in lean and obese male rhesus monkeys, *J. Clin. Endocrinol. Metab.* **67**:1064–1067.

Duckworth, W. C., 1988, Insulin degradation: Mechanisms, products, and significance, *Endocrine Rev.* **9**:319–344.

Duckworth, W. C., and Kitabchi, A. E., 1981, Insulin metabolism and degradation, *Endocrine Rev.* **2**:210–233.

Duckworth, W. C., Heinemann, M. A., and Kitabchi, A. E., 1972, Purification of insulin-specific protease by affinity chromatography, *Proc. Natl. Acad. Sci. USA* **69**:3698–3702.

Duckworth, W. C., Hamel, F. G., Liepnieks, J. J., Peavy, D. E., Ryan, M. P., Hermodson, M. A., and Frank, B. H., 1987, Identification of A-chain cleavage sites in intact insulin produced by insulin protease and isolated hepatocytes, *Biochem. Biophys. Res. Commun.* **147**:615–622.

Duckworth, W. C., Hamel, F. G., Peavy, D. E., Liepnieks, J. J., Ryan, M. P., Hermodson, M. A., and Frank, B. H., 1988, Degradation products of insulin generated by hepatocytes and by insulin protease, *J. Biol. Chem.* **263**:1826–1833.

Duckworth, W. C., Hamel, F. G., Liepnieks, J., Peavy, D., Frank, B., and Rabkin, R., 1989, Insulin degradation products from perfused rat kidney, *Am. J. Physiol.* **256**:E208–E214.

Ferraiolo, B. L., Cronin, M., Bakhit, C., Roth, M., Chestnut, M., and Lyon, R., 1989, The pharmacokinetics and pharmacodynamics of a human relaxin in the mouse pubic symphysis bioassay, *Endocrinology* **125**:2922–2926.

Ferrannini, E., Wahren, J., Faber, O. K., Felig, P., Binder, C., and DeFronzo, R. A., 1983, Splanchnic and renal metabolism of insulin in human subjects: A dose–response study, *Am. J. Physiol.* **244**:E517–E527.

Flueck, J. A., DiBella, F. P., Edis, A. J., Kehrwald, J. M., and Arnaud, C. D., 1977, Immunoheterogeneity of parathyroid hormone in venous effluent serum from hyperfunctioning parathyroid glands, *J. Clin. Invest.* **60**:1367–1375.

Fujita, T., Baba, R. E., Sase, M., Fukushima, M., and Nishii, Y., 1986, Parathyroid hormone-degrading enzyme of high molecular weight in the cytosol of rat renal cortex, *Bone Mineral.* **1**:457–465.

Genuth, S. M., 1972, Metabolic clearance of insulin in man, *Diabetes* **21**:1003–1012.

Given, B. D., Cohen, R. M., Shoelson, S. E., Frank, B. H., Rubenstein, A. H., and Tager, H. S., 1985, Biochemical and clinical implications of proinsulin conversion intermediates, *J. Clin. Invest.* **76**:1398–1405.

Gros, C., Souque, A., Schwartz, J.-C., Duchier, J., Cournot, A., Baumer, P., and Lecomte, J.-M., 1989, Protection of atrial natriuretic factor against degradation: Diuretic and natriuretic responses after *in vivo* inhibition of enkephalinase (EC 3.4.24.11) by acetorphan, *Proc. Natl. Acad. Sci. USA* **86**:7580–7584.

Groves, W. E., Houts, G. E., and Bayse, G. S., 1972, Subcellular distribution of

[125]I-labeled bovine growth hormone in rat liver and kidney, *Biochim. Biophys. Acta* **264**:472–480.

Habener, J. F., Rosenblatt, M., and Potts, J. T., Jr., 1984, Parathyroid hormone: Biochemical aspects of biosynthesis, secretion, action, and metabolism, *Physiol. Rev.* **64**:985–1053.

Hamel, F. G., Peavy, D. E., Ryan, M. P., and Duckworth, W. C., 1986, High performance liquid chromatographic analysis of insulin degradation by rat skeletal muscle insulin protease, *Endocrinology* **118**:328–333.

Hamel, F. G., Posner, B. I., Bergeron, J. J. M., Frank, B. H., and Duckworth, W. C., 1988, Isolation of insulin degradation products from endosomes derived from intact rat liver, *J. Biol. Chem.* **263**:6703–6708.

Harris, R. B., 1989, Processing of pro-hormone precursor proteins, *Arch. Biochem. Biophys.* **275**:315–333.

Harris, R. B., and Wilson, I. B., 1985, Conversion of atriopeptin II to atriopeptin I by atrial dipeptidyl carboxy hydrolase, *Peptides* **6**:393–396.

Hildebrandt, P., Birch, K., Sestoft, L., and Volund, A., 1984, Dose-dependent subcutaneous absorption of porcine, bovine, and human NPH insulins, *Acta Med. Scand.* **215**:69–73.

Hudson, P., John, M., Crawford, R., Haralambidis, J., Scanlon, D., Gorman, J., Tregear, G., Shine, J., and Niall, H., 1984, Relaxin gene expression in human ovaries and the predicted structure of a human preprorelaxin by analysis of cDNA clones, *EMBO J.* **3**:2333–2339.

Johnston, P. D., Burnier, J., Chen, S., Davis, D., Morehead, H., Remington, M., Struble, M., Tregear, G., and Niall, H., 1985, Structure/function studies on human relaxin, in: *Peptides: Structure and Function* (C. M. Deber, V. J. Hruby, and K. D. Kopple, eds.), Pierce Chemical Co., pp. 683–686.

Jorgensen, K. D., Monrad, J. D., Brondum, L., and Dinesen, B., 1988, Pharmacokinetics of biosynthetic and pituitary human growth hormones in rats, *Pharmacol. Toxicol.* **63**:129–134.

Josefsberg, Z., Posner, B. I., Patel, B., and Bergeron, J. J. M., 1979, The uptake of prolactin into female rat liver. Concentration of intact hormone in the Golgi apparatus, *J. Biol. Chem.* **254**:472–480.

Katsube, N., Schwartz, D., and Needleman, P., 1986, Atriopeptin turnover. Quantitative relationship between in vivo changes in plasma levels and atrial content, *J. Pharmacol. Exp. Ther.* **239**:474–479.

Kawamori, R., and Vranic, M., 1977, Mechanism of exercize-induced hypoglycemia in depancreatized dogs maintained on long acting insulin, *J. Clin. Invest.* **59**:331–336.

King, K. A., Courneya, C. A., Tang, C., Wilson, N., and Ledsome, J. R., 1989, Pharmacokinetics of vasopressin and atrial natriuretic peptide in anesthetized rabbits, *Endocrinology* **124**:77–83.

Kobayashi, T., Sawando, S., Itoh, T., Kosaka, K., Hirayama, H., and Kasuya, Y., 1983, The pharmacokinetics of insulin after continuous subcutaneous infusion or bolus subcutaneous injection in diabetic subjects, *Diabetes* **32**:331–336.

Koehn, J. A., Norman, J. A., Jones, B. N., LeSueur, L., Sakane, Y., and Ghai, R. D.,

1987, Degradation of atrial natriuretic factor by kidney cortex membranes. Isolation and characteristics of the primary proteolytic products, *J. Biol. Chem.* **262:**11623–11627.

Koivisto, V. A., and Felig, P., 1980, Alterations in insulin absorption and glycemic control associated with varying insulin injection sites in the diabetic, *Ann. Intern. Med.* **92:**59–61.

Koivisto, V. A., Forthey, S., Hendler, R., and Felig, P., 1981, A rise in ambient temperature augments insulin absorption in diabetic patients, *Metabolism* **30:**402–405.

Kraeger, E. W., and Chisholm, D. J., 1985, Pharmacokinetics of insulin. Implications for continuous subcutaneous insulin infusion therapy, *Clin. Pharmacokinet.* **10:**303–314.

Krieter, P. A., and Trapani, A. J., 1989, Metabolism of atrial natriuretic peptide. Extraction by organs in the rat, *Drug Metab. Disp.* **17:**14–19.

Laragh, J. H., 1985, Atrial natriuretic hormone, the renin–aldosterone axis and blood pressure–electrolyte homeostasis, *N. Engl. J. Med.* **313:**1330–1340.

Lewis, U. J., 1984, Variants of growth hormone and prolactin and their posttranslational modifications, *Annu. Rev. Physiol.* **46:**33–42.

Lewis, U. J., Singh, R. N. P., Vanderlaan, W. P., and Tutwiler, G. F., 1977, Enhancement of the hyperglycemic activity of human growth hormone by enzymic modification, *Endocrinology* **101:**1587–1603.

Lucas, C., Bald, L. N., Martin, M. C., Jaffe, R. B., Drolet, D. W., Mora-Worms, M., Bennett, G., Chen, A. B., and Johnston, P. D., 1989, An enzyme-linked immunosorbent assay to study human relaxin in human pregnancy and in pregnant rhesus monkeys, *J. Endocrinol.* **120:**449–457.

Maack, T., Johnson, V., Kau, S. T., Figueiredo, J., and Sigulem, D., 1979, Renal filtration, transport, and metabolism of low-molecular-weight proteins: A review, *Kidney Int.* **16:**251–270.

Maciag, T., Forand, R., Ilsley, S., Cerundolo, J., Greenlee, R., Kelley, P., and Canalis, E., 1980, The generation of sulfation factor activity by proteolytic modification of growth hormone, *J. Biol. Chem.* **255:**6064–6070.

MacGregor, R. R., Jilka, R. L., and Hamilton, J. W., 1986, Formation and secretion of fragments of parathormone. Identification of cleavage sites, *J. Biol. Chem.* **261:**1929–1934.

MacLennan, A. H., Green, R. C., Grant, P., and Nicolson, R., 1986, Ripening of the human cervix and induction of labor with intracervical purified porcine relaxin, *Obstet. Gynecol.* **68:**598–601.

Martin, K., Hruska, K., Greenwalt, A., Klahr, S., and Slatopolsky, E., 1976, Selective uptake of intact parathyroid hormone by the liver: Differences between hepatic and renal uptake, *J. Clin. Invest.* **58:**781–788.

Martin, K., Frietag, J. J., Conrades, M. B., Hruska, K. A., Klahr, S., and Slatopolsky, E., 1978, Selective uptake of the synthetic amino terminal fragment of bovine parathyroid hormone by isolated perfused bone, *J. Clin. Invest.* **62:**256–261.

Maruyama, M., Fujita, T., and Ohata, M., 1970, Purification and properties of a microsomal endopeptidase from rat kidney preferentially hydrolyzing parathyroid hormone, *Arch. Biochem. Biophys.* **138**:245–253.

Michener, M. L., Gierse, J. K., Seetharam, R., Fok, K. F., Olins, P. O., Mai, M. S., and Needleman, P., 1986, Proteolytic processing of atriopeptin prohormone, *Mol. Pharmacol.* **30**:552–557.

Misbin, R. I., and Almira, E. C., 1989, Degradation of insulin and insulin-like growth factors by enzyme purified from human erythrocytes, *Diabetes* **38**:152–158.

Mittra, I., 1980, A novel "cleaved prolactin" in the rat pituitary: Part I. Biosynthesis, characterization and regulatory control, *Biochem. Biophys. Res. Commun.* **95**:1750–1759.

Mittra, I., 1984, Somatomedins and proteolytic bioactivation of prolactin and growth hormone, *Cell* **38**:347–348.

Mordenti, J., and Chappell, W., 1989, The use of interspecies scaling in toxicokinetics, in: *Toxicokinetics and New Drug Development* (A. Yacobi, J. Skelly, and V. Batra, Eds.), Pergamon Press, Elmsford, N.Y., pp. 42–96.

Mordenti, J., Chen, S. A., Moore, J. A., Ferraiolo, B. L., and Green, J. D., 1991, Interspecies scaling of clearance and volume of distribution data for five therapeutic proteins, *Pharm. Res.* **8**:1351–1359.

Muir, A., Offord, R. E., and Davies, J. G., 1986, The identification of a major product of the degradation of insulin by 'insulin proteinase' (EC 3.4.22.11), *Biochem. J.* **237**:631–637.

Murthy, K. K., Thibault, G., Garcia, R., Gutkowska, J., Genest, J., and Cantin, M., 1986, Degradation of atrial natriuretic factor in the rat, *Biochem. J.* **240**:461–469.

Musah, A. I., Schwabe, C., and Anderson, L. L., 1987, Acute decrease in progesterone and increase in estrogen secretion caused by relaxin during late pregnancy in beef heifers, *Endocrinology* **120**:317–324.

Navalesi, R., Pilo, A., and Ferrannini, E., 1978, Kinetic analysis of plasma insulin disappearance in nonketotic diabetic patients and in normal subjects, *J. Clin. Invest.* **61**:197–208.

O'Byrne, E. M., and Steinetz, B. G., 1976, Radioimmunoassay (RIA) of relaxin in sera of various species using an antiserum to porcine relaxin (39377), *Proc. Soc. Exp. Biol. Med.* **152**:272–276.

O'Byrne, E. M., Brindle, S., Quintavalla, J., Strawinski, C., Tabachnick, M., and Steinetz, B. G., 1982, Tissue distribution of injected ^{125}I-labeled porcine relaxin: Organ uptake, whole body autoradiography and renal concentration of radiometabolites, *Ann. N.Y. Acad. Sci.* **380**:187–197.

Orskov, H., and Christensen, N. J., 1969, Plasma disappearance rate of injected human insulin in juvenile diabetic, maturity-onset diabetic and nondiabetic subjects, *Diabetes* **18**:653–659.

Peavy, D. E., Brunner, M. R., Duckworth, W. C., Hooker, C. S., and Frank, B. H., 1985, Receptor binding and biological potency of several split forms (conversion intermediates) of human proinsulin, *J. Biol. Chem.* **260**:13989–13994.

Pilistine, S. J., and Varandani, P. T., 1986, Degradation of porcine relaxin by glutathione-insulin transhydrogenase and a neutral peptidase, *Mol. Cell. Endocrinol.* **46**:43–52.

Pillai, S., and Zull, J. E., 1986, Production of biologically active fragments of parathyroid hormone by isolated Kupffer cells, *J. Biol. Chem.* **261**:14919–14923.

Pillai, S., Botti, R., and Zull, J. E., 1983, ATP activation of parathyroid hormone cleavage catalyzed by cathepsin D from bovine kidney, *J. Biol. Chem.* **258**:9724–9728.

Poole, G. P., O'Connor, K. J., Lazarus, N. R., and Pogson, C. I., 1982, 125I-labelled insulin degradation by isolated rat hepatocytes: The roles of glutathione-insulin transhydrogenase and insulin-specific protease, *Diabetologia* **23**:49–53.

Posner, B. I., Kahn, M. N., and Bergeron, J. J. M., 1982, Endocytosis of peptide hormones and other ligands, *Endocrine Rev.* **3**:280–298.

Postel-Vinay, M.-C., Kayseer, C., and Desbuquois, B., 1982, Fate of injected human growth hormone in the female rat liver in vivo, *Endocrinology* **111**:244–251.

Powers, C. A., and Hatala, M. A., 1990, Prolactin proteolysis by glandular kallikrien: *In vitro* reaction requirements and cleavage sites, and detection of processed prolactin *in vivo, Endocrinology* **127**:1916–1927.

Raben, M. S., 1958, Treatment of a pituitary dwarf with human growth hormone, *J. Clin. Endocrinol. Metab.* **18**:901–903.

Revers, R. R., Henry, R., Schneiser, L., Kolterman, O., Cohen, R., Bergenstahl, R., Polonsky, K., Jaspan, J., Rubenstein, A. H., Frank, B., Galloway, J., and Olefsky, J. M., 1984, The effect of biosynthetic human proinsulin on carbohydrate metabolism, *Diabetes* **33**:762–770.

Ronco, P., Pollard, H., Galceran, M., Delauche, M., Schwartz, J. C., and Verquist, P., 1988, Distribution of enkephalinase (membrane metalloendopeptidase, E.C. 3.4.24.11) in rat organs—Detection using monoclonal antibody, *Lab. Invest.* **58**:210–217.

Roth, R. A., and Koshland, M. E., 1981, Role of disulfide interchange enzyme in immunoglobulin synthesis, *Biochemistry* **20**:6594–6599.

Roth, R. A., Mesirow, M. L., Yokono, K., and Baba, S., 1984, Degradation of insulin-like growth factors I and II by a human insulin degrading enzyme, *Endocrine Res.* **10**:101–112.

Scheppert, J. M., Hughes, E. F., Postel-Vinay, C., and Hughes, J. P., 1984, Cleavage of growth hormone by rabbit liver plasmalemma enhances binding, *J. Biol. Chem.* **259**:12945–12948.

Schwartz, D., Geller, D. M., Manning, P. T., Siegel, N. R., Fok, K. F., Smith, C. E., and Needleman, P., 1985, Ser-Leu-Arg-Arg-atriopeptin III: The major circulating form of atrial peptide, *Science* **229**:397–400.

Serge, G. V., Naill, H. D., Sauer, R. T., and Potts, J. T., Jr., 1977, Edman degradation of radioiodinated parathyroid hormone: Application to sequence analysis and hormone metabolism *in vivo. Biochemistry* **16**:2417–2427.

Sherwood, O. D., Crnekovic, V. E., Gordon, W. L., and Rutherford, J. E., 1980, Radioimmunoassay of relaxin throughout pregnancy and during parturition in the rat, *Endocrinology* **107**:691–698.

Shii, K., Yokono, K., Baba, S., and Roth, R. A., 1986, Purification and characterization of insulin-degrading enzyme from human erythrocytes, *Diabetes* **35**:675-683.

Shiu, R. P. C., 1980, Processing of prolactin by human breast cancer cells in long term culture, *J. Biol. Chem.* **255**:4278-4282.

Sigel, M. B., Vanderlaan, W. P., Kobrin, M. S., Vanderlaan, E. F., and Thorpe, N. A., 1982, The biological half-life of human growth hormone and a biologically active 20,000-dalton variant in mouse blood, *Endocrin. Res. Commun.* **9**:67-77.

Sinha, Y. N., Gilligan, T. A., Lee, D. W., Hollingsworth, D., and Markoff, E., 1985, Cleaved prolactin: Evidence for its occurrence in human pituitary gland and plasma, *J. Clin. Endocrinol. Metab.* **60**:239-243.

Slovik, D. M., Neer, R. M., and Potts, J. T., 1981, Short-term effects of synthetic human parathyroid hormone (1-34) administration on bone mineral metabolism in osteoporotic patients, *J. Clin. Invest.* **68**:1261-1271.

Slovik, D. M., Rosenthal, D. I., Doppelt, S. H., Potts, J. T., Jr., Daly, M. A., Campbell, J. A., and Neer, R. M., 1986, Restoration of spinal bone in osteoporotic men by treatment with human parathyroid hormone (1-34) and 1,25-dihydroxyvitamin D, *J. Bone Mineral Res.* **1**:377-381.

Sodoyez-Goffaux, F., Sodoyez, J.-C., Koch, M., DeVos, C. J., and Frank, B. H., 1988, Scintigraphic distribution of 123 I labelled proinsulin, split conversion intermediates and insulin in rats, *Diabetologia* **31**:848-854.

Sonnenberg, J. L., Sakane, Y., Jeng, A. Y., Koehn, J. A., Ansell, J. A., Wennogle, L. P., and Ghai, R. D., 1988, Identification of protease 3.4.24.11 as the major atrial natriuretic factor degrading enzyme in the rat kidney, *Peptides* **9**:173-180.

Steiner, D. F., Hallund, O., Rubenstein, A., Sho, S., and Bayless, C., 1968, Isolation and properties of proinsulin, intermediate forms and other minor components from crystalline bovine insulin, *Diabetes* **17**:725-736.

Steinetz, B. G., Beach, V. L., Kroc, R. L., Stasilli, N. R., Nussbaum, R. E., Nemith, P. J., and Dun, R. K., 1960, Bioassay of relaxin using a reference standard: A simple and reliable method utilizing direct measurement of interpubic ligament formation in mice, *Endocrinology* **67**:102-115.

Stentz, F. B., Kitabchi, A. E., Schilling, J. W., Schronk, L. R., and Seyer, J. M., 1989, Identification of insulin intermediates and sites of cleavage of native insulin by insulin protease from human fibroblasts, *J. Biol. Chem.* **264**:20275-20282.

Stephenson, S. L., and Kenny, A. J., 1987, The hydrolysis of a-human atrial natriuretic peptide by pig kidney microvillar membranes is initiated by endopeptidase-24.11, *Biochem. J.* **243**:183-187.

Stoll, R. W., Touber, J. L., Winterscheid, L. C., Ensinck, J. W., and Williams, R. H., 1971, Hypoglycemic activity and immunological half-life of porcine insulin and proinsulin in baboons and swine, *Endocrinology* **88**:714-717.

Striffler, J. S., 1987, Insulin clearance and microsomal glutathione-insulin transhydrogenase in perfused livers of fed and fasted rats, *Diabetes Metab.* **13**:582-590.

Sybertz, E. J., Chiu, P. J. S., Vemulapalli, S., Pitts, B., Foster, C. J., Watkins, R. W., Barnett, A., and Haslanger, M. F., 1989, SCH 39370, a neutral metalloendopeptidase inhibitor, potentiates biological responses to atrial natriuretic factor and lowers blood pressure in desoxycorticosterone acetate-sodium hypertensive rats, *J. Pharmacol. Exp. Ther.* **250**:624–631.

Tager, H. S., Patzelt, C., Assoian, R. K., Chan, S. J., Duguid, J. R., and Steiner, D. F., 1980, Biosynthesis of islet cell hormones, *Ann. N.Y. Acad. Sci.* **343**:133–147.

Tam, C. S., Heersche, J. N. M., Murray, T. M., and Parsons, J. A., 1982, Parathyroid hormone stimulates the bone apposition rate independently of its resorptive action: Differential effects of intermittent and continuous administration, *Endocrinology* **110**:506–512.

Terris, S., and Steiner, D. F., 1975, Binding and degradation of ^{125}I-insulin by rat hepatocytes, *J. Biol. Chem.* **250**:8389–8394.

Tillil, H., Frank, B. H., Pekar, A. H., Broelsch, C., Rubenstein, A. H., and Polonsky, K. S., 1990, Hypoglycemic potency and metabolic clearance rate of intravenously administered human proinsulin and metabolites, *Endocrinology* **127**:2418–2422.

Tregear, G. W., van Rietachoten, J., Greene, E., Keutmann, H. T., Niall, H. D., Reit, B., Parsons, J. A., and Potts, J. T., Jr., 1973, Bovine parathyroid hormone: Minimum chain length of synthetic peptide required for biological activity, *Endocrinology* **93**:1349–1355.

Vanneste, Y., Michel, A., Dimaline, R., Najdovski, T., and Deschodt-Lanckman, M., 1988, Hydrolysis of α-human atrial natriuretic peptide in vitro by human kidney membranes and purified endopeptidase-24.11, *Biochem. J.* **254**:531–537.

Vanneste, Y., Pauwels, S., Lambotte, L., Michel, A., Dimaline, R., and Deschodt-Lanckman, M., 1990, Respective roles of kallikrein and endopeptidase 24.11 in the metabolic pathway of atrial natriuretic peptide in the rat, *Biochem. J.* **269**:801–806.

Varandani, P. T., Shroyer, L. A., and Nafz, M. A., 1972, Sequential degradation of insulin by rat liver homogenates, *Proc. Natl. Acad. Sci. USA* **69**:1681–1684.

Verburg, K. M., Freeman, R. H., Davis, J. O., Villareal, D., and Vari, R. C., 1986, Control of atrial natriuretic factor release in conscious dogs, *Am. J. Physiol.* **251**:R947–R956.

Walton, P. E., Gopinath, R., Burleigh, B. D., and Etherton, T. D., 1989, Administration of recombinant human insulin-like growth factor I to pigs: Determination of circulating half-lives and chromatographic profiles, *Horm. Res.* **31**:138–142.

Weiss, T. J., and Bryant-Greenwood, G. D., 1982, Localization of relaxin binding sites in the rat uterus and cervix by autoradiography, *Biol. Reprod.* **27**:673–679.

Winslow, J., Shih, A., Laramee, G., Bourell, J., Stults, J., and Johnston, P., 1989, Purification and structure of human pregnancy relaxin from corpora lutea, serum and plasma, Program and Abstract Book of the Endocrinology Society 71st Annual Meeting, Seattle (Abstract #889, p. 245).

Wong, V. L. Y., Compton, M. M., and Witorsch, R. J., 1986, Proteolytic modification of rat prolactin by subcellular fractions of the lactating rat mammary gland, *Biochim. Biophys. Acta* **881**:167–174.

Woods, R. L., 1988, Contribution of the kidney to metabolic clearance of atrial natriuretic peptide, *Am. J. Physiol.* **255**:E934–E941.

Yadley, R. A., and Chrambach, A., 1973, Isohormones of human growth hormone. II. Plasmin-catalyzed transformation and increase in prolactin biological activity, *Endocrinology* **93**:858–865.

Yamaguchi, T., Fukase, M., Nishikawa, M., Fujimi, T., and Fujita, T., 1988, Parathyroid hormone degradation by chymotrypsin-like endopeptidase in opossum kidney cell, *Endocrinology* **123**:2812–2817.

Yandle, T. G., Richards, A. M., Nicholls, M. G., Cuneo, R., Espiner, E. A., and Livesey, J. H., 1986, Metabolic clearance rate and plasma half life of a-human atrial natriuretic peptide in man, *Life Sci.* **38**:1827–1833.

Yokono, K., Imamurs, Y., Shii, K., Mizuno, N., Sakai, H., and Baba, S., 1980, Immunochemical studies on the insulin-degrading enzyme from pig and rat skeletal muscle, *Diabetes* **29**:856–859.

Zarrow, M. X., and Money, W. L., 1948, Some studies on the pharmacology of relaxin, *J. Pharmacol. Exp. Ther.* **93**:180–187.

Zull, J. E., and Chuang, J., 1985, Characterization of parathyroid hormone fragments produced by cathepsin D, *J. Biol. Chem.* **260**:1608–1613.

Chapter 5

Pharmacokinetics and Metabolism of Therapeutic Cytokines

Carol A. Gloff and Robert J. Wills

1. TUMOR NECROSIS FACTOR

Tumor necrosis factor (TNF) was first described by Carswell and co-workers (1975) to cause necrosis in experimental tumors. Subsequent studies have shown that TNF is actually two factors, TNF-α and TNF-β; the latter is also known as lymphotoxin (Zahn and Greischel, 1989). Although both of these molecules have demonstrated a variety of biological effects, and both can bind to the same receptor, the majority of published work to date has described TNF-α or cachectin. Consequently, this section will focus exclusively on TNF-α.

TNF-α has been shown to inhibit the activity of lipoprotein lipase (Beutler *et al.,* 1985a) in addition to exerting cytostatic and cytotoxic effects against human tumor cells (Balkwill *et al.,* 1986; Carswell *et al.,* 1975; Helson *et al.,* 1975; Old, 1985). TNF-α is also important in the mediation of inflammation and the regulation of immune effector cells. Due to this variety of functions, it is understandable that TNF-α has moved rapidly through preclinical testing into clinical trials.

Carol A. Gloff • Alkermes, Inc., Cambridge, Massachusetts 02139.   *Robert J. Wills* • R. W. Johnson Pharmaceutical Research Institute, Raritan, New Jersey 08869.

Protein Pharmacokinetics and Metabolism, edited by Bobbe L. Ferraiolo *et al.* Plenum Press, New York, 1992.

1.1. Chemistry

TNF-α is a polypeptide of approximately 17 kDa as a monomer (Aggarwal *et al.*, 1985). The material is nonglycosylated, and usually found in a multimeric form (Selby *et al.*, 1987). The TNF-α sequence is approximately 80% conserved between mouse and man. Kramer and co-workers (1988) have shown that these differences lead to some species preference, although this species specificity is not as pronounced as with the interferons.

1.2. Analytical Techniques

The measurement of TNF-α in plasma has been performed using both immunoassays and bioassays. An ELISA assay technology has been used in the majority of clinical pharmacokinetic studies reported. Most assays utilized a sandwich technique, with two polyclonal antibodies to TNF-α. In some cases, one of the antibodies was linked to horseradish peroxidase (Kimura *et al.*, 1987), while in others biotin was conjugated to the antibody (Wiedenmann *et al.*, 1989). In a few instances, monoclonal antibodies replaced the polyclonals (Selby *et al.*, 1987; Spriggs *et al.*, 1988). In other cases, an RIA (Sheron *et al.*, 1990) or an EIA (Taguchi *et al.*, 1988) was used rather than an ELISA.

In several clinical studies, samples that were positive for TNF-α in an immunoassay were then assayed in a bioassay (Blick *et al.*, 1987; Moritz *et al.*, 1989; Steinmetz *et al.*, 1988). These bioassays involve the measurement of cell cytotoxicity of a sample believed to contain TNF-α. Although more difficult to perform and possibly more variable than the immunoassays, use of a bioassay does guarantee measurement of bioactive material.

Both immunoassays and bioassays have been used to evaluate the pharmacokinetics of TNF-α in animal studies. In addition, TNF-α has been labeled with ^{125}I or ^{3}H, and radioactivity used as a measure of TNF-α present (Bocci *et al.*, 1987; Ferraiolo *et al.*, 1988; Kramer *et al.*, 1988; Pacini *et al.*, 1987; Pessina *et al.*, 1986). The advantages and disadvantages of the use of immunoassays, bioassays, and radioactivity in evaluating protein pharmacokinetics are described elsewhere in this volume.

1.3. Pharmacokinetics

1.3.1. Absorption

1.3.1a. Clinical Studies.   Most clinical studies of TNF-α reported to date have utilized the intravenous route of administration, where "absorption" into the bloodstream is defined as 100%. However, in a few studies, another route of administration was evaluated in addition to the intravenous route. In a study reported by Blick *et al.* (1987), each subject received a single intramuscular and a single intravenous dose; the intramuscular dose was generally 1.33 to 5 times greater than the intravenous dose for each subject. Although serum concentrations could be measured, and pharmacokinetic parameters were readily calculated at intravenous doses of 25 to 100 $\mu g/m^2$, serum concentrations of TNF-α could not be consistently detected after intramuscular administration at doses less than 150 $\mu g/m^2$. In some subjects, serum concentrations of TNF-α were sustained for at least 24 hr after intramuscular dosing. This was in sharp contrast to the rapid clearance from the bloodstream seen after intravenous dosing in the same subjects. Due to the differences in dose levels and the small number of subjects, bioavailability after intramuscular dosing was not calculated.

Chapman *et al.* (1987) performed a study comparing subcutaneous and intravenous dosing. The subcutaneous dose was usually 1.2 to 5 times greater than the corresponding intravenous dose. In this study, subjects were generally escalated through four dose levels for each route of administration. Serum concentrations were generally measurable following intravenous doses of 25 to 200 $\mu g/m^2$. However, TNF-α was only detectable in serum after the highest subcutaneous dose (250 $\mu g/m^2$). This dose was given to only one subject, who demonstrated a peak concentration at 2 hr, but whose serum concentration of TNF-α became undetectable between 8 and 12 hr postdose. No assessment of bioavailability was attempted.

Another clinical study (Taguchi *et al.,* 1988) compared intravenous infusion with intratumoral administration of TNF-α in cancer patients. Each subject received TNF-α by only one of these two routes. Doses used for intratumoral administration were 0.1×10^6 to 2×10^6 units/subject with a specific activity of 3×10^6 units/mg protein. In all, 10 subjects received intratumoral TNF-α, but none demonstrated serum concentrations of TNF-α following this treatment.

1.3.1b. Preclinical Studies.   Studies in laboratory animals provide more complete evidence concerning absorption of TNF-α when administered by a nonintravenous route. However, only one of the studies to date has determined bioavailability, often either because serum sampling is not sufficiently extended to allow an adequate assessment of area-under-the-curve, or because the use of radioactivity as the means of detection creates significant doubts about the identity and structural integrity of the measured material.

Pacini *et al.* (1987) have evaluated the plasma pharmacokinetics of TNF-α in rabbits after intramuscular, subcutaneous, intraperitoneal, and oral administration, using both ^{125}I radioactivity and a bioassay to measure

TNF-α. The dose level was approximately 300 μg TNF-α per rabbit. All routes of administration, including oral, yielded measurable serum concentrations of TNF-α.

Administration via the intraperitoneal, intrastomach wall, and intragut wall routes, in addition to intramuscular, subcutaneous, and intravenous administration, were evaluated by Kojima *et al.* (1988) in rats. These investigators looked for TNF-α in both plasma and lymph, and found that the intraperitoneal route and the two gastrointestinal routes of administration resulted in high lymph/plasma ratios when compared to intravenous, intramuscular, and subcutaneous dosing.

Several groups have studied the absorption of TNF-α in mice. Ferraiolo *et al.* (1988) have compared intravenous and intramuscular dosing, and calculated bioavailability by the intramuscular route to be 12%. Flick and Gifford (1986), using murine TNF-α, found significant absorption after intraperitoneal administration, and a terminal half-life similar to that measured after intravenous administration. Nakano *et al.* (1990) have also measured TNF-α in serum after intraperitoneal administration.

1.3.2. Distribution and Elimination

1.3.2a. Clinical Studies.   Clinical trials of TNF-α administered intravenously as a bolus injection, as a short-term intravenous infusion, or as an intravenous infusion lasting up to 5 days, have been reported. In only one or two of these have the key pharmacokinetic parameters of clearance and volume of distribution been evaluated. However, the terminal half-life has been determined in the majority of these studies, permitting an evaluation of possible differences in apparent pharmacokinetics.

Blick *et al.* (1987) and Chapman *et al.* (1987) have evaluated the pharmacokinetics of TNF-α after intravenous bolus administration. In the former study, over the dose range of 25 to 100 μg/m^2, the terminal half-life remained relatively constant, approximately 14 to 18 min. However, the volume of distribution decreased with increasing dose from 66 to 12 liters, with a concomitant decrease in clearance, as a result of an increase in area-under-the-curve disproportionate to dose. A terminal half-life of 20 min, similar to that obtained by Blick *et al.* (1987), was also measured by Chapman *et al.* (1987) over the dose range of 35 to 100 μg/m^2. This group did not calculate volume of distribution or clearance for their subjects.

Several groups have utilized short-term infusions of TNF-α, ranging from 10 to 60 min. Moritz *et al.* (1989) found that the terminal half-life increased with increasing dose. For doses of 40 to 80 μg/m^2 given over 10 min, the terminal half-life was 11 to 17 min. However, at doses of 200 to 280

μg/m^2, the half-life increased to 54 to 71 min. Data were not available to determine volume of distribution or clearance for these subjects. In another study utilizing infusion times of 10 to 30 min, the terminal half-life was 40 to 50 min at the 100 μg/m^2 dose level (Sheron *et al.,* 1990). Taguchi and co-workers (1988) have evaluated the terminal half-life after doses of 0.5×10^6 to 5×10^6 units/subject (specific activity of 3×10^6 units/mg protein) given over 30 min, and have found this parameter to increase with increasing dose, from 0.5 to 2.4 hr. Kimura *et al.* (1987) utilized a 30-min infusion period, and a dose range of 1×10^5 to 16×10^5 units/m^2 (specific activity of 2.2×10^6 units/mg protein). The terminal half-life increased from approximately 10 to 80 min as the dose level was increased. Once again, volume of distribution and clearance were not determined. Creaven *et al.* (1987), utilizing a short-term intravenous infusion, evaluated TNF-α doses of 1×10^4 to 36×10^4 units/m^2, given as a 1-hr infusion (specific activity of 2.2×10^6 units/mg protein). At doses below 20×10^4 units/m^2, plasma concentrations of TNF-α could not be measured. For higher doses, the terminal half-life appeared to increase with increasing dose, from 0.33 to 0.72 hr, and plasma clearance decreased.

In several studies, infusions of 24 hr or longer have been tested. Wiedenmann *et al.* (1989) were unable to measure TNF-α in plasma after a 24-hr infusion administering doses of 40 to 400 μg/m^2. Steinmetz *et al.* (1988) administered doses of 2×10^5 to 7.5×10^5 units/m^2 (specific activity of 2.3×10^6 units/mg protein) over 24 hr. Plasma concentrations were measurable in a few patients at doses of 3×10^5 units/m^2 or greater, but samples were only collected during the course of the infusion, so pharmacokinetic parameters could not be calculated. A similar study design, with plasma sample collection only during the infusion phase, was used by Spriggs *et al.* (1988). Total doses of 182 to 636 μg/m^2 infused over 24 hr provided measurable plasma TNF-α concentrations, with concentrations generally decreasing after the first 1 to 3 hr of the infusion. A study by Sherman and co-workers (1988) utilized a 5-day continuous infusion, with doses ranging from 5×10^4 to 3×10^5 units/m^2 per day (specific activity of 2.2×10^6 units/mg protein). In no subject were there measurable TNF-α plasma concentrations during the course of the infusion. Finally, a study performed by Selby *et al.* (1987) evaluated the pharmacokinetics of TNF-α when administered as an infusion of unspecified duration. In this study, over a dose range of 9×10^3 to 1.2×10^6 units/m^2 (specific activity of 2.2×10^6 units/mg protein), clearance decreased from 457 to 59.5 ml/min and the volume of distribution decreased from 7.3 to 2.6 liters, while half-life increased from 10.4 to 41.5 min.

In summary, it appears, based upon the available data, that the pharmacokinetics of TNF-α may be dependent upon dose. In the few studies where these pharmacokinetic parameters were calculated, both clearance and volume of distribution decreased with increasing dose. However, these changes may be due to effects of repeated dosing in addition to, or instead of, those caused by increased dose level.

1.3.2b. Preclinical Studies.   A number of preclinical studies have evaluated the tissue distribution of TNF-α. Since this type of information is not available from human studies, the results from animal studies can be very useful in predicting target organs for both efficacy and toxicity.

Ferraiolo et al. (1988) have examined the tissue distribution of TNF-α in mice. Twenty-four hours after intravenous dosing, the highest percentages of the administered radioactive dose were found in liver, kidney, and lung, while at 10 and 240 min after dosing, the majority of the radioactivity was found in liver, kidney, and spleen. Similar results were noted following intramuscular dosing. Palladino and co-workers (1987) have also evaluated the biodistribution of ^{125}I-labeled TNF-α in tumor-bearing and normal mice. When the effects of unbound ^{125}I were taken into account, the distribution of TNF-α was not significantly different from that observed in the previous study. In addition, no significant differences in biodistribution between normal and tumor-bearing mice were noted. Yet another study evaluating the biodistribution of TNF-α in mice (Beutler et al., 1985b) provided results similar to those obtained in the previously mentioned studies.

Rabbits have also been used to evaluate the biodistribution of TNF-α (Pessina et al., 1986). Similar to mice, the largest amounts of TNF-α (measured by ^{125}I) were found in kidney, liver, lung, and spleen up to 90 min after intravenous dosing.

In addition to biodistribution studies in mice and rabbits, the pharmacokinetics of TNF-α have been evaluated in several species. Of particular interest are two studies performed in monkeys. In one study, at doses of approximately 115 to 230 μg/kg given as an intravenous bolus dose, clearance was 0.24 ml/min per kg, while the volume of distribution (β) was 85 ml/kg (Bocci et al., 1987). In another study, performed by Greischel and Zahn (1989), intravenous doses of 22 to 325 μg/kg provided clearance values of 0.2 to 0.3 ml/min per kg, similar to those in the previous study. Values for the volume of distribution (steady state), ranging from 31 to 41 ml/kg, cannot be directly compared to those from the previous study due to the different methods used for the calculation of the volumes of distribution.

Little work has been done to elucidate the elimination mechanisms utilized to remove TNF-α from the body. Ferraiolo et al. (1989) have compared the pharmacokinetics of TNF-α in normal and nephrectomized rats, and have shown a decrease in clearance of a high dose of TNF-α with loss of renal function. The central volume of distribution was decreased in female rats, but not in male rats. With the changes in these parameters, there was a concomitant increase in terminal half-life. Some increase in terminal half-life in nephrectomized rats was also seen at a low dose of TNF-α when compared to the same dose in control rats. A study by Beutler *et al.* (1985b) showed that radioactivity was heavily concentrated in urine after administration of [^{125}I]-TNF-α. However, the ^{125}I was found to no longer be bound to TNF-α, so the percent of the injected dose excreted in urine could not be determined. Finally, Kojima and co-workers (1988) have evaluated the lymphatic transport of TNF-α in rats, and have found that, particularly after administration localized to the gastrointestinal tract or peritoneal cavity, TNF-α can be concentrated in lymph when compared to blood.

1.3.3. Catabolism

Virtually no work has been done to evaluate the catabolism of TNF-α in humans. Beutler *et al.* (1985b) have shown that, within minutes after intravenous administration to mice, the majority of apparent TNF-α in tissues has been degraded to low-molecular-weight species. Pessina *et al.* (1986) have used isolated perfused livers from rabbits and monkeys, and kidneys and lungs from rabbits, to evaluate the catabolism of TNF-α. Livers from both species and rabbit lung were shown to have very little catabolic capability for TNF-α. A higher percentage of TNF-α appears to be catabolized by the rabbit kidney, although the rate of catabolism is not rapid in this tissue.

Ghezzi and co-workers (1986) have considered the effect of TNF-α on the metabolism of other compounds. These studies, using mice and subsequent microsomal assays, showed that TNF-α can cause a marked decrease in activity of cytochrome P450 and other drug-metabolizing enzymes, in the liver and other organs. It appears that although TNF-α is not significantly catabolized by the liver, its presence may affect the metabolism of other molecules.

2. INTERFERONS

2.1. Chemistry

Interferons (IFN) are classified into three major groups which reflect antigenic and structural differentiation. IFN-α (leukocytic interferon) is produced by B lymphocytes, null lymphocytes, and macrophages; its production can be induced by foreign, virus-infected, tumor, or bacterial cells. IFN-

β (fibroblast interferon) is a product of fibroblasts, epithelial cells, and macrophages which can be induced by viral and other nucleic acids. IFN-γ (immune interferon) is a product of activated T lymphocytes stimulated by foreign antigens. An in-depth review of interferon protein structure has been published by Rashidbaigi and Pestka (1988).

2.1.1. Interferon-α

Human IFN-α is a family of more than 15 subtypes, with molecular masses of 17.5–23 kDa. These proteins are composed of 165–166 amino acid residues and show approximately 80% sequence homology. The secondary structure of human IFN-α consists of 55–70% α-helix and less than 16% β-sheet. With the exception of subtypes B and D, these interferons contain four cysteine residues located at positions 1, 29, 99, and 139 which form disulfide bonds (1 to 99 and 29 to 139). Subtype B has a cysteine residue at position 98 instead of 99, and subtype D has an additional cysteine residue at position 86. In general, human IFN-α does not contain N-linked glycosylation sites. Subtype H is an exception, with two potential N-linked glycosylation sites at positions 2 and 78. These interferons tend to be acidic, with isoelectric points of 5.7–7.0.

2.1.2. Interferon-β

Human IFN-β is a single species. The molecular size of the natural glycoprotein is 23 kDa (166 amino acid residues). Twenty one of the original residues are removed during secretion from cells. IFN-β shows 29% structural homology to IFN-α. The secondary structure consists of 36% α-helix and 33% β-sheet. Cysteine residues are located at positions 17, 31, and 141. A potential N-linked glycosylation site exists at position 80. The isoelectric point falls between 6.8 and 7.8.

2.1.3. Interferon-γ

Human IFN-γ is heterogeneous, the molecular size depending on glycosylation and possibly oligomerization. The 20-kDa protein shows N-linked glycosylation at position 28, while the 25-kDa protein is glycosylated at positions 28 and 100. Higher molecular sizes of 50–70 kDa have been reported, suggesting oligomerization. IFN-γ is composed of 143 amino acids; it shows no statistically significant sequence homology to IFN-α or IFN-β.

Cysteine residues occur at positions 1 and 3. IFN-γ is acid-labile and highly basic, with an isoelectric point of 8.6–8.7.

2.2. Analytical Considerations

Interferon doses and concentrations are often defined in units of biological activity; this is derived with respect to the established activity of a reference standard. Since these standards are variable, it is unreasonable to compare or discuss dose and serum concentrations quantitatively for a given family of interferons. This constraint applies to the pharmacokinetic review which follows.

2.3. Pharmacokinetics

The majority of work involving interferons has been done with natural or cloned human interferon preparations. In the 1970s, human testing of interferons was restricted by their limited availability, purity (generally less than 1%), and specific activity (of the order of 10^6 U/mg). Recombinant DNA technology has provided high-purity ($\geq$99%) and high-specific-activity (2×10^8 U/mg) preparations for large-scale human testing (Goeddel et al., 1980). Human leukocyte interferon was the first preparation available. Early phase I studies, intended to establish the safety, tolerance, pharmacokinetics, and efficacy of interferons, targeted a host of viral and oncologic diseases (Gutterman et al., 1982; Hawkins et al., 1984; Krown et al., 1983; Nethersell et al., 1984; Quesada and Gutterman, 1983; Sherwin et al., 1982, 1983; Smith et al., 1983). Although the clinical results were often disappointing, the importance of the route and rate of administration as treatment variables was identified.

Biologic research in animals has been complicated by the species specificity of activity shared to some degree by all interferons (Stewart, 1981), while the activity of most conventional drugs in animals is used as an *in vivo* predictor for similar activity in humans. On the other hand, the pharmacokinetics of conventional drugs commonly demonstrate species specificity. However, the catabolism and excretion of interferons are similar across most species. This is not surprising considering that proteins are endogenous substances and protein processing is a natural function of all species. Therefore, animals may serve as suitable models for the pharmacokinetics of interferons in humans.

2.3.1. Absorption

2.3.1a. Clinical Studies. Oral absorption of intact proteins has met with limited success due to the natural proteolytic capabilities of the gastrointestinal tract. The systemic absorption of interferons from sites other than the gastrointestinal tract has been remarkably good, considering the size of these molecules. Intramuscularly and subcutaneously administered IFN-α (Bornemann *et al.*, 1985; Budd *et al.*, 1984; Gutterman *et al.*, 1982; Hawkins *et al.*, 1984; Omata *et al.*, 1985; Quesada and Gutterman, 1983; Radwanski *et al.*, 1987; Shah *et al.*, 1984; Sherwin *et al.*, 1982; Wells *et al.*, 1988; Wills *et al.*, 1984a) and nonglycosylated IFN-γ (Kurzrock *et al.*, 1985; Thompson *et al.*, 1987) are well absorbed: >80% for IFN-α and 30 to 70% for IFN-γ. These routes exhibit protracted absorption, which results in maximum serum or plasma concentrations occurring after 1 to 8 hr. Concentrations are measurable for 4 to 24 hr after injection for both IFN-α and -γ. Maximum serum concentrations following these routes of administration are at least an order of magnitude less than the highest concentration observed after administration of an equal intravenous dose. The absorption of IFN-β from muscle or skin has not been sufficient to produce serum concentrations much above the limits of assay detection (Billiau *et al.*, 1979; Hawkins *et al.*, 1985; Quesada *et al.*, 1982).

Other routes of administration have been evaluated in clinical studies: inhalation (Kinnula *et al.*, 1989); intralesional (Green *et al.*, 1984); intranasal (Davies *et al.*, 1983; Phillpotts *et al.*, 1984; Samo *et al.*, 1984); intraperitoneal (D'Acquisto *et al.*, 1988); intrathecal (Smith *et al.*, 1982); intraventricular (Jacobs *et al.*, 1986; Smith *et al.*, 1982); and intraocular (Turner *et al.*, 1989). For the most part, these alternative routes were attempts to improve the delivery of interferon to sites not easily accessible via the systemic circulation. These dosing strategies have provided adequate concentrations of interferon in CSF, lymph, nasal mucosa, and peritoneal fluid, but have not led to clinical success, undoubtedly reflecting the lack of understanding of the inherent mechanism of interferon action.

2.3.1b. Preclinical Studies. The similarities between the gastrointestinal tracts of humans and other mammals suggest that oral doses of interferons in laboratory animals should not result in significant interferon concentrations in the systemic circulation. As expected, studies in dogs (Gibson *et al.*, 1985) and monkeys (Wills *et al.*, 1984b) showed no measurable serum interferon concentrations after administration of an oral solution of recombinant IFN-α. However, Yoshikawa *et al.* (1984, 1985) presented preliminary evidence of mixed micelle enhancement of the absorption of IFN-α and -β via the lymphatic system from the rat large intestine, and Bocci *et al.* (1986a) reported enteric absorption of both IFN-α and -β in rats using an absorption enhancer.

Animal models have also been used to evaluate other sites of administration for interferons. Absorption of IFN-α from intramuscular (Cantell *et al.*, 1984; Collins *et al.*, 1985; Gibson *et al.*, 1985; Habif *et al.*, 1975; Naito *et al.*, 1984; Sarkar, 1982; Wills *et al.*, 1984b), subcutaneous (Gibson *et al.*, 1985; Sarkar, 1982), intraperitoneal (Heremans *et al.*, 1980; Sarkar, 1982), intradermal (Bocci *et al.*, 1986b), duodenal (Naito *et al.*, 1984), and rectal (Naito *et al.*, 1984) sites has been reported. In general, IFN-α absorption from these sites is prolonged. Maximum plasma or serum concentrations occur 1–6 hr postinjection, followed by measurable concentrations through 8–24 hr postinjection. The concentration–time profile appears to be independent of the purity or source of IFN-α; i.e., partially purified (Cantell *et al.*, 1984; Habif *et al.*, 1975; Heremans *et al.*, 1980), natural (Naito *et al.*, 1984; Sarkar, 1982), or recombinant (Bocci *et al.*, 1986c; Collins *et al.*, 1985; Wills *et al.*, 1984b). In addition, the disposition profiles were similar across the tested species, including mice (Heremans *et al.*, 1980; Sarkar, 1982), rabbits (Bocci *et al.*, 1986c; Cantell *et al.*, 1984; Naito *et al.*, 1984), dogs (Gibson *et al.*, 1985), and monkeys (Collins *et al.*, 1985; Habif *et al.*, 1975; Wills *et al.*, 1984b). Several of these studies have determined the absolute bioavailability from the intramuscular site to be 42% in dogs (Gibson *et al.*, 1985), and 93% (Wills *et al.*, 1984b) and 56% (Collins *et al.*, 1985) in monkeys.

The absorption of IFN-β from intramuscular (Abreu, 1983; Billiau *et al.*, 1979, 1981; Bocci *et al.*, 1986b; Gomi *et al.*, 1984; Hilfenhaus *et al.*, 1981; Satoh *et al.*, 1984), subcutaneous (Naito *et al.*, 1984), and intraperitoneal (Abreu, 1983) sites is similar to that observed for IFN-α. The concentration–time profiles are also similar to those seen with IFN-α, namely, prolonged absorption with concentrations persisting from 9 to 24 hr after injection. Again, absorption appears to be independent of species and interferon source. Estimates of absolute bioavailability from the intramuscular site can be obtained from published data: 60% in rats (Heremans *et al.*, 1980), 33% (Gomi *et al.*, 1984), and 60% (Satoh *et al.*, 1984) in rabbits, and 43% in monkeys (Gomi *et al.*, 1984). Recently, a bioavailability of 2.2% was reported for IFN-β following intranasal administration to rabbits (Maitani *et al.*, 1986).

The absorption of IFN-γ has not been studied as extensively as that of IFN-α or -β. In relevant papers by Cantell *et al.* (1983, 1984), absorption from intramuscular and subcutaneous sites was observed in rabbits and monkeys after injection of natural human IFN-γ. The resultant profiles were similar to those observed with IFN-α. Concentrations were measurable through 7–9 hr (Cantell *et al.*, 1983) and at least 24 hr (Cantell *et al.*, 1984) postinjection. In contrast, serum concentrations were very low or nondetectable following intramuscular injection of nonglycosylated IFN-γ in monkeys (Weck *et al.*, 1982).

2.3.2. Distribution and Elimination

2.3.2a. Clinical Studies.   The pharmacokinetics of interferons have been evaluated using a variety of dosing regimens varying from intermittent to continuous administration. Although definitive pharmacokinetics have not been reported in all cases, enough information is available to define their basic disposition characteristics. Serum interferon concentrations following intravenous administration decline rapidly in a biexponential manner for IFN-α (Bornemann *et al.*, 1985; Radwanski *et al.*, 1987; Shah *et al.*, 1984; Smith *et al.*, 1985; Wells *et al.*, 1988; Wills and Spiegel, 1985; Wills *et al.*, 1984a) and IFN-β (Grunberg *et al.*, 1987; Hawkins *et al.*, 1985; Liberati *et al.*, 1989; McPherson and Tan, 1980; Sarna *et al.*, 1986). Monoexponential decline has been observed for IFN-γ (Gutterman *et al.*, 1984; Kurzrock *et al.*, 1985, 1986; Vadhan-Raj *et al.*, 1986). Interferon concentrations fall several orders of magnitude over the measurable serum concentration–time course. Terminal half-lives range from 4 to 16 hr for IFN-α, 1 to 2 hr for IFN-β, and 25 to 35 min for IFN-γ. At the doses tested, serum concentrations are generally measurable for between 8 and 24 hr after dosing of IFN-α and up to 4 hr after injection for IFN-β and -γ.

The clearance of IFN-α has been reported to range from 4.9 to 21 liters/hr (Bornemann *et al.*, 1985; Shah *et al.*, 1984; Smith *et al.*, 1985; Wills *et al.*, 1984a) or 24 liters/hr per m^2 (Radwanski *et al.*, 1987). For IFN-β the clearance is much higher, 19–38 liters/hr per m^2 (Sarna *et al.*, 1986), while that of IFN-γ falls between the two at 12.66 to 32.4 liters/hr (Gutterman *et al.*, 1984; Kurzrock *et al.*, 1985).

The volume of distribution is similar for both IFN-α and -γ, ranging from 12 to 40 liters (Bornemann *et al.*, 1985; Gutterman *et al.*, 1984; Kurzrock *et al.*, 1985; Shah *et al.*, 1984; Wills *et al.*, 1984a). Although this volume is not physiological, it is approximately 20 to 60% of body weight. Information regarding IFN-β has not been reported.

The penetration of interferons across the human blood–brain barrier has been actively studied. Partially purified (Smith *et al.*, 1982), natural (Priestman *et al.*, 1982), and recombinant (Jablecki *et al.*, 1983; Martino and Singhakowinta, 1984; Smith *et al.*, 1985) interferons do not readily cross the blood–brain barrier intact after intravenous, intramuscular, or subcutaneous administration. These data must ultimately be reconciled with the occurrence of neurotoxicity, a common side effect of interferon therapy.

Endogenous interferon has been detected in tissues and fluids in a wide variety of seemingly unrelated diseases. For example, interferon has been found in the brain (Salonen, 1983) and CSF (Degre *et al.*, 1976) of patients with multiple sclerosis; in the lesions of patients with recurrent herpes labialis (Overall *et al.*, 1981) and psoriasis (Bjerke *et al.*, 1983); and in the synovial fluid of patients with rheumatoid arthritis (Degre *et al.*, 1983). The clinical relevance of these findings has yet to be determined, although reports by Heremans and Billiau (1989) and Khan *et al.* (1989) shed some light on this issue.

2.3.2b. Preclinical Studies. In laboratory animals, the initial decline in serum concentrations following intravenous administration is rapid for IFN-α, -β and -γ (Abreu, 1983; Bocci *et al.*, 1985a,b; Bohoslawec *et al.*, 1986; Cantell *et al.*, 1983; Collins *et al.*, 1985; Gibson *et al.*, 1985; Gomi *et al.*, 1984; Hilfenhaus *et al.*, 1981; Satoh *et al.*, 1984; Wills *et al.*, 1984b) for virtually all species tested. The half-lives of this initial distribution were on the order of minutes. Serum concentrations initially decrease rapidly, then decline more slowly, with terminal elimination half-lives ranging from minutes [30–40 min for IFN-α in mice (Bohoslawec *et al.*, 1986)] to hours [7–12 hr for IFN-β in rabbits (Gomi *et al.*, 1984; Satoh *et al.*, 1984)], depending on the species selected and the type of interferon (Bocci *et al.*, 1985b; Bohoslawec *et al.*, 1986; Cantell *et al.*, 1983; Collins *et al.*, 1985; Gibson *et al.*, 1985; Gomi *et al.*, 1984; Satoh *et al.*, 1984; Tokazewski-Chen *et al.*, 1983; Wills *et al.*, 1984b).

Both IFN-α and -β have a volume of distribution following intravenous administration that ranges from 20 to 100% of body weight in mice (Bohoslawec *et al.*, 1986), rats (Abreu, 1983), rabbits (Bocci *et al.*, 1985b; Gomi *et al.*, 1984; Satoh *et al.*, 1984), dogs (Gibson *et al.*, 1985), and monkeys (Collins *et al.*, 1985; Wills *et al.*, 1984b), suggesting distribution into a volume that approximates total body water. Similar data for IFN-γ are not available.

The tissue content of the interferons generally parallels that found in serum or plasma. Measurable concentrations or titers of interferons have been demonstrated in brain (Bohoslawec *et al.*, 1986; Heremans *et al.*, 1980), spleen (Abreu, 1983; Billiau *et al.*, 1981; Bohoslawec *et al.*, 1986; Heremans *et al.*, 1980), lung (Abreu, 1983; Billiau *et al.*, 1981; Bohoslawec *et al.*, 1986; Heremans *et al.*, 1980), liver (Bocci *et al.*, 1985a, 1986c; Bohoslawec *et al.*, 1986; Heremans *et al.*, 1980), and kidney (Abreu, 1983; Bocci *et al.*, 1985a; Bohoslawec *et al.*, 1986). For IFN-β and -γ, the amount of interferon determined in specific organs or tissues reflected the amount found in the serum or plasma, suggesting no uptake of these interferons into the sample organs or tissues (Abreu, 1983; Bocci *et al.*, 1985a). The results of tissue distribution studies with IFN-α have been mixed. Heremans *et al.* (1980) reported liver, lung, and spleen uptake of intraperitoneally administered homologous mouse IFN-α in mice. In contrast, Bohoslawec *et al.* (1986) showed no appreciable uptake of murine IFN-α into brain, liver, lung, or spleen following intravenous administration to mice. However, the kidney showed an appreciable uptake of murine IFN-α. Two human IFN-α subtypes, A and D, and one human hybrid, A/D (Bgl), displayed a tissue distribution profile similar to that of biologically active mouse IFN-α (Bohoslawec *et al.*, 1986).

CSF penetration of interferons has been an area of special interest owing to the potential clinical benefit of a central nervous system (CNS) active molecule and to the occurrence of adverse CNS drug reactions with interferons. Penetration into CSF could be achieved only with high intravenous doses of both IFN-α and -β (Abreu, 1983; Collins *et al.*, 1985; Habif *et al.*, 1975; Hilfenhaus *et al.*, 1981; Jablecki *et al.*, 1983). Because interferons do not readily cross the blood–brain barrier, some workers have elected to explore intrathecal or intraventricular administration (Collins *et al.*, 1985; Hilfenhaus *et al.*, 1981). High concentrations of IFN-α and -β in the CSF were obtained, with measurable concentrations persisting through 24–48 hr. Billiau (1981) found that IFN-β equilibrated throughout the CSF pathway after intrathecal injection to monkeys, but interferon was not detected in the substance of the brain upon sacrifice at 3 hr postadministration. Smith and Landel (1987) sought to map the action of interferon in the brain using molecular biology techniques. Partially purified IFN-α was administered to juvenile rhesus monkeys as a 7.5×10^6 IU intravenous dose or a 1×10^6 IU intracisternal dose. Analysis of brain tissue suggested that interferon reached the brain parenchyma from both the vascular and CSF compartments. Thus, the intrathecal or intraventricular routes of administration may provide alternatives to systemic delivery for attaining high interferon concentrations in the CSF. There are no similar data regarding IFN-γ.

2.3.3. Catabolism

2.3.3a. Clinical Studies.   There have been no definitive studies of the catabolism of the interferons in humans, although a number of studies have been performed in laboratory animals. As previously discussed, the catabolism of the three types of interferon, whether native or recombinant, is expected to be similar to the catabolism of other endogenous proteins. This process is likely to be similar across most species. Therefore, it is reasonable to discuss the catabolism of the interferons in animals with the assumption that it is applicable to humans. A good review of the work detailing the physiologic handling of plasma proteins is provided by Maack *et al.* (1979).

2.3.3b. Preclinical Studies. The catabolism of interferons has been the most widely researched area of preclinical interferon pharmacokinetics. Undoubtedly, the most effective and efficient system for catabolizing proteins is the gastrointestinal tract. However, the kidney is the major catabolic organ for circulating low-molecular-mass (<60 kDa) proteins (Ashwell and Morell, 1974). The liver has also been shown to play a role in the catabolism of glycosylated proteins (Bose and Hickman, 1977). Other organs and tissues, such as the lungs and muscle, are thought to play minor roles in protein catabolism (Fishman and Pietra, 1974).

The catabolism of IFN-α has been the most extensively studied of the three types of interferons. In general, IFN-α is filtered through the glomeruli of the kidney followed by luminal endocytosis and proximal tubular reabsorption (Bino *et al.*, 1982a,b; Bocci *et al.*, 1981a,b, 1982b). During reabsorption, IFN-α undergoes proteolytic degradation by lysosomal enzymes (Bino *et al.*, 1982a,b; Bocci *et al.*, 1984; Rosenberg *et al.*, 1985); therefore, negligible amounts are excreted intact in the urine. Several studies have assessed the effect of nephrectomy on the pharmacokinetics of IFN-α. As expected, the clearance was significantly decreased in nephrectomized rabbits (Bocci *et al.*, 1981a) and rats (Tokazewski-Chen *et al.*, 1983). Consistent with the above findings is the fact that the liver has been shown to play a small role in the catabolism of IFN-α (Bocci *et al.*, 1982a, 1983b).

Both IFN-β and -γ undergo renal catabolism (Bocci *et al.*, 1982b, 1983a, 1985a), but to a much smaller extent than IFN-α. The work performed thus far suggests that liver catabolism is the predominant pathway of elimination for IFN-β and -γ (Bocci *et al.*, 1982a, 1985a). Natural IFN-β and -γ are glycoproteins that contain sialic acid groups; they are therefore subject to the liver catabolism characteristic of this class of proteins (Ashwell and Morell, 1974). In one study (Tokazewski-Chen *et al.*, 1983) the clearance of both natural and recombinant IFN-β was unaltered in nephrectomized rats; this finding supports the hypothesis of a nonrenal pathway for catabolism. A detailed review of interferon catabolism is provided by Bocci (1985).

Evidence has been presented that recombinant IFN-α reduces hepatic drug metabolism activity in mice (Parkinson *et al.*, 1982; Secor and Schenker, 1984; Taylor *et al.*, 1985). Further, it has been demonstrated that the greater the antiviral activity of a given interferon in murine cells, the greater its ability to depress hepatic microsomal cytochrome P450 content in the mouse (Parkinson *et al.*, 1982).

REFERENCES

Abreu, S. L., 1983, Pharmacokinetics of rat fibroblast interferon, *J. Pharmacol. Exp. Ther.* **226**:197–220.

Aggarwal, B. B., Henzel, W. J., Moffat, B., Kohr, W. J., and Harkins, R. N., 1985, Primary structure of human lymphotoxin derived from 1788 lymphoblastoid cell line, *J. Biol. Chem.* **260**:2334–2344.

Ashwell, G., and Morell, A. G., 1974, The role of surface carbohydrates in the hepatic recognition and transport of circulating glycoproteins, *Adv. Enz.* **41**:98–128.

Balkwill, F. R., Lee, A., Aldam, G., Moodie, E., Thomas, J. A., Tavernier, J., and Fiers, W., 1986, Human tumor xenografts treated with recombinant human tumor necrosis factor alone or in combination with interferons, *Cancer Res.* **46**:3990–3993.

Beutler, B., Mahoney, J., LeTrang, N., Pekala, P., and Cerami, A., 1985a, Purification of cachectin, a lipoprotein lipase-suppressing hormone secreted by endotoxin-induced RAW 264.7 cells, *J. Exp. Med.* **161**:984–995.

Beutler, B. A., Milsark, I. W., and Cerami, A., 1985b, Cachectin/tumor necrosis factor: Production, distribution, and metabolic fate *in vivo, J. Immunol.* **135**(6):3972–3977.

Billiau, A., 1981, Interferon therapy: Pharmacokinetic and pharmacological aspects, *Arch. Virol.* **67**:121–133.

Billiau, A., de Somer, P., Edy, V. G., de Clercq, E., and Heremans, H., 1979, Human fibroblast interferon for clinical trials: Pharmacokinetics and tolerability in experimental animals and humans, *Antimicrob. Agents Chemother.* **16**:56–63.

Billiau, A., Heremans, H., Ververken, D., Van Damme, J., Carton, H., and de Somer, P., 1981, Tissue distribution of human interferon after exogenous administration in rabbits, monkeys and mice, *Arch. Virol.* **68**:19–25.

Bino, T., Edery, H., Gertler, A., and Rosenberg, H., 1982a, Involvement of the kidney in catabolism of human leukocyte interferon, *J. Gen. Virol.* **59**:39–45.

Bino, T., Madar, Z., Gertler, A., and Rosenberg, H., 1982b, The kidney is the main site of interferon degradation, *J. Interferon Res.* **2**:301–308.

Bjerke, J. R., Linden, J. K., Degre, M., and Matre, R., 1983, Interferon in suction blister fluid from psoriatic lesions, *Br. J. Dermatol.* **108**:295–299.

Blick, M., Sherwin, S. A., Rosenblum, M., and Gutterman, J., 1987, Phase I study of recombinant tumor necrosis factor in cancer patients, *Cancer Res.* **47**:2986–2989.

Bocci, V., 1985, Distribution, catabolism, and pharmacokinetics of interferons, in: *In Vivo and Clinical Fluids* (N. B. Finter and R. K. Oldham, eds.), Vol. 4, Elsevier, Amsterdam, pp. 47–72.

Bocci, V., Pacini, A., Muscettola, M., Paulesu, L., and Pessina, G. P., 1981a, Renal metabolism of rabbit serum interferon, *J. Gen. Virol.* **55**:297–304.

Bocci, V., Pacini, A., Muscettola, M., Paulesu, L., Pessina, G. P., Santiano, M., and Viano, I., 1981b, Renal filtration, absorption and catabolism of human alpha interferon, *J. Interferon Res.* **1**:347–352.

Bocci, V., Pacini, A., Bandinelli, L., Pessina, G. P., Muscettola, M., and Paulesu, L., 1982a, The role of liver in the catabolism of human alpha- and beta-interferon, *J. Gen. Virol.* **60**:397–400.

Bocci, V., Pacini, A., Muscettola, M., Pessina, G. P., Paulesu, L., and Bandinelli, L., 1982b, The kidney is the main site of interferon catabolism, *J. Interferon Res.* **2**:309–314.

Bocci, V., Di Francesco, P., Pacini, A., Pessina, G. P., Rossi, G. B., and Sorrentino, V., 1983a, Renal metabolism of homologous serum interferon, *Antiviral Res.* **3**:53–58.

Bocci, V., Mogensen, K. E., Muscettola, M., Pacini, A., Paulesu, L., Pessina, G. P., and Skiftas, S., 1983b, Degradation of human ^{125}I-interferon alpha by isolated perfused rabbit kidney and liver, *J. Lab. Clin. Med.* **101**:857–863.

Bocci, V., Maunsboch, A. B., and Mogensen, K. E., 1984, Autoradiographic demonstration of human ^{125}I-interferon alpha in lysosomes of rabbit proximal tubule cells, *J. Submicrosc. Cytol.* **16**:753–757.

Bocci, V., Pacini, A., Pessina, G. P., Paulesu, L., Muscettola, M., and Lunghetti, G., 1985a, Catabolic sites of human interferon-gamma, *J. Gen. Virol.* **66**:887–891.

Bocci, V., Pessina, G. P., Pacini, A., Paulesu, L., Muscettola, M., Naldini, A., and Lunghetti, G., 1985b, Pharmacokinetics of human lymphoblastoid interferon in rabbits, *Gen. Pharmacol.* **16**:277–279.

Bocci, V., Corradeschi, F., Naldini, A., and Lencioni, E., 1986a, Enteric absorption of human interferon alpha and beta in the rat, *Int. J. Pharm.* **34**:111–114.

Bocci, V., Muscettola, M., and Naldini, A., 1986b, The lymphatic route. IV. Pharmacokinetics of human recombinant interferon alpha[2] and natural interferon beta administered intradermally in rabbits, *Int. J. Pharm.* **32**:103–110.

Bocci, V., Muscettola, M., Naldini, A., Bianchi, E., and Segre, G., 1986c, The lymphatic route. II. Pharmacokinetics of human recombinant interferon-alpha 2 injected with albumin as a retarder in rabbits, *Gen. Pharmacol.* **17**:93–96.

Bocci, V., Pacini, A., Pessina, G. P., Maioli, E., and Naldini, A., 1987, Studies on tumor necrosis factor (TNF). I. Pharmacokinetics of human recombinant TNF in rabbits and monkeys after intravenous administration, *Gen. Pharmacol.* **18**:343–346.

Bohoslawec, O., Trown, P. W., and Wills, R. J., 1986, Pharmacokinetics and tissue distribution of recombinant human A, D, A/D (Bgl) and I interferons and mouse alpha-interferon in mice, *J. Interferon Res.* **6**:207–213.

Bornemann, L. D., Speigel, H. E., Dziewanowska, Z. E., Krown, S. E., and Colburn, W. A., 1985, Intravenous and intramuscular pharmacokinetics of recombinant leukocyte A interferon, *Eur. J. Clin. Pharmacol.* **28**:469–471.

Bose, S., and Hickman, J., 1977, Role of the carbohydrate moiety in determining the survival of interferon in the circulation, *J. Biol. Chem.* **252**:8336–8337.

Budd, G. T., Bukowski, R. M., Miketo, L., Yen-Lieberman, B., and Proffitt, M. R., 1984, Phase-I trial of ultrapure human leukocyte interferon in human malignancy, *Cancer Chemother. Pharmacol.* **12**:39–42.

Cantell, K., Hirvonen, S., Pyhala, L., De Reus, A., and Schellekens, H., 1983, Circulating interferon in rabbit and monkeys after administration of human gamma interferon by different routes, *J. Gen. Virol.* **64**:1823–1826.

Cantell, K., Fiers, W., Hirvonen, S., and Pyhala, L., 1984, Circulating interferon in rabbits after simultaneous intramuscular administration of human alpha and gamma interferon, *J. Interferon Res.* **4**:291–292.

Carswell, E. A., Old, L. A., Kassel, R. L., Green, S., Fiore, N., and William, B., 1975, An endotoxin-induced serum factor that causes necrosis of tumors, *Proc. Natl. Acad. Sci. USA* **72**:3666–3670.

Chapman, P. B., Lester, T. J., Casper, E. S., Gabrilove, J. L., Wong, G. Y., Kempin, S. J., Gold, P. J., Welt, S., Warren, R. S., Starnes, H. F., Sherwin, S. A., Old, L. J., and Oettgen, H. F., 1987, Clinical pharmacology of recombinant human tumor necrosis factor in patients with advanced cancer, *J. Clin. Oncol.* **5**(12):1942–1951.

Collins, J. M., Riccardi, R., Trown, P., O'Neill, D., and Poplack, D. G., 1985, Plasma and cerebrospinal fluid pharmacokinetics of recombinant interferon alpha A in monkeys: Comparison of intravenous, intramuscular and intraventricular delivery, *Cancer Drug Del.* **2**:247–253.

Creaven, P. J., Plager, J. E., Dupere, S., Huben, R. P., Takita, H., Mittelman, A., and Proefrock, A., 1987, Phase I clinical trial of recombinant human tumor necrosis factor, *Cancer Chemother. Pharmacol.* **20**:137–144.

D'Acquisto, R., Markman, M., Hakes, T., Rubin, S., Hoskins, W., and Lewis, J. L., Jr., 1988, A phase I trial of intraperitoneal recombinant gamma-interferon in advanced ovarian carcinoma, *J. Clin. Oncol.* **6**:689–695.

Davies, H. W., Scott, G. M., Robinson, J. A., Higgins, P. G., Wootton, R., and Tyrrell, D. A., 1983, Comparative intranasal pharmacokinetics of interferon using two spray systems, *J. Interferon Res.* **3**:443–449.

Degre, M., Dahl, H., and Vandvik, B., 1976, Interferon in the serum and cerebrospinal fluid in patients with multiple sclerosis and other neurological disorders, *Acta Neurol. Scand.* **53**:152–160.

Degre, M., Mellbye, O. J., and Clarke-Jensen, O., 1983, Immune interferon in serum and synovial fluid in rheumatoid arthritis and related disorders, *Ann. Rheum. Dis.* **42**:672–676.

Ferraiolo, B. L., Moore, J. A., Crase, D., Gribling, P., Wilking, H., and Baughman, R. A., 1988, Pharmacokinetics and tissue distribution of recombinant human tumor necrosis factor-alpha in mice, *Drug Metab. Dispos.* **16**(2):270–275.

Ferraiolo, B. L., McCabe, J., Hollenbach, S., Hultgren, B., Pitti, R., and Wilking, H., 1989, Pharmacokinetics of recombinant human tumor necrosis factor-alpha in rats. Effects of size and number of doses and nephrectomy, *Drug Metab. Dispos.* **17**(4):369–372.

Fishman, A. P., and Pietra, G. G., 1974, Handling of bioactive materials by the lung, *N. Engl. J. Med.* **291**:884–890.

Flick, D. A., and Gifford, G. E., 1986, Pharmacokinetics of murine tumor necrosis factor, *J. Immunopharm.* **8**(1):89–97.

Ghezzi, P., Saccardo, B., and Bianchi, M., 1986, Recombinant tumor necrosis factor depresses cytochrome P450-dependent microsomal drug metabolism in mice, *Biochem. Biophys. Res. Commun.* **136**:316–321.

Gibson, D. M., Cotler, S., Speigel, H. E., and Colburn, W. A., 1985, Pharmacokinetics of recombinant leukocyte A interferon following various routes and modes of administration to the dog, *J. Interferon Res.* **5**:403–408.

Goeddel, D. V., Yelverton, E., Ullrick, A., Heynecker, H. L., Miozzari, G., Holmes, R., Seeburg, P. H., Tabor, J. M., Gross, M., Familleti, P. C., and Pestka, S., 1980, Human leukocyte interferon produced by *E. coli* is biologically active, *Nature* **287**:411–416.

Gomi, K., Morimoto, M., Inoue, A., Kobayashi, H., Deguchi, T., Hara, T., and Nakamizo, N., 1984, Pharmacokinetics of human recombinant interferon-beta in monkeys and rabbits, *Gann* **75**:292–300.

Green, J. R., Klein, R. J., and Friedman-Kien, A. E., 1984, Intralesional administration of large doses of human leukocyte interferon for the treatment of condyloma acuminata, *J. Infect. Dis.* **150**:612–615.

Greischel, A., and Zahn, G., 1989, Pharmacokinetics of recombinant human tumor necrosis factor alpha in rhesus monkeys after intravenous administration, *J. Pharmacol. Exp. Ther.* **251**:358–361.

Grunberg, S. M., Kempf, R. A., Venturi, C. L., and Mitchell, M. S., 1987, Phase I study of recombinant gamma-interferon given by four-hour infusion, *Cancer Res.* **47**:1174–1178.

Gutterman, J. U., Fine, S., Quesada, J., Horning, S. J., Levine, J. F., Alexanian, R., Bernhardt, L., Kramer, M., Speigel, H., Colburn, W., Trown, P., Merigan, T., and Dziewanowski, F., 1982, Recombinant leukocyte A interferon: Pharmacokinetics, single-dose tolerance, and biologic effects in cancer patients, *Ann. Intern. Med.* **96**:549–556.

Gutterman, J. U., Rosenblum, M. G., Rios, A., Fritsche, H. A., and Quesada, J. R., 1984, Pharmacokinetic study of partially pure gamma-interferon in cancer patients, *Cancer Res.* **44**:4164–4171.

Habif, D. V., Lipton, R., and Cantell, K., 1975, Interferon crosses blood–cerebrospinal fluid barrier in monkeys, *Proc. Soc. Exp. Biol. Med.* **149**:287–289.

Hawkins, M. J., Borden, E. C., Merritt, J. A., Edwards, B. S., Ball, L. A., Grossbard, E., and Simon, K. J., 1984, Comparison of the biological effects of two recombinant human interferon alpha (rA and rD) in humans, *J. Clin. Oncol.* **2**:221–226.

Hawkins, M., Horning, S., Konrad, M., Anderson, S., and Sielaff, K., 1985, Phase I evaluation of a synthetic mutant of beta-interferon, *Cancer Res.* **45**:5914–5920.

Helson, L., Green, S., Carswell, E., and Old, L. J., 1975, Effect of tumour necrosis factor on cultured human melanoma cells, *Nature* **258**:731–732.

Heremans, H., and Billiau, A., 1989, The potential role of interferons and interferon antagonists in inflammatory disease, *Drugs* **38**:957–972.

Heremans, H., Billiau, A., and DeSorner, P., 1980, Interferon in experimental viral infection in mice: Tissue interferon levels resulting from virus infection and from exogenous interferon therapy, *Infect. Immun.* **30**:513–522.

Hilfenhaus, J., Damm, H., Hofstaetter, T., Mauler, R., Ronneberger, H., and Weinmann, E., 1981, Pharmacokinetics of human interferon-beta in monkeys, *J. Interferon Res.* **1**:427–436.

Jablecki, C. K., Poplack, D., Howell, S., Kingsbury, D., and Cantell, K., 1983, Highdose intravenous infusions of interferon, *Neurology* **33**:141–142.

Jacobs, L., Herndon, R., Freeman, A., Cuetter, A., Smith, W. A., Salazar, A. H., Ruse, P. A., Jose Rowica, R., Husam, F., Ekes, R., and O'Malley, J. A., 1986, Multicenter double-blind study of effect of intrathecally administered natural fibroblast interferon on exacerbations of multiple sclerosis, *Lancet* **2**:1411–1413.

Khan, N. U. D., Pulford, K. A. F., Farquharson, M. A., Howatson, A., Stewart, C., Jackson, R., McNicol, A. M., and Foulis, A. K., 1989, The distribution of immunoreactive interferon-alpha in normal human tissues, *Immunology* **66**:201–206.

Kimura, K., Taguchi, T., Urushizaki, I., Ohno, R., Abe, O., Furue, H., Hattori, T., Ichihashi, H., Inoguchi, K., Majima, H., Niitani, H., Ota, K., Saito, T., Suga, S., Suzuoki, Y., Wakui, A., Yamada, K., and the A-TNF Cooperative Study Group, 1987, Phase I study of recombinant human tumor necrosis factor, *Cancer Chemother. Pharmacol.* **20**:223–229.

Kinnula, V., Mattson, K., and Cantell, K., 1989, Pharmacokinetics and toxicity of inhaled human interferon-alpha in patients with lung cancer, *J. Interferon Res.* **9**:419–423.

Kojima, K., Takahashi, T., and Nakanishi, Y., 1988, Lymphatic transport of recombinant human tumor necrosis factor in rats, *J. Pharmacobio-Dyn.* **11**:700–706.

Kramer, S. M., Aggarwal, B. B., Essalu, T. E., McCabe, S. M., Ferraiolo, B. L., Figari, I. S., and Palladino, M. A., Jr., 1988, Characterization of the *in vitro* and *in vivo* species preference of human and murine tumor necrosis factor-alpha, *Cancer Res.* **48**:920–925.

Krown, S. E., Real, F. X., Cunningham-Rundles, S., Myskowski, P. L., Koziner, B., Fein, S., Mittleman, A., Oettgen, H. F., and Safai, B., 1983, Preliminary observations on the effect of recombinant leukocyte A interferon in homosexual men with Kaposi's sarcoma, *N. Engl. J. Med.* **308**:1071–1076.

Kurzrock, R., Rosenblum, M. G., Sherwin, S. A., Rios, A., Talpaz, M., Quesada, J. R., and Gutterman, J. U., 1985, Pharmacokinetics, single-dose tolerance, and biological activity of recombinant gamma-interferon in cancer patients, *Cancer Res.* **45**:2866–2872.

Kurzrock, R., Quesada, J. R., Rosenblum, M. G., Sherwin, S. A., and Gutterman, J. U., 1986, Phase I study of iv administered recombinant gamma interferon in cancer patients, *Cancer Treat. Rep.* **70**:1357–1364.

Liberati, A. M., Biscottini, B., Fizzotti, M., Schippa, M., de Angelis, V., Senatore, M., Vittori, O., Teggia, L., Natali, K., Palmisano, L., and Canali, S., 1989, A phase I study of human natural interferon-beta in cancer patients, *J. Interferon Res.* **9**:339–348.

Maack, T., Johnson, V., Kan, S. T., Figueiredo, J., and Sigulem, D., 1979, Renal filtration, transport and metabolism of low-molecular weight proteins: A review, *Kidney Int.* **16**:251–270.

McPherson, T. A., and Tan, Y. H., 1980, Phase I pharmacotoxicology study of human fibroblast interferon in human cancers, *J. Natl. Cancer Inst.* **65**:75–79.

Maitani, Y., Igawa, T., Machida, Y., and Nagai, T., 1986, Intranasal administration of beta-interferon in rabbits, *Drug Design Del.* **1**:65–70.

Martino, S., and Singhakowinta, A., 1984, Serial interferon alpha-2 levels in serum and cerebrospinal fluid, *Cancer Treat. Rep.* **68**:1057–1058.

Moritz, T., Niederle, N., Baumann, J., May, D., Kurschel, E., Osieka, R., Kempeni, J., Schlick, E., and Schmidt, C. G., 1989, Phase I study of recombinant human tumor necrosis factor alpha in advanced malignant disease, *Cancer Immunol. Immunother.* **29**:144–150.

Naito, S., Tanaka, S., Mizuno, M., and Kawashima, H., 1984, Concentrations of human interferons alpha and beta in rabbit body fluids, *Int. J. Pharm.* **18**:117–125.

Nakano, Y., Onozuka, K., Terada, Y., Shinomiya, H., and Nakano, M., 1990, Protective effect of recombinant tumor necrosis factor-alpha in murine salmonellosis, *J. Immunol.* **144**(5):1935–1941.

Nethersell, A., Smedley, H., Katrak, M., Wheeler, T., and Sikora, K., 1984, Recombinant interferon in advanced breast cancer, *Br. J. Cancer* **49**:615–620.

Old, L. J., 1985, Tumour necrosis factor (TNF), *Science* **230**:630–632.

Omata, M., Imazeki, F., Yokosuka, O., Ito, Y., Uchiumi, K., Mori, J., and Okuda, K., 1985, Recombinant leukocyte A interferon treatment in patients with chronic hepatitis B virus infection, *Gastroenterology* **88**:870–880.

Overall, J. C., Spruance, S. L., and Green, J. H., 1981, Viral-induced leukocyte interferon in vesicle fluid from lesions of recurrent herpes labialis, *J. Infect. Dis.* **143**:543–547.

Pacini, A., Maioli, E., Bocci, V., and Pessina, G. P., 1987, Studies on tumor necrosis factor (TNF). III. Plasma disappearance curves after intramuscular, subcutaneous, intraperitoneal and oral administration of human recombinant TNF, *Cancer Drug Del.* **4**(1):17–23.

Palladino, M. A., Jr., Shalaby, M. R., Kramer, S. M., Ferraiolo, B. L., Baughman, R. A., Deleo, A. B., Crase, D., Marafino, B., Aggarwal, B. B., Figari, I. S., Liggitt, D., and Patton, J. S., 1987, Characterization of the antitumor activities of human tumor necrosis factor-alpha and the comparison with other cytokines: Induction of tumor-specific immunity, *J. Immunol.* **138**(11):4023–4032.

Parkinson, A., Lasker, J., Kramer, M. J., Huang, M.-T., Thomas, P. E., Ryan, D. E., Reik, L. M., Norman, R. L., Levin, W., and Conney, A. H., 1982, Effects of three recombinant human leukocyte interferons on drug metabolism in mice, *Drug Metab. Dispos.* **10**:579–585.

Pessina, G. P., Pacini, A., Bocci, V., Maioli, E., and Naldini, A., 1986, Studies on tumor necrosis factor (TNF): II. Metabolic fate and distribution of human recombinant TNF, *Lymphokine Res.* **6**:35–44.

Phillpotts, R. J., Davies, H. W., Willman, J., Tyrell, D. A. J., and Higgins, P. G., 1984, Pharmacokinetics of intranasally applied medication during a cold, *Antiviral Res.* **4**:71–74.

Priestman, T. J., Johnston, M., and Whiteman, P. D., 1982, Preliminary observations on the pharmacokinetics of human lymphoblastoid interferon given by intramuscular injection, *Clin. Oncol.* **8**:265–269.

Quesada, J. R., and Gutterman, J. U., 1983, Clinical study of recombinant DNA-produced leukocyte interferon (clone A) in an intermittent schedule in cancer patients, *J. Natl. Cancer Inst.* **70**:1041–1046.

Quesada, J. R., Gutterman, J. U., and Hersh, E. M., 1982, Clinical and immunological study of beta interferon by intramuscular route in patients with metastatic breast cancer, *J. Interferon Res.* **2**:593–599.

Radwanski, E., Perenthesis, G., Jacobs, S., Oden, E., Affrime, M., Symchowicz, S., and Zampaglione, N., 1987, Pharmacokinetics of interferon alpha-2b in healthy volunteers, *J. Clin. Pharmacol.* **27**:432–435.

Rashidbaigi, A., and Pestka, S., 1988, Interferons: Protein structure, in: *The Interferon System* (S. Baron, G. J. Dianzani, W. R. Stanton, and W. R. Fleischmann, Jr., eds.), University of Texas Press, Austin, pp. 149–168.

Rosenberg, H., Madar, F., Gertler, A., Rubinstein, M., and Bino, T., 1985, The fate of ^{125}I-labeled human leukocyte-derived alpha interferon in the rat, *J. Interferon Res.* **5**:121–127.

Salonen, R., 1983, CSF and serum interferon in multiple sclerosis: Longitudinal study, *Neurology* **33**:1604–1606.

Samo, T. C., Greenberg, S. B., Palmer, J. M., Couch, R. B., Harmon, M. W., and Johnson, P., 1984, Intranasally applied recombinant leukocyte A interferon in normal volunteers: II. Determination of minimal effective and tolerable dose, *J. Infect. Dis.* **150**:181–188.

Sarkar, F. H., 1982, Pharmacokinetic comparison of leukocyte and Escherichia coli-derived human interferon type alpha, *Antiviral Res.* **2**:103–106.

Sarna, G., Pertcheck, M., Figlin, R., and Ardalan, B., 1986, Phase I study of recombinant ser 17 interferon in the treatment of cancer, *Cancer Treat. Rep.* **70**:1365–1372.

Satoh, Y., Kasama, K., Kajita, A., Shinizer, H., and Ida, N., 1984, Different pharmacokinetics between natural and recombinant human interferon beta in rabbits, *J. Interferon Res.* **4**:411–422.

Secor, J., and Schenker, S., 1984, Effect of recombinant alpha-interferon on in vivo and in vitro markers of drug metabolism in mice, *Hepatology* **4**:1081.

Selby, P., Hobbs, S., Viner, C., Jackson, E., Jones, A., Newell, D., Calvert, A. H., McElwain, T., Fearon, K., Humphreys, J., and Shiga, T., 1987, Tumour necrosis factor in man: Clinical and biological observations, *Br. J. Cancer* **56**:803–808.

Shah, I., Band, J., Samson, M., Young, J., Robinson, R., Bailey, R., Lerner, A. M., and Prasad, A. S., 1984, Pharmacokinetics and tolerance of intravenous and intramuscular recombinant alpha-2 interferon in patients with malignancies, *Am. J. Hematol.* **17**:363–371.

Sherman, M. L., Spriggs, D. R., Arthur, K. A., Imamura, K., Frei, E., III, and Kufe, D. W., 1988, Recombinant human tumor necrosis factor administered as a five-day continuous infusion in cancer patients: Phase I toxicity and effects on lipid metabolism, *J. Clin. Oncol.* **6**:344–350.

Sheron, N., Lau, J. N., Hofmann, J., Williams, R., and Alexander, G. J. M., 1990, Dose-dependent increase in plasma interleukin-6 after recombinant tumour necrosis factor infusion in humans, *Clin. Exp. Immunol.* **82**:427–428.

Sherwin, S. A., Knost, J. A., Fein, S., Abrams, P. G., Foon, K. A., Ochs, J. J., Schoenberger, C., Maluish, A. E., and Oldham, R. K., 1982, A multiple-dose phase I trial of recombinant leukocyte A interferon in cancer patients, *J. Am. Med. Assoc.* **248**:2461–2466.

Sherwin, S. A., Mayer, D., Ochs, J. J., Abrams, P. G., Knost, J. A., Foon, K. A., Fein, S., and Oldham, R. K., 1983, Recombinant leukocyte A interferon in advanced breast cancer, *Ann. Intern. Med.* **98**:598–602.

Smith, C. I., Weissberg, J., Bernhardt, L., Gregory, P. B., Robinson, W. S., and Merigan, T. C., 1983, Acute Dane particle suppression with recombinant leukocyte A interferon in chronic hepatitis beta virus infection, *J. Infect. Dis.* **148**:907–913.

Smith, R. A., and Landel, C. P., 1987, Mapping the action of interferon on primate brain, in: *The Biology of the Interferon System 1986* (K. Cantell and H. Schellekens, eds.), Nijhoff, The Hague, pp. 563–566.

Smith, R. A., Kingsbury, D., Alksne, J., James, H., and Cantell, K., 1982, Distribution of interferon in cerebrospinal fluid after systemic, intrathecal, and intraventricular administration, *Ann. Neurol.* **12**:81.

Smith, R. A., Norris, F., Palmer, D., Bernhardt, L., and Wills, R. J., 1985, Distribution of alpha interferon in serum and cerebrospinal fluid after systemic administration, *Clin. Pharmacol. Ther.* **37**:85–88.

Spriggs, D. R., Sherman, M. L., Michie, H., Arthur, K. A., Imamura, K., Wilmore, D., Frei, E., III, and Kufe, D. W., 1988, Recombinant human tumor necrosis factor administered as a 24-hour intravenous infusion. A phase I and pharmacologic study, *J. Natl. Cancer Inst.* **80**:1039–1044.

Steinmetz, T., Schaadt, M., Gahl, R., Schenk, V., Diehl, V., and Pfreundschuh, M., 1988, Phase I study of 24-hour continuous intravenous infusion of recombinant human tumor necrosis factor, *J. Biol. Respir. Mod.* **7**:417–423.

Stewart, W. E., II, 1981, *The Interferon System,* Springer-Verlag, Berlin.

Taguchi, T., and PT-50 Study Group, 1988, Phase I study of recombinant human tumor necrosis factor (rHu-TNF: PT-050), *Cancer Detect. Prevent.* **12**:561–572.

Taylor, G., Marafino, B. J., Jr., Moore, J. A., Gurley, V., and Blaschke, T. F., 1985, Interferon reduces hepatic drug metabolism in vivo in mice, *Drug Metab. Dispos.* **13**:459–463.

Thompson, J. A., Cox, W. W., Lindgren, C. G., Collins, C., Neraas, K. A., Bonnem, E. M., and Fefer, A., 1987, Subcutaneous recombinant gamma interferon in cancer patients: Toxicity, pharmacokinetics and immunomodulatory effects, *Cancer Immunol. Immunother.* **25**:47–53.

Tokazewski-Chen, S. A., Marafino, B. J., Jr., and Stebbing, N., 1983, Effects of nephrectomy on the pharmacokinetics of various cloned human interferons in rats, *J. Pharmacol. Exp. Ther.* **227**:9–15.

Turner, R. B., Durcan, F. J., Albrecht, J. K., and Crandall, A. S., 1989, Safety and tolerance of ocular administration of recombinant alpha interferons, *Antimicrob. Agents Chemother.* **33**:396–397.

Vadhan-Raj, S., Nathan, C. F., Sherwin, S. A., Oettgen, H. F., and Krown, S. E., 1986, Phase I trial of recombinant interferon gamma by 1-hr iv infusion, *Cancer Treat. Rep.* **70**:609–614.

Weck, P. K., Shalaby, M. R., Apperson, S., Gray, P. W., and Goeddel, D. V., 1982, Comparative biological properties of human alpha, beta and gamma IFN's derived from bacteria, Abstracts of the Third International Congress for Interferon Research, Miami, Florida.

Wells, R. J., Weck, P. K., Baehner, R. L., Krivit, W., Raney, R. B., Ortega, J. A., Bernstein, I. O., Lampkin, B., Whisnant, J. K., Sather, H. N., and Hammond, G. D., 1988, Interferon in children with recurrent acute lymphocytic leukemia: A Phase I study of pharmacokinetics and tolerance, *J. Interferon Res.* **8**:309–318.

Wiedenmann, B., Reichardt, P., Rath, U., Theilmann, L., Schule, B., Ho, A. D., Schlick, E., Kempeni, J., Hunstein, W., and Kommerell, B., 1989, Phase-I trial of intravenous continuous infusion of tumor necrosis factor in advanced metastatic carcinomas, *J. Cancer Res. Clin. Oncol.* **115**:189–192.

Wills, R. J., and Spiegel, H. E., 1985, Continuous intravenous infusion pharmacokinetics of interferon to patients with leukemia, *J. Clin. Pharmacol.* **25**:616–619.

Wills, R. J., Dennis, S., Spiegel, H. E., Gibson, D. M., and Nadler, P. I., 1984a, Interferon kinetics and adverse reactions after intravenous, intramuscular and subcutaneous injection, *Clin. Pharmacol. Ther.* **35**:722–727.

Wills, R. J., Spiegel, H. E., and Soike, K. F., 1984b, Pharmacokinetics of recombinant leukocyte A interferon following iv infusion and bolus, im, and po administration to African green monkeys, *J. Interferon Res.* **4**:399–409.

Yoshikawa, H., Takada, K., Muranishi, S., Satoh, Y., and Naruse, N., 1984, A method to potentiate enteral absorption of interferon and selective delivery into lymphatics, *J. Pharmaco. Dyn.* **7**:59–62.

Yoshikawa, H., Takada, K., Satoh, Y., Naruse, N., and Muranishi, S., 1985, Potentiation of enteral absorption of human interferon alpha and selective transfer into lymphatics in rats, *Pharm. Res.* **2**:249–250.

Zahn, G., and Greischel, A., 1989, Pharmacokinetics of tumor necrosis factor alpha after intravenous administration in rats, *Arzneim. Forsch./Drug Res.* **39**(II):1180–1182.

Chapter 6

Pharmacokinetics and Metabolism of Cardiovascular Therapeutic Proteins

Paul A. Cossum and Robert A. Baughman, Jr.

1. PROTEINS OF THE BLOOD COAGULATION PATHWAY

The blood coagulation pathway consists of a series of inactive and active enzymes (Fig. 1). In certain individuals, some of the elements, or factors, of this cascade may be inoperative or have reduced activity, and in some cases may be missing. Such cases give rise to the bleeding disorder hemophilia. There are three categories of patients with the deficiency: patients with normal amounts of factor that has reduced clotting activity; patients with factor and activity equally reduced; patients in whom the factor and its activity are undetectable. The existence of at least two forms of hemophilia was suggested by the results of experiments performed by Pavlovsky (1947) in which the mixing of the blood of two patients classified as hemophilics caused a correction of the clotting times of each blood sample. The more common hemophilia A, or classical hemophilia, and hemophilia B, or Christmas disease, occur as a result of factor VIII (FVIII) and factor IX (FIX) deficiency, respectively. These two diseases are X-linked recessive traits in which males are affected. Patients with severe hemophilia A or B have undetectable (less than 1% of normal) concentrations of FVIII or FIX and suffer from recur-

Paul A. Cossum • Department of Preclinical Development, Isis Pharmaceuticals, Carlsbad, California 92008.   *Robert A. Baughman, Jr.* • Research and Development, Emisphere Technologies, Hawthorne, New York 10532.

Protein Pharmacokinetics and Metabolism, edited by Bobbe L. Ferraiolo *et al.* Plenum Press, New York, 1992.

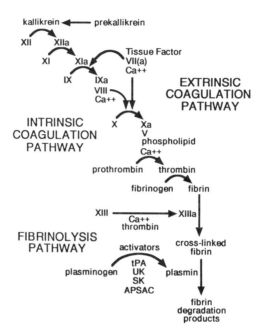

Figure 1. Simplified scheme depicting the relationship of the intrinsic and extrinsic coagulation pathways and the fibrinolysis pathway. Roman numerals refer to individual coagulation factors, and tPA, UK, SK, and APSAC refer to tissue plasminogen activator, urokinase, streptokinase, and anisoylated plasminogen streptokinase activator complex, respectively.

rent spontaneous hemarthroses and retroperitoneal bleeding. A related disorder is von Willebrand's disease which is caused by a lack of von Willebrand's factor (and also FVIII). von Willebrand's factor is essential for platelet aggregation. Other recognized hemophilias result from deficiencies in factors V, VII, X, XI, and XIII.

Several of the proteins involved in the blood coagulation pathway (Fig. 1) have been used for many years as replacement therapy in patients suffering from hemophilia. Until the 1950s and 1960s the only replacement therapy available for patients with classical hemophilia A was blood and fresh frozen plasma. The development of factor concentrates, processed from human plasma and suitable for clinical use, greatly advanced the treatment of hemophilic patients. The benefits of this form of treatment continued to be seen until the 1980s when the emergence of the acquired immune deficiency syndrome (AIDS) devastated the hemophilic community. Although this coincided with the development of recombinant DNA techniques that produce coagulation factors free from any risk of human immunodeficiency virus (HIV) infection, such factors are not yet approved for use. Current therapy uses heat-treated plasma concentrates that have undergone additional purification steps.

For the purpose of this chapter, only factors III, VII, VIII, IX, and XIII will be reviewed since these have been administered either to man or to other animals. Most of the pharmacokinetic data gathered on the coagulation factors emanate from studies using factor concentrates, i.e., factors extracted from human plasma. These procedures resulted in concentrates containing most or all of the coagulation factors. Consequently, it is possible that the clearance of any one factor may have been influenced by the concomitant administration of the other factors. Data have also been gathered using single factors purified from plasma. Recombinant DNA technology has provided at least two of the principal coagulation factors suitable for clinical use. The reader is referred to earlier references on the pharmacokinetics of the clotting factors (Owen and Bowie, 1975; Williams, 1983).

2. FACTOR VIII

2.1. Background

FVIII activity purified from plasma is associated with multiple polypeptide chains having molecular masses ranging from 80 kDa to 210 kDa (Eaton *et al.,* 1986). The amino acid sequence deduced from cDNA predicts that the molecule is originally synthesized as a single chain precursor with a molecular mass of about 300 kDa (Wood *et al.,* 1984; Toole *et al.,* 1984). In plasma it is noncovalently bound to von Willebrand's factor (vWF) which, apart from stabilizing FVIII, functions in platelet adhesion reactions. Of all the replacement therapies involving coagulation factors, FVIII has probably been used the most. Hemophilia A is an X-linked recessive genetic disorder, affecting about 1 in 10,000 males, that is due to a deficient or defective FVIII molecule. FVIII has also been used in the treatment of von Willebrand's disease, a platelet function deficit occurring with a concomitant reduction in FVIII activity. FVIII concentrates derived from plasma were introduced into clinical practice in the 1960s. This facilitated a marked improvement in the quality of life and the life expectancy of hemophilia A patients.

2.2. Endogenous Distribution

When immunohistochemical techniques have been used to study the localization of FVIII in selected human tissues, positive staining was obtained in the endothelial cells of liver sinusoids, veins and arteries, lung and spleen (Kadhom *et al.,* 1988). The localization of antigens in tissues and cells does not, however, distinguish the sites of synthesis from those of storage. The detection of FVIII mRNA may give a better indication of the sites of synthesis. Wion *et al.* (1985) detected FVIII mRNA in a range of differentiated cell types: isolated hepatocytes, spleen, lymph nodes, and kidney, but not in white blood cells or cultured endothelial cells. It is probable, therefore, that the liver, as for the vitamin K-dependent enzymes, is a primary site of FVIII production. Other strong evidence confirming the role of the liver in FVIII synthesis is that at least one patient's hemophilia was alleviated (at least temporarily) after receiving a liver transplant (Lewis *et al.,* 1985).

2.3. Pharmacokinetics

There are many commercial sources of FVIII concentrates, each one potentially differing from the next in potency (Ratnoff, 1986). Moreover, the methods used for the quantitation of the decay in plasma biological or immunological activity and for estimating pharmacokinetic parameters have been diverse. The reader is referred to reviews which compare FVIII pharmacokinetic methodologies (Messori *et al.,* 1987), and reported pharmacokinetic data (Kjellman, 1984).

2.3.1. Model-Dependent Pharmacokinetics

One confounding feature of the pharmacokinetics of FVIII is that plasma decay curves can be monophasic or biphasic in nature, although most are biphasic. When model-dependent methods have been used to estimate parameters, the initial and terminal half-lives, respectively, for humans were: 1.3–2.0 and 10.4–15.6 hr (Morfini *et al.,* 1984), 5.2 and 13.0 hr (Rousell *et al.,* 1989), 2.4 and 15.1 hr (Over *et al.,* 1978). Kjellman (1984) lists the elimination half-lives from several literature studies of various FVIII preparations; they vary from approximately 11 to 15 hr. Another occurrence making it difficult to compare various literature reports of the pharmacokinetics of FVIII is that some authors calculate half-lives based on plasma FVIII biological activities while other authors use plasma FVIII antigen concentrations to estimate half-lives. Hellstern *et al.* (1987) highlight this problem by reporting that the elimination half-life of an FVIII preparation was 3.8, 2.3, or 24.0 hr when estimated using plasma data generated by an IRMA, an ELISA, and a biological activity assay, respectively. In contrast to these data, McLellan *et al.* (1982) found, using an activity assay different from that used by Hellstern *et al.* (1987), similar half-lives for plasma antigen and plasma activity.

2.3.2. Model-Independent Pharmacokinetics

Model-independent methods have also been used to describe the pharmacokinetics of FVIII in hemophilia A patients administered a variety of FVIII concentrate preparations (Matucci *et al.,* 1985; Longo *et al.,* 1986; Noe *et al.,* 1986; Messori *et al.,* 1988). In these studies, plasma clearance values were in the range 3.3–5.0 ml/hr per kg, although there was often considerable patient-to-patient variability. This variability prompted Matucci *et al.* (1985) to suggest that there is a need to individualize FVIII dosage. These same authors reported that there was evidence of an inverse correlation between the clearance of FVIII and the patient's age (Matucci *et al.,* 1985). Data from the study of Messori *et al.* (1988) indicated that FVIII clearance was nonlinear, i.e., the clearance increased at higher doses. The authors, however, discounted this unusual finding. Since bleeding (Matucci *et al.,* 1985) is known to increase the consumption (clearance) of the deficient factor in hemophilics, the increase in clearance with increasing dose may reflect the fact that patients with more serious bleeding episodes received larger doses of FVIII. In the above studies, mean residence time (MRT) values were in the range 13.2–16.6 hr. The MRT values are equivalent to a half-life of approximately 9–11 hr. Messori *et al.* (1988) also analyzed their data using compartment analysis and estimated the half-life to be 10.3–12.2 hr. Surgery is one situation which may lead to an alteration in the half-life of FVIII in some hemophilia patients. In most cases, surgery has no apparent effect on the pharmacokinetics of FVIII, but in a small number of patients the postoperative half-life may be shortened to 6 hr. Since the volume of distribution of FVIII in these patients is apparently similar to that of unaffected patients, the clearance of FVIII must be altered. These patients represent a group who could be at risk for postoperative bleeding due to inadequate plasma concentrations of FVIII and so careful monitoring of postoperative plasma FVIII concentrations is encouraged (Messori *et al.,* 1987) in order that increased dosages may be administered if necessary.

The *in vivo* recovery is a distribution parameter that has been widely used in characterizing the pharmacokinetics of FVIII (Kjellman, 1984) and other coagulation factors. It is calculated as the ratio of plasma FVIII activity/concentration measured immediately following the end of dosing, and the theoretically expected activity/concentration. It is necessary to estimate the patient's plasma volume in order to calculate the latter value. The calculation of recovery is fraught with several uncertainties (Kjellman, 1984; Messori *et al.*, 1987) to the extent that large errors in the estimation are almost unavoidable. The calculation was originally introduced as a measure of the potency of FVIII products. In practice, however, it can be regarded as the inverse of the volume of distribution. The *in vivo* recovery may be of some value when compartmental kinetic analysis is performed, but the more relevant steady-state volume of distribution, which is determined in noncompartmental analysis, is preferred. In the studies discussed above (Matucci *et al.*, 1985; Longo *et al.*, 1986; Noe *et al.*, 1986; Messori *et al.*, 1988), the steady-state volume of distribution was 49–58 ml/kg. These values indicate that the factor was mostly distributed in the vascular space.

As mentioned above, reliance on plasma as the source of FVIII resulted in the exposure of hemophilia patients not only to alloantigens, but to hepatitis viruses and, more recently, to HIV. Proteins produced by recombinant DNA technology, apart from being produced in relatively large quantities, are free from such viruses. With the cloning and expression of the FVIII gene (Toole *et al.*, 1984; Wood *et al.*, 1984), this factor is now in clinical trials. In a study comparing the pharmacokinetics of recombinant human FVIII (rhFVIII) with plasma-derived human FVIII in hemophilia patients, White *et al.* (1989) showed the plasma decay curves to be biphasic; the elimination half-lives were 15.3 and 16.1 hr for the plasma-derived and recombinant factor, respectively. Another study of rhFVIII pharmacokinetics was performed by Schwartz *et al.* (1990). Model-independent pharmacokinetic analysis gave the clearance (2.5 ml/hr per kg), MRT (21.2 hr), volume of distribution at steady state (50 ml/kg), and elimination half-life (15.8 hr) after administration of a single dose of rhFVIII. There were no substantial differences in these values after multiple doses of rhFVIII. Nor were there any biologically significant differences in the values when compared to those of a plasma-derived FVIII measured in the same study. Both studies (White *et al.*, 1989; Schwartz *et al.*, 1990) demonstrated the equivalence of rhFVIII and plasma-derived FVIII in terms of efficacy, pharmacokinetics, and safety. The incidence of generation of inhibitors/antibodies to the rhFVIIIs was no greater than for plasma-derived FVIIIs, as determined from historic data.

2.4. Metabolism

Several enzymes are capable of degrading FVIII *in vitro.* Thrombin converted the 80- to 210-kDa proteins of FVIII to polypeptides of 73, 50, and 43 kDa. Maximum coagulant activity of thrombin-activated FVIII was correlated with the generation of these polypeptides. Activated protein C

(APC) and FXa also cleaved FVIII; proteolysis by APC inactivated FVIII, and proteolysis by FXa initially activated, but subsequently inactivated, FVIII (Eaton *et al.*, 1986). Plasmin, trypsin, and α-chymotrypsin all degraded functional and antigenic FVIII reactivity when added to fresh human plasma at room temperature (Rick *et al.*, 1985). Apart from providing a clue as to the *in vivo* regulation of FVIII concentrations, this study also shows that *in vitro* processing of FVIII may occur due to endogenous enzymes after blood collection. Indeed, loss of FVIII procoagulant activity following blood collection was a major problem when the only source of FVIII was plasma. Rock *et al.* (1983) investigated the ability of a series of protease inhibitors to stabilize FVIII activity in human plasma. Heparin offered the most protection. Calcium chelators caused a loss of factor activity that could be reversed upon the addition of calcium.

There were no differences in the decay of blood radioactivity (which would also have indicated the decay of the FVIII/von Willebrand's complex) when [^{125}I]-FVIII was administered to normal humans and hemophilia A patients (Over *et al.*, 1978). The higher-molecular-weight forms of the FVIII/von Willebrand's complex (i.e., those binding to platelets) disappeared more rapidly than the lower-molecular-weight forms (those remaining in the cryo-supernatant plasma). While the overall elimination half-life of plasma radioactivity was about 19 hr, the half-life of the highest-molecular-weight species was only 12 hr and that of the lowest-molecular-weight species was 27 hr. The longer survival of the low-molecular-weight forms may have been a result of their continued generation from the higher-molecular-weight forms. A similar study was performed in patients with von Willebrand's disease (Over *et al.*, 1981). The disappearance of radioactivity in the cryosupernatant fraction paralleled the decrease in coagulant activity, indicating that in von Willebrand's disease patients, the procoagulant activity of FVIII is associated with the low-molecular-weight form of von Willebrand's factor. Overall, the elimination half-lives for the FVIII and subfractions were similar to those in normals and hemophilia A patients. The urinary recovery of radioactivity (in the form of free ^{125}I) was 70 and 83% of the infused dose 66 hr after administration into normal and hemophilia A patients, or von Willebrand's disease patients, respectively.

2.5. Antibodies

One clinical feature of FVIII (and other factor) treatment is the development of antibodies to the factor. This may occur in as many as 20% of hemophilia A patients, although most reports cite about a 10% incidence

(Gill, 1984). The presence of either neutralizing or nonneutralizing antibodies to a protein may affect the efficacy and disposition of that protein (Working and Cossum, 1991). The most serious outcome of such an occurrence can be complete failure of FVIII therapy. This may occur with a concomitant faster plasma clearance of the factor (Allain, 1984; Nilsson *et al.*, 1990). Longo *et al.* (1986) reported that the values for plasma clearance, MRT, and steady-state volume of distribution in one patient in their series were 10.3 ml/hr per kg, 9.9 hr, and 102 ml/kg. These compared with 3.8 ml/hr per kg, 15.8 hr, and 56 ml/kg for the mean values of the patients without inhibitor. One treatment of such patients comprises the combined administration of FVIII, cyclophosphamide, and high-dose intravenous IgG (Nilsson *et al.*, 1988). Following this regimen, complexes of the FVIII and antibodies still formed, but the antibodies differed in specificity, lacked coagulation inhibitory activity, and did not enhance the rate of elimination of FVIII (Nilsson *et al.*, 1990).

Antibody generation to coagulation factors also impacts their preclinical development. Hemophilic dogs are often used to test the preclinical efficacy of these factors. Not surprisingly, Littlewood and Barrowcliffe (1987) found that all four of their hemophilic dogs generated antibodies to human FVIII. Therefore, unexpectedly reduced activity in dogs could mistakenly be interpreted as lack of efficacy of the molecule or formulation for use in humans.

3. FACTOR IX

3.1. Background

Human FIX is a vitamin K-dependent single-chain glycoprotein of approximately 57 kDa. It contains approximately 17% carbohydrate, so that the size of the polypeptide chain is approximately 48 kDa. FIX is activated by the action of FXIa in the intrinsic pathway or by the FVII/tissue factor complex in the extrinsic pathway (Fig. 1).

3.2. Pharmacokinetics

Aggeler (1961) reported that the postinfusion plasma activity data for FIX in humans could be described using a biexponential equation. Since then, a variety of disappearance rates for the two phases have been reported for humans and other animals. This variation probably reflects several features: the majority of infusion studies since Aggeler (1961) have relied on fewer postinfusion points with the assumption of monoexponential disappearance plasma curves, and most recently model-independent techniques have been used; the quality of earlier activity and plasma concentration assays; and the relative impurity of the administered preparations.

3.2.1. Model-Dependent Pharmacokinetics

The values for the distribution and elimination half-lives of FIX in patients with hemophilia B have been reported as 0.8–6.4 and 23–33 hr, respectively (Kohler et al., 1988; Smith and Thompson, 1981; Zauber and Levin, 1977). Distribution and elimination half-lives in hemophilic dogs were 1.3 and 31 hr, respectively. Distribution and elimination half-lives for normal baboons were 1.2 and 15.8 hr, respectively (Thompson et al., 1980). Kohler et al. (1988) found that different assays of FIX activity/concentration provided substantially different results, thus pointing out the need for a uniform system of assaying this coagulation factor.

3.2.2. Model-Independent Pharmacokinetics

Model-independent techniques have been used by Noe et al. (1986) and Longo et al. (1987) to describe the disposition of FIX in hemophilia B patients. The mean plasma clearance was 3.4 ml/hr per kg in patients receiving FIX infusions of 20–167 U/hr; the infusion rate was linearly related to the steady-state plasma FIX activity. Longo et al. (1987) reported a clearance of 5 ml/hr per kg and a steady-state volume of distribution of 100 ml/kg. These clearance values are consistent with the value 5.0 ml/hr per kg reported by Hoag et al. (1969).

The steady-state volume of distribution of FIX (100 ml/kg) and recovery from plasma (60%) (Longo et al., 1987) were approximately twice and half, respectively, that for FVIII (Longo et al., 1986). Several other reports (Barrowcliffe et al., 1973; Bidwell et al., 1967; Zauber and Levin, 1977; Kohler et al., 1988) show a low plasma recovery of FIX after intravenous administration; however, the recovery of most of the other infused factors is quantitative (Biggs and Denson, 1963; Loeliger and Hensen, 1964; Longo et al., 1986). The rapid distribution phase following intravenous administration of FIX, and the associated larger volume of distribution and lower plasma recovery, when compared to FVIII, may be related to the binding of FIX to endothelium. Extending in vitro studies in which they showed that cultured endothelium has specific binding sites for FIX (Stern et al., 1983), Stern et al. (1987) infused the bovine factor into normal baboons. They found that the infused bovine factor was capable of displacing baboon FIX from the putative vessel wall binding sites, resulting in a dose-related increase in baboon plasma factor antigen. Presumably, following intravenous administration to hemophilia B patients who have greatly reduced circulating concentrations of the factor, it would rapidly distribute to the pool of unoccupied vascular receptors. The liver may also be a site of binding of FIX (see below).

3.3. Metabolism

The fate of administered FIX is activation to FIXa and/or degradation. Activation of human FIX to FIXa by FXIa in the intrinsic pathway (Fig. 1) requires two proteolytic cleavages; the development of FIXa activity parallels the appearance of a 28-kDa protein. Activation in the extrinsic pathway apparently occurs via the same two cleavages as occur in the intrinsic pathway (Bertina and Veltkamp, 1981).

After intravenous administration of ^{125}I-labeled FIX to hemophilia B patients (Smith and Thompson, 1981) and normal baboons (Thompson *et al.*, 1980), no high-molecular-weight form of radioactivity was detected in urine, indicating that only free labeled iodine had been excreted. FIX is presumably degraded by the normal pathways of protein catabolism and the constituent amino acids are reutilized. Most of the radioactivity from a dose of ^{125}I-labeled FIX was found in the liver of baboons (Stern *et al.*, 1987) and mice (Fuchs *et al.*, 1984). The liver accounted for 86% of the dose in mice, and 68% of the dose in baboons, 2 and 30 min after administration, respectively. The same pattern of distribution of radioactivity was seen after administration of labeled FIXa in mice (Fuchs *et al.*, 1984). The liver may be another site of binding that helps explain the volume of distribution of FIX (see above). There is evidence that enzymatic cleavage of sialic acid from FIX reduces its clotting ability *in vitro* (Chavin and Weidner, 1984). It is not known if sialic acid removal occurs as part of the *in vivo* processing of FIX. Similarly, little is known of how FIX is transported in the blood. It is known, however, that FIXa complexes with antithrombin III, but not α_2-macroglobulin (Fuchs *et al.*, 1984).

3.4. Antibodies

Although 8–20% of patients with hemophilia A develop antibodies to FVIII, and a lesser percentage of hemophilia B patients develop antibodies to

FIX (Goodnight *et al.,* 1979), little is known of the capacity of the antibodies to affect the disposition of these two coagulation factors, particularly FIX. Goodnight *et al.* (1979) identified one patient who had circulating concentrations of an inhibitor to FIX and in whom the clearance of the infused factor was markedly reduced. Because plasma concentrations of FIX were measured by an activity assay, and not an immunoassay, it is likely that this was truly a faster clearance of the drug.

4. FACTOR VII/VIIa

4.1. Background

The single-chain glycoprotein zymogen FVII binds to tissue factor to be activated to the two-chain FVIIa; the complex is then presumably capable of activating factors IX and X (Fig. 1). The molecular mass of human FVII is about 48 kDa (Broze and Majerus, 1980). The zymogen shows substantial sequence homology with other vitamin K-related coagulation proteins. Less than 10% of the plasma FVII needs to be converted to FVIIa to initiate activation of FIX (Rao *et al.,* 1986). Severe hemostatic defects are observed, therefore, only in patients with very low concentrations of FVII; this disease is very rare. Due to its location in the extrinsic coagulation pathway, infusions of FVII are administered as a bypass therapy in patients with neutralizing antibodies to FVIII and FIX.

4.2. Pharmacokinetics

4.2.1. Factor VII

Generally, infusions of FVII in deficient patients have resulted in biexponential plasma decay curves of activity, although monoexponential curves have also been observed. Elucidation of the disposition profile of FVII has not been optimized; infrequent blood sampling at early times following infusion of the drug has been characteristic. When the frequency of blood sampling has allowed estimation of biexponential pharmacokinetic parameters, the half-lives of the distribution and elimination phases were approximately 0.5 and 4.0 hr (Hoag *et al.,* 1960), respectively. Half-lives associated with monoexponential declines in plasma activity concentrations were 3.5 hr (Mariani *et al.,* 1978), 4 hr (Kuzel *et al.,* 1988), and 6.6 hr (Kohler *et al.,*

1989). The volume of distribution was 2.5 liters and recovery from plasma was 133% (Kohler *et al.*, 1989).

4.2.2. Factor VIIa

Patients with severe hemophilia A and hemophilia B, complicated by high titers of neutralizing antibodies against FVIII and FIX, respectively, represent a serious therapeutic problem. Because FVIIa should bypass the intrinsic coagulation pathway and cause the generation of thrombin by the extrinsic pathway (Fig. 1), patients with severe hemophilia complicated by antibodies should benefit from FVIIa infusions. Plasma FVIIa activity had a half-life of approximately 2 hr when an FIX concentrate containing FVIIa was infused into nonbleeding hemophilia A and B patients (Seligsohn *et al.*, 1979). The persistence of FVIIa in the blood in this study suggested that the factor might have a hemostatic effect in these patients. FVIIa, purified from human plasma, was first conclusively shown to be a treatment for hemophilia patients with high-titer neutralizing antibodies by Hedner and Kisiel (1983). Subsequently, FVIIa produced by recombinant DNA technology was shown to be equipotent to the natural factor (Hedner *et al.*, 1988). In dogs with hemophilia A, the half-life of plasma FVIIa activity and antigen concentration after intravenous boluses of 50–220 μg/kg was approximately 2.8 hr. This finding was very similar to that for normal dogs. A mean plasma recovery of 34% was found for hemophilia A dogs and a recovery of 44% for hemophilia B dogs, indicating either an extravascular distribution or intravascular binding of the factor (Brinkhous *et al.*, 1989).

5. FACTOR XIII

5.1. Background

Fibrin-stabilizing factor (FXIII) circulates in blood as a four-chain precursor of 320 kDa (Loewy *et al.*, 1961). Subunit A carries the active site of the molecule which is activated by thrombin and calcium to form FXIIIa. This activated factor induces lysine-glutamine covalent cross-linking between adjacent fibrin molecules. In addition to its fibrin-stabilizing effects, FXIII may also cross-link fibrin to fibronectin and other adhesive proteins and promote platelet–vessel wall interactions (Bloom, 1990). Most patients with a congenital deficiency lack immunologically identifiable subunit A.

Inherited FXIII deficiency results in a serious hemorrhagic diathesis with a high risk of early death from intracranial hemorrhage.

5.2. Pharmacokinetics

FXIII derived from various sources has been used as replacement therapy. The pharmacokinetic parameters reported have been variable; this may be related to the source of the factor as well as the frequency of blood sampling and the assay methodology (Fear *et al.*, 1983). The terminal half-life following a bolus injection of human placental FXIII concentrate was 9–10 days when plasma activity and subunit A concentrations were measured (Fear *et al.*, 1983). An initial distribution phase was not quantitated. The observed 10-day half-life is supported by results in a patient who had been treated for 14 years, initially with plasma and then with FXIII concentrate. The patient only experienced bleeding episodes if the period between doses was greater than 6 weeks. A 10-day half-life would result in FXIII plasma concentrations at 6 weeks being about 2% of normal concentrations following the appropriate dose. This FXIII concentration would allow complete dimerization of fibrin. No bleeding was seen in two infants who were administered plasma cryoprecipitate FXIII prophylactically every 4 weeks, and in whom FXIII plasma concentrations had been shown to decay with a half-life of about 7 days (Daly and Haddon, 1988). Plasma concentrations of a pasteurized placental preparation of FXIII followed a biexponential decay with an initial half-life of 4 hr and a terminal half-life of about 11 days in two factor-deficient patients (Rodeghiero *et al.*, 1991). The plasma recovery was 50%, which suggested a distribution volume about twice the plasma volume. There have been no reports of antibodies interfering with FXIII replacement therapy.

6. TISSUE FACTOR (FACTOR III)

6.1. Background

Tissue factor (FIII, thromboplastin, TF) is a cell membrane-associated lipoprotein that serves as the receptor and essential cofactor for FVII and FVIIa. It has an amino acid content of 263 residues and a molecular mass of about 45 kDa; it contains approximately 33% carbohydrate (Fisher *et al.*,

1987). Its potential as a therapeutic agent lies in its possible ability to bypass FVIII, leading to coagulation in bleeding patients who have neutralizing antibodies to FVIII and in whom infusions of FVIII are of little value.

6.2. Pharmacokinetics

The pharmacokinetics of canine tissue factors, derived from brain, lungs, arteries, and veins, were investigated in four normal dogs (Gonmori and Takeda, 1975). Each of the factors (30–50 μg per dog) was labeled with ^{125}I and injected intravenously. Blood was sampled for up to 28 hr and the plasma was precipitated by trichloroacetic acid (TCA). The elimination of TCA-precipitable plasma radioactivity was described, in all cases, by biexponential equations. Initial half-lives were 0.60, 0.74, 0.57, and 0.87 hr, and the terminal half-lives were 7.4, 8.1, 14.6, and 24.3 hr, for the factors derived from brain, artery, lung, and vein, respectively. It is likely that the actual terminal half-lives were longer than the reported values since blood was only sampled for 28 hr.

More recently, the pharmacokinetics of recombinant human tissue factor apoprotein (rhTF) in normal and hemophilic dogs have been reported (Cossum et al., 1990). rhTF was administered as intravenous bolus doses of 5, 10, or 100 μg/kg in normal dogs, and 5, 10, or 25 μg/kg in hemophilic dogs. Blood was collected for up to 5 days and plasma rhTF concentrations were determined by ELISA. Plasma immunoreactivity decayed in a biexponential manner in all animals. In normal dogs, initial half-lives ranged from 1.1 to 2.0 hr and terminal half-lives ranged from 41 to 60 hr. The steady-state volumes of distribution were 98–137 ml/kg and plasma clearance ranged from 1.2 to 2.4 ml/hr per kg. There were no apparent dose-related differences in any of the pharmacokinetic parameters. Parameters for the hemophilic dogs were very similar to those for the normal dogs. TF has not been administered to humans.

6.3. Metabolism/Inactivation

The pharmacologic effects of TF may be attenuated by endogenous inhibitors. The presence of a serum inhibitor of TF was first suggested by experiments in which intravenous injections of extracts of the factor, preincubated with human serum, failed to kill mice. Injections of extracts which were not preincubated with serum were lethal (Schneider, 1947). Using *in vitro* assays, Broze and Miletich (1987) showed progressive inhibition of TF activity upon its addition to human plasma or serum. The agent responsible for the neutralization of TF reacted with the FVII(a)–Ca^{2+}–TF complex. TF was not destroyed per se, since functional activity could be restored following chelation of Ca^{2+} and by dilution.

At least two identified endogenous agents can inhibit TF activity. Lipo-protein-associated coagulation inhibitor (LACI), also known as extrinsic pathway inhibitor (EPI) (Sandset et al., 1991), contains three tandemly re-peated Kunitz-type serine protease inhibitory domains. One hypothesis to explain the mechanism by which LACI inhibits FVIIa/TF activity involves some of the generated FXa binding by its active site to the second Kunitz domain of LACI, leading to inhibition of the FXa activity. The FXa/LACI complex then binds to, and inhibits, the VIIa/TF complex, probably by forming a Xa/LACI/VIIa/TF quaternary complex (Girard et al., 1989). Sandset et al. (1991) reported that infusion of an antibody to EPI rendered rabbits susceptible to the disseminated intravascular coagulation effects of a dose of TF that had little effect in uncompromised animals. The inhibitory function of EPI/LACI in vivo may protect against the thrombotic complica-tions of infections and other coagulation-inducing phenomena incurred dur-ing normal living. Another inhibitory agent of TF activity, although proba-bly of less physiological importance, is purported to be apolipoprotein A-II (Carson, 1987). It is thought that this lipoprotein inhibits TF activity by altering the lipids around TF, thus reducing its ability to functionally asso-ciate with FVIIa.

The presence of these inhibitors may make interpretation of pharmaco-kinetic–pharmacodynamic relationship data more difficult since the biologi-cal half-life of TF may be different from the half-life calculated from plasma immunoreactivity data.

7. FIBRINOLYTIC THERAPEUTIC PROTEINS

In normal hemostasis a tightly controlled and complex series of cellular and biochemical interactions function to arrest bleeding at the site(s) of acute vessel injury, ultimately forming a thrombus. Typically, the thrombotic event produces a clot that is attached to the vessel or heart wall without obstructing the lumen or, in the case of the microvasculature, the clotting may be such that the vessel is completely occluded. Essential to this cascade of events is the ultimate dissolution of that thrombus and, as vascular repair occurs, the return of normal blood flow. It is the fibrinolytic pathway (Fig. 1) that initiates the dissolution of the thrombus.

Certain thrombotic events, however, fall outside the protection of "nor-mal" hemostasis, and result in frank medical emergencies (e.g., total or subtotal occlusion of a coronary artery resulting in myocardial infarction). The capacity of the fibrinolytic pathway to lyse the clot in these events is often overwhelmed, first by the size and location of the thrombus, and ultimately by the speed at which necrosis ensues in the surrounding tissue. Specific proteins of the exogenous fibrinolytic pathway, the plasminogen activators, promote the dissolution of a thrombus by activating the conversion of plasminogen to plasmin. The previous section focused on the pathologies associated with inoperative, reduced, or missing coagulation proteins, and those therapeutic proteins of the blood coagulation pathways that have been produced to overcome these deficiencies. This section will describe the pharmacokinetics of the plasminogen activators—tissue plasminogen activator (tPA), streptokinase (SK), anisoylated streptokinase, urokinase (uPA), and single-chain urokinase (scuPA)—that promote the dissolution of a thrombus by activating the conversion of plasminogen to plasmin.

8. TISSUE PLASMINOGEN ACTIVATOR

8.1. Background

The identification of tPA was made in the 1940s (Christensen and MacLeod, 1945; Astrup and Permin, 1947), but it was not until the purification and characterization of plasminogen activator from human uterine tissue (Rijken *et al.*, 1979), and subsequently from the Bowes melanoma cell line (mtPA) (Rijken and Collen, 1981), that sufficient quantities could be obtained for adequate studies. With the advances in recombinant DNA technologies, tPA was subsequently cloned (Pennica *et al.*, 1983) and expressed in a mammalian cell line. Native tPA is a serine protease, synthesized by vascular endothelial cells as a single polypeptide chain, composed of 527 amino acids. The molecule is glycosylated, with roughly 7–9% of its 64,000 molecular weight consisting of carbohydrates. There are three glycosylation sites (amino acids 117, 184, and 448) where oligosaccharides are bound to asparagine (Pohl *et al.*, 1984). There is a protease-sensitive peptide bond (Arg_{275}–Ile_{276}) that is cleaved by plasmin during fibrinolysis. Hydrolysis of this peptide bond converts tPA from a one-chain into a two-chain molecule (A and B chains) (Ranby *et al.*, 1989a). The A chain is composed of 275 amino acids and possesses the fibrin-binding region, while the B chain (252 amino acids) contains the protease-active site, as well as the binding sites for plasminogen activator inhibitor-1 (PAI-1). There is no net loss of protein mass since the N-terminal A chain remains connected by a single disulfide bridge to the B chain.

8.2. Analytical Issues

Data generated from a number of analytical techniques will be presented. The most commonly used methods are labeling with ^{125}I, activity assays, and immunoassays. Unfortunately, if all three methods were used with the same biological samples, differing results would be obtained. Each method has its place in characterizing tPA, but the shortcomings of each should be understood prior to initiating any studies. The use of total radioactivity and TCA precipitates has utility in studying tPA's disposition with early sampling *only*. After a few minutes, small peptide fragments or free ^{125}I will obscure further characterization with this technique. Activity assays have traditionally yielded lower plasma concentrations in pharmacokinetic studies than with immunoassays. In most cases, activity assays are laborious and difficult to standardize between laboratories. However, they are essential to fully characterizing the fate of tPA in biological fluids. For most pharmacokinetic studies, the ELISA, with its high sensitivity, has the greatest utility. The exact nature of the detected material is rarely characterized, but by choosing antibodies (monoclonal and/or polyclonal) to known epitopes (Holvoet *et al.*, 1985; Schleef *et al.*, 1986; Fong *et al.*, 1988; Ranby *et al.*, 1989b) the assay can be made specific enough to detect free tPA and/or tPA–inhibitor complexes.

Mohler *et al.* (1986) were able to prevent *in vitro* artifacts due to continued rtPA activation of plasminogen to plasmin, by adding D-Phe-Pro-Arg-chloromethylketone (PPACK) to biological samples. This procedure provided for more consistent measures of fibrinogen, fibrin(ogen) degradation products and α_2-antiplasmin, as well as a significant increase in the recovery of immunoreactive (ELISA) rtPA. The use of PPACK was further studied with similar results by Siefried and Tanswell (1987). This inhibitor has been incorporated into a number of rtPA ELISA assays (Hotchkiss *et al.*, 1988; Siefried *et al.*, 1988, 1989; Tanswell *et al.*, 1989; Tebbe *et al.*, 1989).

8.3. Pharmacokinetics

8.3.1. Preclinical Pharmacokinetics

The euglobulin fibrinolytic activity (EFA) in the plasma of animals and a few human subjects provided the first tPA pharmacokinetic data. In rabbits receiving mtPA as a bolus, the EFA declined with a half-life of 2–3 min (Korninger *et al.*, 1981). When [^{125}I]mtPA was administered to rabbits, the half-life of radioactivity (2.5 min) was similar to the EFA half-life over the first 20 min, and up to 60% of the radioactive dose was found in the liver. There was no significant loss of radioactivity in the plasma when mtPA was administered to hepatectomized rabbits. This finding suggested that over the first 20 min, ^{125}I radioactivity was associated with unchanged mtPA and that the initial clearance of the molecule was via the liver.

The disposition of radioactivity following [^{125}I]mtPA administration to normal rabbits was described by a three-compartment model by Nilsson *et al.* (1985). The half-lives of the three compartments were assigned to physiologic events: mtPA uptake by the liver and other eliminating organs; elimination of mtPA and formation and elimination of mtPA–inhibitor complexes; and a complex function of mtPA elimination, formation of complexes and mtPA metabolites, and the generation of free ^{125}I. Although it is inappropriate to assign physiologic events to mathematical compartments, the authors, nevertheless, present a reasonable hypothesis to explain tPA disposition. Rats were injected with 20 mg [^{125}I]mtPA and the distribution of radioactivity was determined at 3, 15, 60, 360, and 4320 min. Initially there was a rapid accumulation of radioactivity in the liver (58% of the dose), which decreased to 5% of the dose after 60 min.

In a series of experiments, Fuchs *et al.* (1985) administered [^{125}I]mtPA as an intravenous bolus to mice. The total radioactivity in plasma postdosing was equal to that obtained following the precipitation of plasma with TCA, again demonstrating that the activity measured (EFA or radioactivity) probably reflects intact mtPA. The multicompartment disposition of mtPA was also supported by the TCA-precipitable data generated in this study. The clearance of [^{125}I]mtPA from the plasma was unchanged in the presence of 1 mg asialoorosomucoid or a 1000-fold excess of unlabeled mtPA, leading the authors to conclude that plasminogen activator clearance is not carbohydrate mediated and the clearance is nonsaturable. The conclusions are in disagreement with subsequent work with mtPA disposition.

The presence of one- and two-chain (C1 and C2) forms of tPA was identified with mtPA early on by SDS–PAGE. But as additional work was conducted in the presence of aprotinin (a protease inhibitor) with human blood vessel perfusate (Binder *et al.*, 1979), and porcine heart and human uterus extracts (Rijken *et al.*, 1979; Wallen *et al.*, 1981), only one-chain tPA was found. In a fibrin clot model (Rijken *et al.*, 1982), C1 and C2 mtPA were shown to have the same activity and, when dissolved in plasma, the two tPA forms had similar activities toward preformed clots containing [^{125}I]fibrin. Unlike other serine proteases, the one- and two-chain forms appear to possess comparable activity.

Collen *et al.* (1984) compared rtPA from a mammalian cell line to mtPA. rtPA and mtPA caused a very similar dose-related degree of fibrinolysis, and bolus injections resulted in rapid and similar disappearances of EFA and radioactivity from the plasma ($t_{1/2}$ of 3 min).

Recombinant tPA was initially produced by a roller bottle process; this was changed to a suspension culture process to increase yield. The percentage of one-chain material in the final product was 1–5% and 70–80% for the two processes, respectively. It was unfortunate (and incorrect) that the roller bottle product was referred to as two-chain material and suspension culture product termed one-chain rtPA—because the materials from the two processes were not exclusively one-chain or two-chain, but rather different ratios of C1 and C2 rtPA and because this designation implied that the only differences between the two materials were related to the "chainedness" of the molecule. The two materials exhibited moderately different half-lives and clearances in studies where roller bottle and suspension culture rtPAs were administered intravenously by bolus injection or 60-min infusion to cynomolgous monkeys (Baughman, unpublished observation). Suspension culture material generated lower plasma concentrations than the roller bottle rtPA, indicating an apparent process-related difference in immunoreactive rtPA concentrations. The disposition of the material from both processes was described by a two-compartment model over the 45 min of sampling. The half-lives for roller bottle and suspension culture rtPA were 3.6 ± 0.5 and 2.9 ± 0.4 min for the initial, α phase, and 17.4 ± 3.8 and 13.3 ± 6.4 min for the terminal, β phase, respectively. The variability in the pharmacokinetic parameters was considerable; however, the clearance rate estimated for the suspension culture rtPA was 30–40% greater than that of the roller bottle material. This difference would eventually be described in the clinical setting by Garabedian et al. (1987).

In characterizing the roller bottle and suspension culture rtPAs, the question focused on the one-chain/two-chain issue and its effect on tPA's pharmacokinetics. Recombinant tPA (suspension culture, 79.5% one-chain = C1) was subjected to limited plasmin proteolysis generating material that was >95% two-chain (C2). C1 and C2 tPAs were administered as an intravenous bolus to adult rhesus monkeys (Baughman, 1987). Plasma concentrations and the resulting pharmacokinetic parameter estimates were not statistically significantly different. Thus, the differences observed in previous roller bottle/suspension culture comparisons that were attributed solely to the presence of one-chain or two-chain material were not observed when C1 and C2 rtPA (derived from C1 rtPA) were compared. It is concluded that the differences between the roller bottle and suspension culture rtPA were not due to "chainedness" but rather to some other process-related difference.

It is very important that the characterization of material be well established prior to initiating kinetic or metabolic studies with a recombinant protein before comparisons are made between materials. With three glycosylation sites, 17 disulfide bridges, and a primary sequence of 527 amino acids, minor perturbations in the rtPA molecule may be expected to have a significant impact on its disposition. Hotchkiss *et al.* (1988) characterized major alterations in rtPA disposition caused by alterations in the molecule's carbohydrate structure.

Fong and Lynn (1986) reported saturable extraction of a C2 rtPA in an isolated, perfused rat liver preparation (1 μg/ml). The dose-dependent disposition of the same material in anesthetized dogs was reported simultaneously by Fong *et al.* (1988) and Kopia *et al.* (1988). In this model, the disposition of C2 rtPA was characterized during a 90-min infusion (0.5, 1, 2, 4, and 8 μg/kg per min) and 30 min postdosing. Most notable was that steady-state plasma concentrations were linear up to 4 μg/kg per min, but clearance decreased by roughly 25% at the highest dose.

The most definitive preclinical pharmacokinetic study to date was conducted by Tanswell *et al.* (1990). Increasing doses of rtPA were administered to rats, rabbits, and marmosets, and in an isolated perfused rat liver preparation. Plasma concentrations (ELISA) were characterized by a three-compartment model. The corresponding half-lives were $t_{1/2\alpha}$ = 1.1–2.4 min, $t_{1/2\beta}$ = 10–40 min, and $t_{1/2\gamma}$ = 1–1.7 hr. The majority of the area under the plasma concentration–time curve (65–77%) was found in the α phase, with only 7–10% in the γ phase. The initial volume of distribution (V_1) was 46–91 ml/kg, which corresponds to the plasma volume. They further demonstrated that the *in vivo* disposition of rtPA at high doses (well above doses used in the clinical setting) was saturable, with a K_m = 12–15 μg/ml and a V_{max} of 200–350 μg/ml per hr, which is pertinent to animal studies at high doses.

8.3.2. Clinical Pharmacokinetics

8.3.2a. Pharmacokinetics in Normal Subjects. Four studies have described rtPA pharmacokinetics in normal subjects. Verstraete *et al.* (1985) were the first to show the multicompartment biexponential disposition of rtPA. This resulted in estimates for $t_{1/2\alpha}$ of 6.2 min and $t_{1/2\beta}$ of 55 min. Although material from the roller process was used in this study, the authors demonstrated the dependency of the plasma concentrations on infusion rate, and the large contribution of the initial α phase to the overall disposition.

Baughman (1987) described the pharmacokinetics of three suspension culture lots of rtPA in groups of nine subjects. The subjects received 0.25 mg/kg as a 10-min infusion, with frequent blood sampling (into PPACK) during and after the infusion. Plasma rtPA concentrations were measured by ELISA, and the resulting data were analyzed by both compartment-dependent and independent methods. Alpha phase half-lives ranged from

3.9 to 4.2 min, and $t_{1/2\beta}$ ranged from 29 to 36 min. V_1 was 4.3 liters and clearance ranged from 630 to 730 ml/min.

Siefried *et al.* (1988) administered suspension culture rtPA to eight normal volunteers at 0.25 mg/kg over 30 min. Plasma antigen was measured by ELISA (with and without PPACK), and by activity assay on both fibrin plates and in a chromogenic assay. The highest concentrations were measured in the ELISA with PPACK-treated plasma. The data were analyzed by a two-compartment model, and yielded the following mean results: C_{max} = 973 ng/ml, CL = 687 ml/min, $t_{1/2\alpha}$ = 3.3 min, $t_{1/2\beta}$ = 26 min, and V_1 = 3.9 liters. Most notable in this study was that linear regressions of the fibrin plate and chromogenic assays versus the ELISA yielded excellent correlations ($R > 0.96$), but the slopes were 0.76 and 0.64. This indicates that 25–35% of rtPA antigen in thawed plasma samples is not detected in activity assays, which is due, at least in part, to the *in vitro* binding of tPA to inhibitors.

Tanswell *et al.* (1989) reported the pharmacokinetic parameters from normal subjects (N = 6/group) receiving two doses (0.25 and 0.5 mg/kg) over 30 min. The clearance was 620 ± 70 ml/min, V_1 = 4.4 ± 0.6 liters, and the α and β half-lives were 4.4 and 40 min, respectively. More importantly, dose linearity was established over this dosage range.

8.3.2b. **Pharmacokinetics in Myocardial Infarction Patients.** Few studies have evaluated in detail the kinetics of the marketed form (suspension culture) of tPA in patients. The pharmacokinetics of roller bottle-derived and suspension culture rtPA were compared in patients being treated for myocardial infarction (MI). All patients received an IV bolus of 10% of the calculated total dose followed by a 90-min constant rate infusion of 4–11 μg/kg per min. Plateau plasma concentrations ranged from 0.52 to 1.8 μg/ml, and were linearly correlated with the infusion rate. However, suspension culture rtPA resulted in plasma concentrations that were typically 30–35% lower than those obtained with roller bottle material. For those subjects receiving suspension culture product, their α and β half-lives were 3.6–4.6 and 39–53 min, respectively. V_1 was 3.8 to 6.6 liters, and the plasma clearance was 520–1000 ml/min.

A more detailed study has been reported by Siefried *et al.* (1989) in which the recommended dosage regimen was utilized in 12 MI patients and plasma was sampled out to 72 hr. These patients received 10 mg infused within 2 min, followed by 50 mg in 1 hr, and 30 mg in 1.5 h. It was found for all patients that a three-compartment model ($t_{1/2\alpha}$ = 3.6 ± 0.9 min, $t_{1/2\beta}$ = 16 ± 5.4 min, $t_{1/2\gamma}$ = 3.7 ± 1.4 hr) yielded significantly better fits of the data than the two-compartment model. However, terminal phase tPA concentrations approached background levels (5 ng/ml) and >97% of the AUC was contained within the α and β phases. However, the observed clearance of 383 $\pm$ 74 ml/min was appreciably lower than that observed in previous studies in healthy volunteers. Whether this is due to the impaired blood flow during an infarction and/or the older age of this patient population cannot be discerned from the data. Baughman (1987) reported that the clearance and V_1 were lower in eight elderly patients with thrombo-occlusive disease than in healthy volunteers (549 versus 680 ml/min, and 3.8 versus 4.3 liters, respectively).

Finally, in a study evaluating a single 50-mg rtPA bolus in MI patients (Tebbe *et al.*, 1989), the kinetic parameters derived following the bolus were comparable to parameters obtained from patients receiving an infusion. Since the C_{max} was well in excess of what is observed during the infusion, the authors conclude that tPA clearance is not saturated during a myocardial infarction.

8.4. Metabolism

There is a wide body of evidence that tPA is subject to hepatic extraction. Korninger *et al.* (1981) and Nilsson *et al.* (1984) demonstrated that intravenously injected tPA is rapidly cleared from the circulation by the liver. The half-life of tPA in hepatectomized rabbits was increased 20-fold (Nilsson *et al.*, 1985). Bakhit *et al.* (1987) confirmed that [125I]rtPA was taken up by the liver as the free enzyme and was not complexed to protease inhibitors. The interaction *in vivo* of 125I-labeled tPA with rat liver and various liver cell types was also characterized by Kuiper *et al.* (1988). They reported that [125I]tPA was rapidly cleared from the circulation by the liver, and then promptly degraded in lysosomes. The intrahepatic recognition site(s) for tPA were determined by subfractionation of the liver, with 54.5% in parenchymal cells, 39.5% in endothelial cells, and 6% in Kupffer cells. The association of tPA with the parenchymal cells was not mediated by a carbohydrate-specific receptor; however, the association with endothelial cells could be inhibited 80% by ovalbumin (a mannose-terminated glycoprotein). The authors concluded that the uptake of tPA by the liver is mainly mediated by two recognition systems—a specific tPA site on parenchymal cells and via mannose receptors on endothelial cells.

Smedsrod *et al.* (1988) demonstrated that tPA binds to a mannose receptor in rat liver endothelial cells. As tPA contains a mannose-rich sugar at position 117, this position would be expected to contribute significantly to tPA removal. Hotchkiss *et al.* (1988) showed that removing the sugar at this position effectively reduced tPA clearance twofold *in vivo*. Bakhit *et al.*

(1987) described a tPA-specific receptor that functions independently of the carbohydrate receptor(s), probably in parenchymal cells. To date, this receptor has not been fully characterized.

The active site of tPA is probably not involved in its clearance from the circulation (Mohler *et al.*, 1986), and data from Smedsrod *et al.* (1988) would suggest that B-chain sites contribute little to tPA's catabolism. Metabolism studies are now focusing on the N-terminus as an additional contributor to tPA clearance.

9. STREPTOKINASE AND ANISOYLATED STREPTOKINASE

9.1. Background

Streptokinase (SK) is an extracellular protein produced by Lancefield group C strains of β-hemolytic streptococci. It has a molecular weight of 47,000 and a primary structure of 414 amino acids; the NH_2-terminal 230 amino acids show homology with trypsin-type serine-proteases (Jackson and Tang, 1982). However, SK does not possess an active-site serine. "Thus, streptokinase cannot cleave peptide bonds in proteins, has no esterase or amidase activities against synthetic substrates. . . . In fact, streptokinase is not an enzyme by any of the criteria used in defining a protein as an enzyme" (Reddy, 1988). SK acts by combining with plasminogen to form an activator complex (the active site is within the modified plasminogen) that then combines with an additional plasminogen and converts the latter into plasmin. It is this plasmin which then lyses the fibrin in the thrombus. SK has been reported to exist in four modified forms, differing in molecular mass: SK_a, 40 kDa; SK_b, 36 kDa; SK_c, 31 kDa; and SK_d, 26 kDa. The SK_b form is the major one in humans (Markus *et al.*, 1976) and the SK_d form is found in dogs and rabbits (Reddy, 1976; Siefring and Castellino, 1976). Plasminogens from different species show variability in their response to SK. For example, plasminogens from mouse, rat, sheep, pig, and cow do not form a complex with SK; rabbits require a large excess of SK (Siefring and Castellino, 1976) and dog plasminogen forms a complex with equimolar amounts of SK (Reddy, 1976). The biological activity of SK depends upon the reaction of SK with plasminogen forming the SK–plasmin complex, which can bind directly to fibrin.

SK was first introduced as a thrombolytic agent in the 1950s, but fell into disfavor due to a number of factors. Although effective, it was administered via intracoronary infusion. This route of administration has a number of disadvantages; it is costly and requires considerable expertise and catheterization facilities. It has also been associated with coronary perforation. Comparable doses used intravenously did not approach the efficacy of the intracoronary route. As increased doses (1.5 million IU) were used intravenously, reperfusion rates between 35 and 55% were typically observed. This dose resulted in a relatively high incidence of systemic fibrinogenolysis. To circumvent these problems, an essential serine (residue 740) in the catalytic center of the plasminogen light B chain is acylated. This masking of the catalytic center does not interfere with the high capacity of the complex to bind fibrin, as the fibrin binding sites are in a different part of the molecule (Ferres *et al.,* 1987). This anisoylated complex (*p*-anisoylated plasminogen streptokinase activator complex, APSAC) is inactive, as the formation of plasmin is delayed. Deacylation begins immediately upon injection, and ultimately a sufficient portion of the complex is bound to fibrin at the time of deacylation. The plasmin formed via this pathway is protected from circulating antiplasmins and is available for fibrin degradation. Since the rate of plasmin formation is retarded, circulating deacylated complexes are rapidly neutralized, at least until sufficient plasminogen is deacylated to overwhelm the concentration of antiplasmins.

Standring *et al.* (1988) measured the rate of loss of fibrinolytic activity in plasma of both SK–plasmin(ogen) and APSAC on fibrin plates after euglobulin fractionation. The total potential fibrinolytic activity of APSAC was longer than that of SK–plasmin(ogen). Additionally, the loss of SK function in APSAC in plasma was much slower. This finding supports the earlier work by Fears *et al.* (1987), who observed that the acyl group protected the complex from neutralization by plasmin inhibitors in plasma.

9.2. Pharmacokinetics

The pharmacokinetics of SK remain poorly defined, primarily because no direct assay of the protein exists. In addition, the relatively rapid clearance of SK activity contributes to the poor kinetic characterization. Staniforth *et al.* (1983) reported an SK activity half-life of 25 min. The majority of the kinetic work has been conducted with APSAC, probably in an effort to correlate rate of acylation with activity clearance rates. Esmail *et al.* (1984) were able to demonstrate a correlation between deacylation rate constants and clearance in rabbits. Been *et al.* (1986) reported an APSAC clearance half-life for fibrinolytic activity of 87.5 ± 5.0 min. However, this work could not differentiate between APSAC, metabolites, and other components that contribute to total fibrinolytic activity.

The chromogenic substrate S-2251 is split by both plasmin and the SK–plasminogen complex. Grierson and Bjornsson (1987) developed a plasma SK activity assay based on the rate of amidolysis of S-2251. The range in plasma SK activity following an intravenous infusion of 500,000 or 250,000 U over 30 min in five patients with MI and six patients with venous thromboembolism was considerable. Two healthy subjects receiving 100,000 U over 15 min had tenfold differences in C_{max} and clearance. After terminating the infusion, plasma SK activity declined monoexponentially in all patients and subjects studied. Unfortunately, the immense variability in plasma concentrations and essentially an order-of-magnitude difference in some parameters does little to adequately characterize SK disposition. The mean half-life of 82 min for all individuals (range 34–116 min) agreed with a previously described estimate for SK half-life (83 min), which was based on radioactivity measurements following dosing with iodinated SK (Fletcher *et al.,* 1959). Col *et al.* (1989) reported activity half-lives of 184 and 169 min following SK doses of 1.5 million and 500,000 U, respectively. A 30-U dose of APSAC resulted in a comparable half-life of 188 min. Again, the sampling was very limited, and this finding was generated with a mean of three data points per treatment.

Perhaps the most complete SK/APSAC pharmacokinetic study done to date was reported by Gemmill *et al.* (1991). The authors measured plasma concentrations using a functional bioassay. Patients presenting with acute MI received either 1.5 million U of SK infused over 60 min (*n* = 12), or 30 U of APSAC administered over 5 min (*n* = 12). The model-independent volume of distribution was similar between treatments (5.9 and 5.7 liters for APSAC and SK, respectively). The clearance and terminal half-life for APSAC were 3.87 liters/hr and 1.16 hr, respectively, and 7.08 liters/hr and 0.61 hr for SK.

10. UROKINASE AND SINGLE-CHAIN UROKINASE

10.1. Background

The urokinase precursor, pro-uPA, contains 411 amino acids and has a molecular mass of approximately 54 kDa. The single-chain entity (scuPA) can be found in urine (Stump *et al.,* 1986a) and is produced in various cell cultures (Stump *et al.,* 1986b). Plasmin and kallikrein cleave scuPA at residue 158, producing two-chain uPA (tcuPA) or, as it is also called, high-molecular-weight uPA (HMW uPA) (Ichinose *et al.,* 1986), that is held together by a single disulfide bridge. The complete primary amino acid sequence of HMW uPA was described by Gunzler *et al.* and Steffens *et al.* in 1982. HMW uPA and scuPA hydrolyze the Arg_{560}–Val_{561} bond of plasminogen. Likewise, plasmin and thrombin can inactivate HMW uPA by cleaving the Phe_{157}–Lys_{158} bond. Further, plasmin can hydrolyze the Lys_{134}–Lys_{135} bond of HMW uPA producing a 33-kDa, two-chain protein, termed low-molecular-weight urokinase (LMW uPA). Collen *et al.* (1986) reported that scuPA converts plasminogen to plasmin, even in the presence of aprotinin and α_2-macroglobulin. It has been reported (Pannell and Gurewich, 1986) that scuPA generates up to a fourfold better thrombolytic response than does HMW uPA. It has been postulated that the presence of fibrin in the systemic circulation induces scuPA to exhibit thrombolytic activity without initiating fibrinogenolysis.

10.2. Pharmacokinetics

Van der Werf *et al.* (1987) infused recombinant scuPA in dogs with 1-hr-old clots in the LAD coronary artery at rates of 2, 4, 8, and 20 μg/kg per min. A linear correlation was observed between the infusion rate, and the plateau scuPA concentrations in blood. At the highest infusion rate the blood concentration reached 2.5 ± 0.45 μg/ml; this concentration was not associated with systemic fibrinolytic activation.

The pharmacokinetics of three molecular forms of uPA were studied in rabbits (Stump *et al.*, 1987). Urinary scuPA (M_r = 54 kDa), cellular scuPA from cultured human cells (54 kDa), and LMW uPA (32 kDa) were injected via IV bolus into rabbits at varying doses, together with 0.5 μCi/kg of ^{125}I-labeled material. There was a considerable difference in peak concentrations observed following dosing; however, in each case, the plasma disposition (as determined by ELISA) was described by a two-compartment model. Disposition rate constants, volume estimates, and plasma clearance were essentially identical between the two groups. The radioactivity was recovered in the liver and kidney of the rabbits, but in squirrel monkeys dosed with labeled and unlabeled cellular scuPA, the clearance was primarily via the liver. Steady-state plasma concentrations were proportional to the administered dose, and the clearances determined at steady state were comparable to those determined from the bolus injection.

Additional studies are now ongoing to examine the effect of conjugating scuPA with monoclonal antibodies specific for cross-linked fibrin. Preliminary data suggest that the thrombolytic potency of scuPA can be enhanced by targeting fibrin (Collen *et al.*, 1990).

Van der Werf *et al.* (1986) treated 17 MI patients with heparin and scuPA. The material used in this study was produced via expression of the cloned full-length human scuPA gene in *E. coli.* Eight patients receiving a 10-mg bolus followed by 30 mg over 1 hr had plateau concentrations of 3.1 ± 1.3 μg/ml. In nine patients receiving the 10-mg bolus with 60 mg scuPA over 1 hr, the plateau concentrations were 7.6 ± 3.4 μg/ml. Plasma antigen concentrations declined biexponentially when the infusion was terminated, resulting in identical α and β half-lives of 7.9 and 48 min, respectively, in both groups.

11. CONCLUSIONS

The coagulation factors are a group of relatively high-molecular-weight proteins, some of which have been used as replacement therapy to treat bleeding disorders. The range of terminal half-lives reported for those factors that have been administered to patients (factors VII, VIII, IX, and XIII) is approximately 2 hr to 10 days. These values are probably related to the intrinsic turnover of the factors in the normal coagulation pathway. Progress has been, and is being, made in the treatment of hemophilia both by improving the existing therapies (e.g., using highly purified plasma extracts of factors VIII and IX, and by investigation of the usefulness of recombinant factors); in addition, investigating the potential of coagulation factors that have previously not been considered as therapies (factor VIIa and tissue factor). Reliance on replacement therapies may, in the not too distant future, be reduced, due to the development of the ability to replace the defective gene encoding the affected coagulation factor.

Despite considerable clinical data regarding the plasminogen activators, there remains a paucity of pharmacokinetic data for most of these agents. The increased use of thrombolytic therapy would be expected to increase the kinetic as well as the metabolic study of these proteins. Advances in analytical methodologies may also aid in the study of agents that currently are only poorly characterized (e.g., streptokinase). There is, however, a greater understanding of the importance of well-characterized pharmacokinetic profiles early in the development of these agents; the recognition and implementation of this strategy should continue to move potent new therapeutic proteins into the clinical arena.

ACKNOWLEDGMENT

On April 19, 1991, at the age of 42, Dr. Adair Hotchkiss died suddenly from sarcoidosis. Although Adair's scientific career was relatively brief, his accomplishments were many. Most notable were his experiments on fibrinolysis and thrombolytic models carried out during his seven years at Genentech, Inc. Adair was a scientist whose quality of work and achievements deserve the acclaim and appreciation of his colleagues and peers. It is with sadness at our loss and great respect for him that we dedicate this chapter to his memory.

REFERENCES

Aggeler, P. M., 1961, Physiological basis for transfusion therapy in hemorrhagic disorders, *Transfusion* **1**:71–74.

Allain, J. P., 1984, Principles of in vivo recovery and survival studies, *Scand. J. Haematol.* **33**(Suppl. 40):161–165.

Astrup, T., and Permin, P. M., 1947, Fibrinolysis in animal organism, *Nature* **159**:681–682.

Bakhit, C., Lewis, D., Billings, R., and Malfroy, B., 1987, Cellular catabolism of recombinant tissue-type plasminogen activator, *J. Biol. Chem.* **262**:8716–8720.

Barrowcliffe, T. W., Stableforth, R., and Dormandy, K. M., 1973, Small scale preparation and clinical use of factor IX prothrombin complex, *Vox Sang.* **25**:426–441.

Baughman R. A., 1987, Pharmacokinetics of tissue plasminogen activator, in: *Tissue Plasminogen Activator in Thrombolytic Therapy* (B. E. Sobel, D. Collen, and E. B. Grossbard, eds.), Dekker, New York, pp. 41–53.

Been, M., deBono, D. P., Muir, A. L., Boulton, F. E., Fears, R., Standring, R., and Ferres, H., 1986, Clinical effects and kinetic properties of intravenous anistreplase–anisoylated plasminogen–streptokinase activator complex (BRL26921) in acute myocardial infarction, *Int. J. Cardiol.* **11**:53–61.

Bertina, R. M., and Veltkamp, J. J., 1981, Physiology and biochemistry of factor IX, in: *Haemostasis and Thrombosis* (A. L. Bloom and D. P. Thomas, eds.), Churchill Livingstone, Edinburgh, pp. 98–110.

Bidwell, E., Booth, J. M., Dike, G. W. R., and Denson, K. W. E., 1967, The preparation for therapeutic use of a concentrate of factor IX containing also factors II, VII and X, *Br. J. Haematol.* **13**:586–590.

Biggs, R., and Denson, K. W. E., 1963, The fate of prothrombin and factors VII, IX and X transfused to patients deficient in these factors, *Br. J. Haematol.* **9**:532–547.

Binder, B. R., Spragg, J., and Austen, K. F., 1979, Purification and characterization of human vascular plasminogen activator derived from blood vessel perfusates, *J. Biol. Chem.* **254**:1998–2003.

Bloom, A. L., 1990, Physiology of blood coagulation, *Haemostasis* **20**(Suppl.): 14–29.

Brinkhous, K. M., Hedner, U., Garris, J. B., Diness, V., and Read, M. S., 1989, Effect of recombinant factor VIIa on the hemostatic defect in dogs with hemophilia A, hemophilia B, and von Willebrand disease, *Proc. Natl. Acad. Sci. USA* **86**:1382–1386.

Broze, G. J., and Majerus, P. W., 1980, Purification and properties of human coagulation factor VII, *J. Biol. Chem.* **255**:1242–1247.

Broze, G. J., and Miletich, J. P., 1987, Characterization of the inhibition of tissue factor in serum, *Blood* **69**:150–155.

Carson, S. D., 1987, Tissue factor (coagulation factor III) inhibition by apolipoprotein A-II, *J. Biol. Chem.* **262**:718–721.

Chavin, S. I., and Weidner, S. M., 1984, Blood clotting factor IX. Loss of activity after cleavage of sialic acid residues, *J. Biol. Chem.* **259**:3387–3390.

Christensen, L. R., and MacLeod, C. M., 1945, Proteolytic enzyme of serum: Characterization, activation, and reaction with inhibitors, *J. Gen. Physiol.* **28**:559–583.

Col, J. J., Col-DeBeys, S. M., Renkin, J. P., LaVenne-Pardonge, E. M., Bachy, J. L., and Morian, M. H., 1989, Pharmacokinetics, thrombolytic efficacy and hemorrhagic risk of different streptokinase regimens in heparin-treated acute myocardial infarction, *Am. J. Cardiol.* **63**:1185–1192.

Collen, D., Stassen, J. M., Marafino, B. J., Builder, S., DeCock, F., Ogez, J., Tajiri, D., Pennica, D., Bennett, W. F., Salwa, J., and Hoyng, C. F., 1984, Biological properties of human tissue-type plasminogen activator obtained by expression of recombinant DNA in mammalian cells, *J. Pharmacol. Exp. Ther.* **231**:146–152.

Collen, D., Zamarron, C., Lijnen, H. R., and Hoylaerts, M., 1986, Activation of plasminogen pro-urokinase. II. Kinetics, *J. Biol. Chem.* **261**:1259–1266.

Cossum, P., Littlewood, J., Ferraiolo, B., Green, J., and Bunting, S., 1990, Recombinant human tissue factor (rhTF) pharmacokinetics and effects in normal and hemophiliac dogs, *Pharm. Res.* 7(Suppl.):S-45 (abstract).

Daly, H. M., and Haddon, M. E., 1988, Clinical experience with a pasteurized human plasma concentrate in factor XIII deficiency, *Thromb. Haemostas.* **59**:171–174.

Eaton, D., Rodriguez, H., and Vehar, G. A., 1986, Proteolytic processing of human factor VIII. Correlation of specific cleavages by thrombin, factor Xa, and activated protein C with activation and inactivation of factor VIII coagulant activity, *Biochemistry* **25**:505–512.

Esmail, A. F., Dupe, R. J., English, P. D., and Smith, R. A. G., 1984, Pharmacokinetic and pharmacodynamic comparison of acylated streptokinase plasminogen complexes with different deacylation rate constant, *Haemostasis* **14**:84.

Fear, J. D., Miloszewski, K. J. A., and Losowsky, M. S., 1983, The half-life of factor XIII in the management of inherited deficiency, *Thromb. Haemostas.* **49**:102–105.

Fears, R., Ferres, H., and Standring, R., 1987, The protective effect of acylation on the stability of anisoylated plasminogen streptokinase activator complex in human plasma, *Drugs* 33(Suppl. 3):57–63.

Ferres, H., Hibbs, M., and Smith, R. A. G., 1987, Deacylation studies in vitro on anisoylated plasminogen streptokinase activator complex, *Drugs* 33(Suppl. 3):80–82.

Fisher, K. L., Gorman, C., Vehar, G., O'Brien, D. P., and Lawn, R. M., 1987, Cloning and expression of tissue factor cDNA, *Thromb. Res.* **48**:89–99.

Fletcher, A. D., Alkjaersig, N., and Sherry, S., 1959, The clearance of heterologous protein from the circulation of normal and immunized man, *J. Clin. Invest.* **37**:1306–1315.

Fong, K.-L., and Lynn, R. K., 1986, Disposition and metabolism of tissue-type plasminogen activator (tPA) in the isolate perfused rat liver, *Pharmacologist* **28**:117.

Fong, K.-L., Crysler, C. S., Mico, B. A., Boyle, K. E., Kopia, G. A., Kopaciewicz, L., and Lynn, R. K., 1988, Dose-dependent pharmacokinetics of recombinant tissue-type plasminogen activator in anesthetized dogs following intravenous infusion, *Drug Metab. Dispos.* **16**:201–206.

Fuchs, H. E., Trapp, H. G., Griffith, M. J., Roberts, H. R., and Pizzo, S. V., 1984, Regulation of factor IXa in vitro in human and mouse plasma and in vivo in the mouse, *J. Clin. Invest.* **73**:1696–1703.

Fuchs, H. E., Berger, H., and Pizzo, S. V., 1985, Catabolism of human tissue plasminogen activator in mice, *Blood* **65**:539–544.

Garabedian, H. D., Gold, H. K., Leinbacj, R. C., Johns, J. A., Yasuda, T., Kanke, M., and Collen, D., 1987, Comparative properties of two clinical preparations of recombinant tissue-type plasminogen activator in patients with acute myocardial infarction, *J. Am. Coll. Cardiol.* **9**:599–607.

Gemmill, J. D., Hogg, K. J., Burns, J. M., Rae, A. P., Dunn, F. G., Fears, R., Ferres, H., Standring, R., Greenwood, H., Pierce, D., and Hills, W. S., 1991, A comparison of the pharmacokinetic properties of streptokinase and anistreplase in acute myocardial infarction, *Br. J. Clin. Pharmacol.* **31**:143–147.

Gill, F. M., 1984, The natural history of factor VIII inhibitors in patients with hemophilia A, *Prog. Clin. Biol. Res.* **150**:19–24.

Girard, T. J., Warren, L. A., Novotny, W. F., Likert, K. M., Brown, S. G., Miletich, J. P., and Broze, G. J., 1989, Functional significance of the Kunitz-type inhibitory domains of lipoprotein-associated coagulation inhibitor, *Nature* **338**:518–520.

Gonmori, H., and Takeda, Y., 1975, Properties of canine tissue thromboplastin from brain, lung, arteries and veins, *Am. J. Physiol.* **229**:618–626.

Goodnight, S. H., Britell, C. W., Wuepper, K. D., and Osterud, B., 1979, Circulating factor IX antigen–inhibitor complexes in hemophilia B following infusion of a factor IX concentrate, *Blood* **53**:93–103.

Grierson, D. S., and Bjornsson, T. D., 1987, Pharmacokinetics of streptokinase in patients based on amidolytic activator complex activity, *Clin. Pharmacol. Ther.* **41**:304–313.

Gunzler, W. A., Steffens, G. J., Otting, F., Kim, S. M., Frankus, E., and Flohe, L., 1982, The primary structure of high molecular mass urokinase from human urine. The complete amino acid sequence of the A chain, *Hoppe-Seylers Z. Physiol. Chem.* **363**:1155–1165.

Hedner, U., and Kisiel, W., 1983, Use of human factor VIIa in the treatment of two hemophilia A patients with high-titer inhibitors, *J. Clin. Invest.* **71**:1836–1841.

Hedner, U., Glazer, S., Pingel, K., Alberts, K. A., Blomback, M., Schulman, S., and Johnsson, H., 1988, Successful use of recombinant factor VIIa in patients with severe haemophilia A during synovectomy, *Lancet* **2:**1193.

Hellstern, P., Miyashita, C., Kohler, M., von Blohn, G., Kiehl, R., Biro, G., Schwerdt, H., and Wenzel, E., 1987, Measurement of factor VIII procoagulant antigen in normal subjects and in hemophilia A patients by an immunoradiometric assay and by an enzyme-linked immunosorbent assay, *Haemostasis* **17:**173–181.

Hoag, M. S., Aggeler, P. M., and Powell, A. H., 1960, Disappearance rate of concentrated proconvertin extracts in congenital and acquired hypoconvertinemia, *J. Clin. Invest.* **39:**554–563.

Hoag, M. S., Johnson, F. F., Robinson, J. A., and Aggeler, P. M., 1969, Treatment of hemophilia B with a new clotting-factor concentrate, *N. Engl. J. Med.* **280:**581–583.

Holvoet, P., Cleemput, H., and Collen, D., 1985, Assay of human tissue-type plasminogen activator (t-PA) with an enzyme-linked immunosorbent assay (ELISA) based on three murine monoclonal antibodies to t-PA, *Thromb. Haemostas.* **54:**684–687.

Hotchkiss, A., Refino, C. J., Leonard, C. K., O'Connor, J. V., Crowley, C., McCabe, J., Tate, K., Nakamura, G., Powers, D., Levinson, A., Mohler, M., and Spellman, M., 1988, The influence of carbohydrate structure on the clearance of recombinant tissue-type plasminogen activator, *Thromb. Haemostas.* **60:** 255–261.

Ichinose, A., Fujikawa, K., and Suyama, T., 1986, The activation of prourokinase by plasma kallikrein and its inactivation by thrombin, *J. Biol. Chem.* **261:** 3486–3489.

Jackson, K. W., and Tang, J., 1982, Complete amino acid sequence of streptokinase and its homology with serine proteases, *Biochemistry* **21:**6620–6625.

Kadhom, N., Wolfrom, C., Gautier, M., Allain, J. P., and Frommel, D., 1988, Factor VIII procoagulant antigen in human tissues, *Thromb. Haemostas.* **59:**289–294.

Kjellman, H., 1984, Calculations of factor VIII in vivo recovery and half-life, *Scand. J. Haematol.* **33**(Suppl. 40):165–174.

Kohler, M., Seifreid, E., Hellstern, P., Pindur, G., Miyashita, C., Morsdorf, S., Fasco, F., and Wenzel, E., 1988, In vivo recovery and half-life time of a steam-treated factor IX concentrate in hemophilia B patients, *Blut* **57:**341–345.

Kohler, M., Hellstern, P., Pindur, G., Wenzel, E., and von Blohm, G., 1989, Factor VII half-life after transfusion of a steam-treated prothrombin complex concentrate in a patient with homozygous factor VII deficiency, *Vox Sang.* **56:**200–201.

Kopia, G. A., Kopaciewicz, L. J., Fong, K.-L., Crysler, C. S., Boyle, K., and Ruffolo, R. R., 1988, Evaluation of the acute hemodynamic effects and pharmacokinetics of coronary thrombolysis produced by intravenous tissue-type plasminogen activator in the anesthetized dog, *J. Cardiovasc. Pharmacol.* **12:**308–316.

Korninger, C., Stassen, J. M., and Collen, D., 1981, Turnover of human extrinsic (tissue-type) plasminogen activator in rabbits, *Thromb. Haemostas.* **46:** 658–661.

Kuiper, J., Otter, M., Rijken, D. C., and van Berkel, T. J. C., 1988, Characterization of the interaction in vivo of tissue-type plasminogen activator with liver cells, *J. Biol. Chem.* **263**:18220–18224.

Kuzel, T., Green, D., Stulberg, S. D., and Baron, J., 1988, Arthropathy and surgery in congenital factor VII deficiency, *Am. J. Med.* **84**:771–774.

Lewis, J. H., Bontempo, F. A., Spero, J. A., Ragni, M. V., and Starzi, T. E., 1985, Liver transplantation in a hemophiliac, *N. Engl. J. Med.* **312**:1189–1192.

Littlewood, J. D., and Barrowcliffe, T. W., 1987, The development and characterization of antibodies to human factor VIII in haemophilic dogs, *Thromb. Haemostas.* **57**:314–321.

Loeliger, E. A., and Hensen, A., 1964, On the turnover of factors II, VII, IX, X under pathological conditions, *Thromb. Diath. Haemorrh.* **13**(Suppl.):95.

Loewy, A. G., Dahlberg, A., Dunathan, D., Kriel, R., and Wolfinger, H. L., 1961, Fibrinases. II. Some physical properties, *J. Biol. Chem.* **236**:2634–2643.

Longo, G., Matucci, M., Messori, A., Morfini, M., and Rossi-Ferrini, P., 1986, Pharmacokinetics of a new heat-treated concentrate of factor VIII estimated by model-independent methods, *Thromb. Res.* **42**:471–476.

Longo, G., Cinotti, S., Filimberti, E., Giustarini, G., Messori, A., Morfini, M., and Rossi-Ferrini, P., 1987, Single-dose pharmacokinetics of factor IX evaluated by model-independent methods, *Eur. J. Haematol.* **39**:426–433.

Markus, G., Evers, J. L., Hobika, J. H., 1976, Activator activities of the transient forms of the human plasminogen-streptokinase complex during its proteolytic conversion to the stable activator complex, *J. Biol. Chem.* **251**:6495–6504.

McLellan, D. S., Pelly, C., McLellan, H. G., Jones, P., and Aronstam, A., 1982, The in vivo survival characteristics of factor VIII procoagulant antigen (VIII:CAg) in haemophilia A subjects, *Thromb. Res.* **25**:33–39.

Mariani, G., Mannucci, P. M., Mazzucconi, M. G., and Capitanio, A., 1978, Treatment of congenital factor VII deficiency with a new concentrate, *Thromb. Haemostas.* **39**:675–682.

Matucci, M., Messori, A., Donati-Cori, G., Longo, G., Vannini, S., Morfini, M., Tendi, E., and Rossi-Ferrini, P. L., 1985, Kinetic evaluation of four factor VIII concentrates by model-independent methods, *Scand. J. Haematol.* **34**:22–28.

Messori, A., Longo, G., Matucci, M., Morfini, M., and Rossi-Ferrini, P. L., 1987, Clinical pharmacokinetics of factor VIII in patients with classic hemophilia, *Clin. Pharmacokin.* **13**:365–380.

Messori, A., Longo, G., Morfini, M., Cinotti, S., Filimberti, E., Giustarini, G., and Rossi-Ferrini, P., 1988, Multi-variate analysis of factors governing the pharmacokinetics of exogenous factor VIII in haemophiliacs, *Eur. J. Clin. Pharmacol.* **35**:663–668.

Mohler, M. A., Refino, C. J., Chen, S. A., Chen, A. B., and Hotchkiss, A. J., 1986, D-Phe-Pro-Arg-chloromethylketone: its potential use in inhibiting the formation of in vitro artifacts in blood collected during tissue-type plasminogen activator thrombolytic therapy, *Thromb. Haemostas.* **56**:160–164.

Morfini, M., Longo, G., Matucci, M., Vannini, S., Messori, A., Filimberti, E., Duminuco, M., Avanzi, G., and Rossi-Ferrini, P., 1984, Cryoprecipitate and factor

VIII commercial concentrates: In vitro characteristics and in vivo compartmental analysis, *Ric. Clin. Lab.* **14**:681–691.

Nilsson, I. M., Berntorp, E., and Zettervall, O., 1988, Induction of immune tolerance in patients with hemophilia and antibodies to factor VIII by combined treatment with intravenous IgG, cyclophosphamide and factor VIII, *N. Engl. J. Med.* **318**:947–949.

Nilsson, I. M., Berntorp, E., Zettervall, O., and Dahlback, B., 1990, Noncoagulation inhibitory factor VIII antibodies after induction of tolerance to factor VIII in hemophilia A patients, *Blood* **75**:378–383.

Nilsson, S., Wallen, P., and Mellbring, G., 1984, In vivo metabolism of human tissue-type plasminogen activator, *Scand. J. Haematol.* **33**:49–53.

Nilsson, S., Einarsson, M., Ekvarn, L., Haggroth, L., and Mattson, C., 1985, Turnover of tissue plasminogen activator in normal and hepatectomized rabbits, *Thromb. Res.* **39**:511–521.

Noe, D. A., Bell, W. R., Ness, P. M., and Levin, J., 1986, Plasma clearance rates of coagulant factors VIII and IX in factor-deficient individuals, *Blood* **67**:969–972.

Over, J., Sixma, J. J., Doucet-de Brune, M., Trieschnigg, M. M., Vlooswijk, R. A., Beeser-Visser, N. H., and Bouma, B. N., 1978, Survival of [125]iodine-labelled factor VIII in normals and patients with classic hemophilia, *J. Clin. Invest.* **62**:223–234.

Over, J., Sixma, J. J., Bouma, B. N., Bolhuis, P. A., Vlooswijk, R. A., and Beeser-Visser, N. H., 1981, Survival of iodine-125-labeled factor VIII in patients with von Willebrand's disease, *J. Lab. Clin. Med.* **97**:332–344.

Owen, C. A., and Bowie, W. J., 1975, Infusion therapy in hemophilia A and B, in: *Handbook of Hemophilia* (K. M. Brinkhous and H. C. Hemker, eds.), Excerpta Medica, Amsterdam, pp. 449–463.

Pannell, R., and Gurewich, V., 1986, Pro-urokinase: A study of its stability in plasma and of a mechanism for its selective fibrinolytic effect, *Blood* **67**:1215–1223.

Pavlovsky, A., 1947, Contribution to the pathogenesis of hemophilia, *Blood* **2**:185–191.

Pennica, D., Holmes, W. E., Kohr, W. J., Harkins, R. N., Vehar, G. A., Ward, C. A., Bennett, W. F., Yelverton, E., Seeburg, H. L., Heyneker, H. L., Goeddel, D. V., and Collen, D., 1983, Cloning and expression of human tissue-type plasminogen activator cDNA in *E. coli, Nature* **301**:214–221.

Pohl, G., Kalstrom, M., Bergsdorf, N., Wallen, P., and Jornvall, H., 1984, Tissue plasminogen activator: peptide analyses confirm an indirectly derived amino acid sequence; identify the active site serine residue, establish glycosylation sites and localize variant differences, *Biochemistry* **23**:3701–3707.

Ranby, M., Bergesdorf, N., and Nilsson, T., 1989a, Enzymatic properties of the one- and two-chain form of tissue plasminogen activator, *Thromb. Res.* **27**:175–183.

Ranby, M., Nguyen, G., Scarabin, P. Y., and Samama, M., 1989b, Immunoreactivity of tissue plasminogen activator and its inhibitor complexes: Biochemical and multicenter validation of a two-site immunosorbent assay, *Thromb. Haemostas.* **61**:409–414.

Rao, L. V. M., Rapaport, S. I., and Bajaj, S. P., 1986, Activation of human factor VII in the initiation of tissue factor-dependent coagulation, *Blood* **68**:685–691.

Ratnoff, O. D., 1986, Factor VIII concentrates, *J. Am. Med. Assoc.* **255**:325–326.

Reddy, K. N. N., 1976, Kinetics of active center formation in dog plasminogen by streptokinase and activity of a modified streptokinase, *J. Biol. Chem.* **251**:3913–3920.

Reddy, K. N. N., 1988, Streptokinase—Biochemistry and clinical application, *Enzyme* **40**:79–89.

Rick, M. E., Popovsky, M. A., and Krizek, D. M., 1985, Degradation of factor VIII coagulant antigen by proteolytic enzymes, *Br. J. Haematol.* **61**:477–486.

Rijken, D. C., and Collen, D., 1981, Purification and characterization of the plasminogen activator secreted by human melanoma cells in culture, *J. Biol. Chem.* **256**:7035–7041.

Rijken, D. C., Wijngaards, G., Zaal-DeJong, M., and Welbergen, J., 1979, Purification and partial characterization of plasminogen activator from human uterine tissue, *Biochim. Biophys. Acta* **580**:140.

Rijken, D. C., Hoylaerts, M., and Collen, D., 1982, Fibrinolytic properties of one-chain and two-chain human extrinsic (tissue-type) plasminogen activator, *J. Biol. Chem.* **257**:2920–2925.

Rock, G. A., Cruickshank, W. H., Tackaberry, E. S., Ganz, P. R., and Palmer, D. S., 1983, Stability of VIII:C in plasma: The dependence on protease activity and calcium, *Thromb. Res.* **29**:521–535.

Rodeghiero, F., Tosetto, A., DiBona, E., and Castaman, G., 1991, Clinical pharmacokinetics of a placenta-derived factor XIII concentrate in type I and type II factor XIII deficiency, *Am. J. Hematol.* **36**:30–34.

Rousell, R. H., Kasper, C. K., and Schwartz, R. S., 1989, The pharmacology of a new pasteurized antihemophilic factor concentrate derived from human blood plasma, *Transfusion* **29**:208–212.

Sandset, P. M., Warn-Cramer, B. J., Rao, L. V., Maki, S. L., and Rapaport, S. I., 1991, Depletion of extrinsic pathway inhibitor (EPI) sensitizes rabbits to disseminated intravascular coagulation induced with tissue factor: Evidence supporting a physiologic role for EPI as a natural anticoagulant, *Proc. Natl. Acad. Sci. USA* **88**:708–712.

Schleef, R. R., Wagner, N. V., Sinha, M., and Loskutoff, D. J., 1986, A monoclonal antibody that does not recognize tissue-type plasminogen activator bound to its naturally occurring inhibitor, *Thromb. Haemostas.* **56**:328–332.

Schneider, C. L., 1947, The active principle of placental toxin: thromboplastin; its inactivator in blood: antithromboplastin, *Am. J. Physiol.* **149**:123–129.

Schwartz, R. S., Abildgaard, C. F., Aledort, L. M., Arkin, S., Bloom, A. L., Brackmann, H. H., Brettler, D. R., Fukui, H., Hilgartner, M. W., Inwood, M. J., Kasper, C. K., Kernoff, P. B., Levine, P. H., Lusher, J. M., Mannucci, P. M., Scharrer, I., MacKenzie, M. A., Pancham, N., Kuo, H. S., and Allred, R. U., 1990, Human recombinant DNA-derived antihemophilic factor (factor VIII) in the treatment of hemophilia A, *N. Engl. J. Med.* **323**:1800–1805.

Seligsohn, U., Kasper, C. K., Osterud, B., and Rapaport, S. I., 1979, Activated factor VII: Presence in factor IX concentrates and persistence in the circulation after infusion, *Blood* **53**:828–837.

Siefried, E., and Tanswell, P., 1987, Comparison of specific antibody, D-Phe-Pro-Arg-chloromethylketone and aprotinin for prevention of in vitro effects of recombinant tissue-type plasminogen activator on haemostasis parameters, *Thromb. Haemostas.* **58**:921–926.

Siefried, E., Tanswell, P., Rijken, D. C., Barrett-Bergshoeff, M. M., Su, C. A., and Kluft, C., 1988, Pharmacokinetics of antigen and activity of recombinant tissue-type plasminogen activator after infusion in healthy volunteers, *Arzneim. Forsch.* **38**:418–422.

Siefried, E., Tanswell, P., Ellbruck, D., Haerer, W., and Schmidt, A., 1989, Pharmacokinetics and haemostatic status during consecutive infusion of recombinant tissue-type plasminogen activator in patients with acute myocardial infarction, *Thromb. Haemostas.* **61**:497–501.

Siefring, G. E., and Castellino, F. J., 1976, Interaction of streptokinase with plasminogen: Isolation and characterization of a streptokinase degradation product, *J. Biol. Chem.* **257**:3913–3920.

Smedsrod, B., Einarsson, M., and Pertoft, H., 1988, Tissue plasminogen activator is endocytosed by mannose and galactose receptors of rat liver cells, *Thromb. Haemostas.* **59**:480–484.

Smith, K. J., and Thompson, A. R., 1981, Labeled factor IX kinetics in patients with hemophilia-B, *Blood* **58**:625–629.

Standring, R., Fears, R., and Ferres, H., 1988, The protective effect of acylation on the stability of APSAC (Eminase) in human plasma, *Fibrinolysis* **2**:157–163.

Staniforth, D. H., Smith, R. A. G., and Hibbs, M., 1983, Streptokinase and anisoylated streptokinase plasminogen complex—Their action on haemostasis in human volunteers, *Eur. J. Clin. Pharmacol.* **24**:751–756.

Steffens, G. J., Gunzler, W. A., Otting, F., Frankus, E., and Flohe, L., 1982, The complete amino acid sequence of low molecular mass urokinase from human urine, *Hoppe-Seylers Z. Physiol. Chem.* **363**:1043–1058.

Stern, D. M., Drillings, M., Nossel, H. L., Hurlet-Jensen, A., LaGamma, K., and Owen, J., 1983, Binding of factors IX and IXa to cultured vascular endothelial cells, *Proc. Natl. Acad. Sci. USA* **80**:4119–4123.

Stern, D. M., Knitter, G., Kisiel, W., and Nawroth, P. P., 1987, In vivo evidence of intravascular binding sites for coagulation factor IX, *Br. J. Haematol.* **66**:227–232.

Stump, D. C., Thienpont, M., and Collen, D., 1986a, Urokinase-related proteins in human urine. Isolation and characterization of single-chain urokinase (prourokinase) and urokinase–inhibitor complex, *J. Biol. Chem.* **261**:1267–1273.

Stump, D. C., Lijnen, H. R., and Collen, D., 1986b, Purification and characterization of single-chain urokinase-type plasminogen activator from human cell cultures, *J. Biol. Chem.* **261**:1274–1278.

Stump, D. C., Kieckens, L., De Cock, F., and Collen, D., 1987, Pharmacokinetics of single-chain forms of urokinase-type plasminogen activator, *J. Pharmacol. Exp. Ther.* **242**:245–250.

Tanswell, P., Seifried, E., Su, P. C., Feuerer, W., and Rijken, D. C., 1989, Pharmaco-kinetics and systemic effects of tissue-type plasminogen activator in normal subjects, *Clin. Pharmacol. Ther.* **46**:155–162.

Tanswell, P., Heinzel, G., Greischel, A., and Krause, J., 1990, Nonlinear pharmacoki-netics of tissue-type plasminogen activator in three animal species and isolated perfused rat liver, *J. Pharmacol. Exp. Ther.* **255**:318–324.

Tebbe, U., Tanswell, P., Seifried, E., Feuerer, W., Scholz, K. H., and Herrmann, K. S., 1989, Single-bolus injection of recombinant tissue-type plasminogen acti-vator in acute myocardial infarction, *Am. J. Cardiol.* **64**:448–453.

Thompson, A. R., Forrey, A. W., Gentry, P. A., Smith, K. J., and Harker, L. A., 1980, Human factor IX in animals: Kinetics from isolated, radiolabelled protein and platelet destruction following crude concentrate infusions, *Br. J. Haematol.* **45**:329–342.

Toole, J. J., Knopf, J. L., Wozney, J. M., Sultzman, L. A., Buecker, J. L., Pittman, D. D., Kaufman, R. J., Brown, E., Shoemaker, C., Orr, E. C., Amphlett, G. W., Foster, W. B., Coe, M. L., Knutson, G. J., Foss, D. W., and Hewick, R. M., 1984, Molecular cloning of a cDNA encoding human antihemophilic factor, *Nature* **312**:342–347.

Van der Werf, F., Vanhaecke, J., de Geest, H., Verstraete, M., and Collen, D., 1986, Coronary thrombolysis with recombinant single-chain urokinase-type plasmin-ogen activator in patients with acute myocardial infarction, *Circulation* **74**:1066–1070.

Van der Werf, F., Jang, I. K., and Collen, D., 1987, Thrombolysis with recombinant human single-chain urokinase-type plasminogen activator (rscu-PA): Dose-re-sponse in dogs with coronary artery thrombosis, *J. Cardiovasc. Pharmacol.* **9**:91–93.

Verstraete, M., Bounameaux, H., de Cock, F., Van der Werf, F., and Collen, D., 1985, Pharmacokinetics and systemic fibrinolytic effects of recombinant human tissue-type plasminogen activator (rt-PA) in humans, *J. Pharmacol. Exp. Ther.* **235**:506–512.

Wallen, P., Ranby, M., Bergsdorf, N., and Kok, P., 1981, Purification and character-ization of tissue plasminogen activator: On the occurrence of two different forms and the enzymatic properties, in: *Progress in Fibrinolysis,* Volume 5 (J. P. Da-vidson, I. M. Nilsson, and B. Astedt, eds.), Churchill Livingstone, Edinburgh, pp. 16–23.

White, G. C., McMillan, C. W., Kingdon, H. S., and Shoemaker, C. B., 1989, Use of recombinant antihemophilic factor in the treatment of two patients with classic hemophilia, *Lancet* **320**:166–170.

Williams, W. J., 1983, Life span of plasma coagulation factors, in: *Hematology* (W. J. Williams, ed.), McGraw–Hill, New York, pp. 1230–1237.

Wion, K. L., Kelly, D., Summerfield, J. A., Tuddenham, E. G., and Lawn, R. M., 1985, Distribution of factor VIII mRNA and antigen in human liver and other tissues, *Nature* **317**:726–729.

Wood, W. I., Capon, D. J., Simonsen, C. C., Eaton, D. L., Gitschier, J., Keyt, B., Seeburg, P. H., Smith, D. H., Hollingshead, P., Wion, K. L., Delwart, E., Tuddenham, E. G., Vehar, G. A., and Lawn, R. M., 1984, Expression of active human factor VIII from recombinant DNA clones, *Nature* **312**:330–337.

Working, P. K., and Cossum, P. A., 1991, Clinical and preclinical studies with recombinant human proteins: The effect of antibody production, in: *Peptides, Peptoids and Proteins: Proceedings of the 5th Pittsburgh Pharmacodynamics Conference* (P. Gazonne, ed.), Harvey Wilkes Books, Cincinnati.

Zauber, N. P., and Levin, J., 1977, Factor IX levels in patients with hemophilia B (Christmas disease) following transfusion with concentrates of factor IX or fresh frozen plasma (FFP), *Medicine* **56**:213–224.

Chapter 7

Pharmacokinetics and Metabolism of Hematopoietic Proteins

John B. Stoudemire

1. INTRODUCTION

Hematopoietic growth factors or colony-stimulating factors (CSFs) are a group of glycoproteins regulating the survival, proliferation, and differentiation of hematopoietic progenitor cells as well as the function and activation of the mature cells. These factors include macrophage-CSF (M-CSF, also known as CSF-1), granulocyte-CSF (G-CSF), granulocyte macrophage-CSF (GM-CSF), multipotential CSF or interleukin-3 (IL-3), and erythropoietin (Epo) (Metcalf, 1985, 1986; Clark and Kamen, 1987; Sieff, 1987). The term CSF was derived from *in vitro* studies showing that these factors stimulated the growth of colonies of bone marrow progenitor cells in a semisolid agar medium. The factors were named according to the progenitor population that they stimulated. A schema of the actions of the CSFs is shown in Fig. 1. The CSFs are produced by a variety of cells and range in molecular mass from 14 to 90 kDa (Table I). These factors have been purified, cloned, and produced through recombinant DNA techniques. The recombinant factors have been shown to have biologic properties and actions that are similar to the naturally occurring factors. The availability of quantities of the recombinant factors has resulted in their introduction into clinical trials and into the market.

John B. Stoudemire • Cytel Corporation, San Diego, California 92121.

Protein Pharmacokinetics and Metabolism, edited by Bobbe L. Ferraiolo *et al.* Plenum Press, New York, 1992.

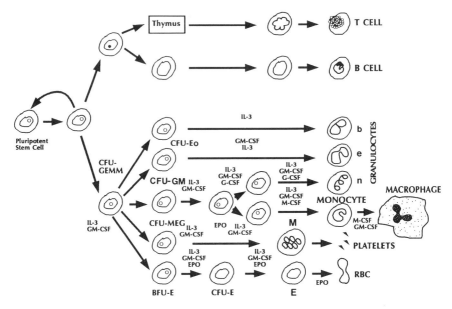

Figure 1. Interactions of the CSFs with hematopoietic cells. Progenitor cells identified in *in vitro* culture systems are CFU-GEMM (colony-forming unit, granulocyte-erthrocyte-monocyte-megakaryocyte), CFU-MEG (CFU-megakaryocyte), CFU-Eo (CFU-eosinophil), CFU-GM (CFU-granulocyte/monocyte), CFU-E (CFU-erythroid), and BFU-E (burst-forming unit-erythroid); n, neutrophil; e, eosinophil; b, basophil; m, monocyte/macrophage; E, erythrocyte; and M, megakaryocyte.

The pharmacokinetics of these factors have been evaluated in animals and in man. Early studies, however, utilized impure or poorly characterized fractions and the interpretation of these results is difficult. With the availability of recombinant proteins, these limitations were overcome and definitive studies have been completed. In general, the biologic effects of these factors can be quantitated by measuring the effects on the target hematopoietic cell population. This has provided a basis for relating the pharmacokinetic profile of the factor to the pharmacodynamic response. As clinical trials and the therapeutic applications of these factors are expanded, the knowledge and understanding of the pharmacokinetics of the drug will be critical to guide the clinician in the choice of dosing routes and schedule.

The majority of the literature on CSFs deals with descriptions of the cloning, purification, and identification of the factors and the *in vitro* and *in vivo* biologic effects. There have been, however, relatively few references to the pharmacokinetics of these factors. When pharmacokinetic parameters are reported, these reports are often limited to a brief indication of a calcu-

Table I
Characteristics of the Human CSFs

Name	Size (kDa)	Cellular sources	Hematopoietic lineages found in colonies[a]
Epo	34–39	Kidney	E
G-CSF	18–22	Monocytes Fibroblasts Endothelial cells	n
GM-CSF	14–35	T cells Endothelial cells Fibroblasts Monocytes	n, m, e, E, M
IL-3 (multi-CSF)	14–28	T cells	n, m, e, b, E, M
M-CSF	45(monomer) × 2	Monocytes Fibroblasts Endothelial cells	m

[a] Key: n, neutrophil; e, eosinophil; b, basophil; m, monocyte/macrophage; E, erythrocyte; M, megakaryocyte

lated plasma half-life. The methods used to calculate the pharmacokinetic parameters can be variable. Assuming that the first sample taken after administration of the dose is equivalent to 100% of the injected dose can lead to erroneous conclusions. If a significant quantity of the protein has been equilibrated or cleared by the time that first sample is taken, the resulting references to that value will be invalid. Another potential difficulty in the interpretation of these reports relates to the quantitation of the protein. There are as yet no standards for the assay of CSFs in serum/plasma or tissues. The methods currently in use include immunoassays (RIA or ELISA) that quantitate immunoreactive protein, bioassays that quantitate bioactive species, or isotopic techniques that rely upon the quantitation of radiolabeled species. The qualitative and/or quantitative results from these assays may not be consistent and careful interpretation of the data is warranted. In addition, there may be methods or assays which are laboratory specific. This leads to difficulties in the comparison of the results of one study to another.

This review will summarize the preclinical and clinical pharmacokinetics of Epo, G-CSF, GM-CSF, and M-CSF.

2. ERYTHROPOIETIN

Epo was the first CSF to be identified (Krumdieck, 1943; Erslev *et al.,* 1953). Epo is a sialoglycoprotein with a molecular mass of 34–39 kDa. This hormone is primarily produced in the kidney with some production also occurring in the liver (Bondurant and Koury, 1986; Koury *et al.*, 1988; Paul *et al.*, 1984; Zanjani *et al.*, 1974). Epo is a lineage-specific factor that stimulates the proliferation and differentiation of erythroid lineage precursor cells [colony-forming unit erythroid (CFU-E) and burst-forming unit erythroid (BFU-E)] to mature erythrocytes (Goldwasser and Kung, 1968). Stimulation of these cells by Epo results in increases in the hematocrit and hemoglobin levels. The cDNA for Epo has been cloned and expressed in Chinese hamster ovary cells (Jacobs *et al.*, 1985; Lin *et al.*, 1985; Powell *et al.*, 1986).

2.1. Epo Preclinical Pharmacokinetics

Early studies on the pharmacokinetics of Epo were performed with impure and poorly characterized preparations and Epo concentrations were determined by imprecise bioassays. Consequently, the results of these studies in both animals and humans showed wide variability. The techniques used in these studies have included administration of Epo-rich plasma, measurement of Epo concentration following stimulation by hypoxia, and administration of recombinant Epo. In early studies, hypoxia was induced in animals and Epo pharmacokinetics were estimated by monitoring the decline in the serum concentrations of the endogenous Epo produced in response. Using this technique, several groups showed a monoexponential disappearance of Epo from the plasma of rats with a half-life of 3–5 hr and a volume of distribution equal to the plasma volume (Reissmann *et al.*, 1965; Stohlman and Howard, 1962). Reissmann *et al.* (1965) injected exogenous Epo from the serum of anemic rabbits into rats; biexponential clearance was noted with a calculated volume of distribution equal to twice the plasma volume. Weintraub *et al.* (1964) administered a single intravenous bolus injection of purified sheep Epo to dogs and demonstrated biphasic clearance from the plasma, with estimated α and β half-lives of 20–45 min and 9–10.5 hr, respectively. Urinary recovery of the Epo was approximately 2–5% of the injected dose, suggesting an extrarenal mechanism of clearance. In a study using ^{125}I-labeled human Epo administered to rabbits, Roh *et al.* (1972a) also reported biphasic plasma clearance curves with half-lives similar to those of the previous studies.

Studies in rabbits demonstrated that Epo is readily absorbed from the peritoneal cavity (Bargman *et al.*, 1990), indicating that this might be a suitable route for patients on chronic ambulatory peritoneal dialysis (CAPD). rhEpo administered intraperitoneally to uremic rabbits without dialysate was almost completely absorbed ($\sim$98%) as compared to less absorption ($\sim$60%) when the protein was administered with dialysate or followed by dialysate. Contrasting results have been reported in human studies.

Kampf *et al.* (1989) compared the kinetics of rhEpo following subcutaneous, intraperitoneal, and intravenous administration. The mean plasma half-lives were 8.4, 15, and 11 hr, respectively. The bioavailability of the subcutaneous and intraperitoneal doses was estimated to be 49 and 6.8%, with maximum plasma concentrations observed at 18 and 14 hr, respectively. The mean plasma residence times (MRT) of the subcutaneous and intraperitoneal rhEpo were 33 and 28 hr, compared to 11 hr for intravenous delivery. The plasma concentration–time curves show that elimination following subcutaneous or intraperitoneal administration is probably absorption rate limited. Although these data suggest that the subcutaneous and intraperitoneal routes would be ineffective relative to intravenous administration, clinical studies have demonstrated that to obtain the same therapeutic effect achieved with intravenous dosing, smaller subcutaneous doses ($\sim$50% of the intravenous dose) are required. Therefore, bioavailability alone is not a sufficient criterion for predicting therapeutic application. These data also suggest that sustained plasma concentrations are more important than peak plasma concentrations. The essential pharmacokinetic parameter for predicting the pharmacodynamic response of rhEpo should be the MRT in the plasma.

2.1.1. Epo Hepatic Catabolism

A number of investigators have examined the role of the liver in the metabolism of Epo. Burke and Morse (1962) showed that the Epo activity of blood perfusates was decreased following perfusion through rat livers and that carbon tetrachloride pretreatment would abolish this capability. Additional early evidence of the sites of clearance of Epo came from Roh *et al.* (1972b), who examined the clearance of Epo in a liver perfusion system in which liver function was either impaired by microsomal enzyme inhibitors or enhanced by phenobarbital. Clearance was increased in the phenobarbital-treated livers as compared to controls and decreased in the impaired livers.

The plasma half-life of endogenous Epo showed a small but significant increase in rats pretreated with a hepatotoxic agent (*d*-galactosamine-HCl). Nephrectomy resulted in a prolongation of the half-life (Dinkelaar *et al.*, 1981a,b). In liver perfusion studies conducted by these investigators, there was no change in Epo levels during a 4-hr perfusion. Overall, these studies indicated that hepatic metabolism of Epo was minimal.

2.1.2. Epo Renal Catabolism

The majority of animal studies have also indicated a limited role for the kidney in Epo metabolism. Naets and Wittek (1974) evaluated the plasma clearance of endogenous and exogenous Epo in normal and nephrectomized rats, and rats with ligated or sectioned ureters. With the exogenous and endogenous Epo, the plasma $t_{1/2}$ in controls and the ureteral section or ligation animals was 1.5 to 1.9 and 3.8 hr, respectively. The $t_{1/2}$ increased to 12.2 and 8.3 hr after nephrectomy, suggesting that the kidney was involved in Epo metabolism. In a study in rabbits, the plasma half-life of rhEpo was also significantly longer in nephrectomized animals as compared to the controls, 5.1 versus 3.0 hr (Brown *et al.*, 1990). The pharmacokinetic parameters of native unlabeled Epo and radiolabeled rhEpo were determined in intact and anephric dogs (Fu *et al.*, 1988). The plasma concentration–time curves were biexponential for both preparations. The mean $t_{1/2}$ α of native Epo in intact dogs was 75 ± 21 min as compared to 23 ± 5 min for the rhEpo. There were, however, no significant differences in the apparent volume of distribution (approximately twice the estimated volume), elimination half-life (9 hr), or clearance rates (0.011 liter/kg per hr) for the two preparations. Clearance was decreased approximately 30% in the anephric dog, supporting the conclusion that although the kidney contributes to Epo elimination, the majority of Epo clearance occurs through nonrenal mechanisms.

A number of animal studies have shown that impaired renal function does not significantly affect Epo clearance. Emmanouel *et al.* (1984) examined the metabolism of native, human Epo in the rat. The clearance of Epo from the plasma was biexponential with an initial $t_{1/2}$ of 0.9 hr and a terminal $t_{1/2}$ of approximately 3.5 hr in normal rats; the terminal $t_{1/2}$ was prolonged to 4.4 hr in rats with ligated renal pedicles. They concluded that renal degradation made a limited contribution (approximately 8%) to the metabolism of the protein. Steinberg *et al.* (1986) reported an elimination half-life of approximately 85 min in both normal and anephric rats receiving Epo-rich plasma. The clearances of Epo-rich plasma in anephric and normal mice were also similar, showing a biphasic plasma concentration decline, with initial and terminal half-lives of 15 min and 7 hr, respectively (Miller, 1985). Similar observations were made for human urinary Epo, rabbit Epo-rich plasma, and rhEpo in rats (Stohlman and Howard, 1962; Reissmann *et al.*, 1965; Scigalla *et al.*, 1987). Epo kinetics in normal and uremic sheep showed a $t_{1/2}$ of approximately 9 hr that was independent of renal function (Mladenovic *et al.*, 1985). The half-life, however, was shortened by multiple infusions. Disparity in the conclusions from these studies may be attributed to differences in assay techniques, test article preparations, or methods of calculating pharmacokinetic parameters.

2.1.3. Epo Erythroid Cell Catabolism

Several studies in animals have evaluated the role of the erythroid cells in Epo metabolism. Naets and Wittek (1969) compared endogenous Epo clearance in rats with aplasia, hyperplasia, and hypoplasia of the erythroid marrow. The observed plasma $t_{1/2}$ of Epo in the controls was 128 min; no significant differences in half-life were noted in the other groups, suggesting that the erythroid cells do not contribute to Epo clearance. These investigators (Naets and Wittek, 1968) made similar observation in dogs, showing that erythroid hyperplasia did not increase catabolism of Epo.

2.1.4. Influence of Carbohydrate Residues on Epo Pharmacokinetics

The carbohydrate residues of glycoproteins can be important in the expression of their *in vitro* and *in vivo* activities. Native Epo and rhEpo produced in mammalian cells are sialated glycoproteins. Approximately 40% of the molecular weight of rhEpo is contributed by carbohydrates, including one *O*-linked and three *N*-linked oligosaccharides (Sasaki *et al.*, 1987; Davis and Arakawa, 1987). It has been shown that asialo-Epo is more active *in vitro* than sialated Epo. Although the sialic acid residues are not necessary for the *in vitro* activity of the protein, these residues are necessary for *in vivo* activity (Dordal *et al.*, 1985; Goldwasser *et al.*, 1974; Lukowsky and Painter, 1972; Takeuchi *et al.*, 1989). The loss of *in vivo* activity can be explained by the rapid clearance of the desialated form from the circulation. Desialation exposes a galactose residue which is recognized by galactose receptors in the liver. The presence of sialic acid residues attached to complex *N*-linked oligosaccharides is required for the survival of glycoproteins in the circulation (Morell *et al.*, 1968, 1971; Ashwell and Kawasaki, 1978). Removal of the sialic acid exposes galactose residues, facilitating the binding and uptake of the glycoprotein by galactose binding receptors in the liver.

In a rat liver perfusion system, desialated native Epo was cleared more rapidly than intact Epo (Briggs *et al.*, 1974). The plasma clearance and biodistribution of rhEpo and desialated rhEpo were also compared in the rat (Spivak and Hogans, 1989). The plasma concentration–time curve of intact rhEpo was biexponential, with an initial $t_{1/2}$ of 53 min and an elimination $t_{1/2}$ of 180 min. The desialated species, however, was cleared rapidly with a $t_{1/2}$ of 2 min. The sialated species showed limited localization at 30 min in the kidney and bone marrow, while the desialated form was cleared to the liver. Similar studies were carried out by Fukuda *et al.* (1989a,b). Intact rhEpo was cleared from the plasma of rats with an approximate half-life of 2 hr, while desialated rhEpo was cleared within 6 min, with 85% of the injected dose

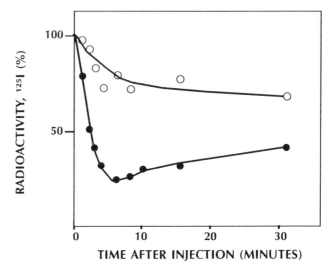

Figure 2. Plasma radioactivity versus time curves of native and desialated rhEpo in the rat.
[125]I-labeled recombinant erythropoietin was injected intravenously into rats before (○) and after
(●) sialidase treatment. (Reproduced with permission from Fukuda *et al.,* 1989a).

being recovered in the liver at 30 min (Fig. 2). A subpopulation of rhEpo
molecules with more than three *N*-acetyllactosaminyl repeats were isolated,
and the clearance of these species was compared to that of the unfractionated
population which primarily contained species with a single additional *N*-
acetyllactosamine unit. The three *N*-acetyllactosaminyl repeat species were
found to clear more rapidly than those having less than two repeats. Presumably, the rapid clearance was mediated by galactose binding receptors in
the liver.

In a recent study, Wasley *et al.* (1991) examined the effects of carbohydrate additions on the biologic activities of rhEpo. rhEpo was produced with
altered carbohydrate composition by expression in a mutant Chinese hamster ovary (CHO) cell line. Fully glycosylated Epo was secreted in the presence of galactose (Gal) and *N*-acetylgalactosamine (GalNAc). Omission of
the GalNAc yielded a species which lacked *O*-linked carbohydrate, while
omission of Gal or both Gal and GalNAc yielded species with incompletely
processed *N*-linked oligosaccharides. The species without *O*-linked carbohydrate showed normal *in vitro* and *in vivo* biological activity and *in vivo*
plasma clearance. The species with incompletely processed *N*-linked oligosaccharides exhibited normal *in vitro* activity and *in vivo* plasma clearance,
but lacked *in vivo* activity.

2.2. Epo Clinical Pharmacokinetics

The availability of rhEpo and the development of sensitive and accurate methods for its detection (ELISA and RIA) have enabled more consistent and reliable pharmacokinetic evaluations to be performed. The majority of the studies in humans have shown that the plasma concentration–time curve of intravenously administered rhEpo is best described by a biexponential function, with a plasma half-life from 3 to 16 hr. The pharmacokinetics of rhEpo have been extensively investigated in a variety of therapeutic settings in patients with impaired renal function and anemia, and those undergoing CAPD or hemodialysis. These comparisons have been useful in determining the effects of disease states and therapeutic regimens on the pharmacokinetics and the pharmacodynamics of rhEpo. Dosing regimens have included single or multiple doses administered by intravenous, subcutaneous, or intraperitoneal routes.

2.2.1. Single-Dose Clinical Studies with Epo

In normal volunteers receiving [^{125}I]Epo by intravenous bolus injection, the serum radioactivity–time curve was best described by a tri-exponential function with estimated half-lives of 1.5, 3.5, and 26.3 hr, respectively (Coles et al., 1990). The material recovered from serum in the final phase was not immunoreactive and may have represented metabolic products. There was no significant excretion of the protein into the urine. The pharmacokinetic profile of rhEpo was studied following a single intravenous dose in patients with various degrees of renal function (Kindler et al., 1989). These studies showed that the Epo was cleared with a half-life between 6.5 and 12.7 hr in all groups. The volume of distribution was approximately equal to the plasma volume. Renal clearance accounted for less than 3% of the total clearance. These results were in agreement with animal studies which demonstrated that Epo is cleared by nonrenal mechanisms.

2.2.2. Multiple-Dose Clinical Studies with Epo

Several studies have investigated the effects of increasing doses and multiple doses on the pharmacokinetics of rhEpo. Doses of rhEpo ranging from 10 to 1000 U/kg were given as single intravenous bolus injections to normal men (Flaharty et al., 1988, 1990a). As the dose was increased, the serum half-life increased from 4.4 to 11 hr, and clearance decreased from 15 ml/hr to 4 ml/hr. The volume of distribution was approximately 90 ml/kg at the lowest dose, and 40–60 ml/kg at the higher doses. Total urinary excretion accounted for less than 5% of the dose at all dose levels.

Multiple doses of rhEpo were administered intravenously or subcutaneously at doses of 150 or 300 U/kg to normal male volunteers on Days 1, 4, 6, 8, and 10 (McMahon *et al.*, 1990). Following intravenous injection, the serum concentration–time curve was monoexponential on Days 1 and 10. A trend toward increasing rate of clearance was noted after multiple doses, while a trend toward a decreased rate of clearance was seen as the dose was increased.

In a study performed by Muirhead *et al.* (1988), chronic renal failure patients received multiple doses of rhEpo by intravenous bolus injection at doses of 50, 100, or 200 U/kg. On the first day of dosing, the serum half-lives were 8.6 ± 1.2, 8.8 ± 1.8, and 10.6 ± 5.2 hr, respectively. A decrease in half-life was seen following 22 days of dosing, with estimated half-lives of 7.7 ± 1.7, 7.7 ± 1.1, and 6.7 ± 1.7 hr. The volume of distribution was increased on Day 22, suggesting the recruitment of tissue binding sites.

Lim *et al.* (1989) administered rhEpo intravenously to predialysis patients at doses of 50, 100, or 150 IU/kg. Pharmacokinetic evaluations were performed on the first day of dosing and on Day 54. There were no apparent dose-related pharmacokinetic effects. In some cases the plasma concentration–time curves were biexponential at the beginning of the study and monoexponential at the end of the study (Fig. 3). On the first day of the study, the serum half-life was 7.7 hr and the clearance was 6.6 ml/min. On Day 54 the half-life was decreased to 4.6 hr and the clearance was increased to 9.6 ml/min. The volume of distribution was equal to the plasma volume throughout this study. The increased rate of clearance was not attributable to antibodies as there were no anti-Epo antibodies detected. There was a dose-dependent rise in hematocrit during the study period, and the shortened half-life could be the result of increased uptake and metabolism by an expanded pool of Epo-receptor-positive cells induced during the course of therapy.

rhEpo pharmacokinetics have been compared in hemodialysis patients prior to therapy and following treatment and the establishment of a maintenance regimen. In one study (Cotes *et al.*, 1989), pharmacokinetic evaluations were performed at the initiation of therapy and from 96 to 378 days after initiation of therapy. The plasma concentration versus time curve was best described by a monoexponential function with an estimated half-life of 2.3–7.3 hr (at initiation) and 3.2–5.2 hr (96–378 days). In a second study (Neumayer *et al.*, 1989), the initial plasma half-life and clearance were 9.5 ± 3.9 hr and 6.4 ± 1.2 ml/min, respectively. Following 3 months of therapy, the plasma half-life was decreased to 5.6 ± 2.4 hr and the clearance was increased to 9.0 ± 4.4 ml/min. The pharmacokinetics of a single dose of

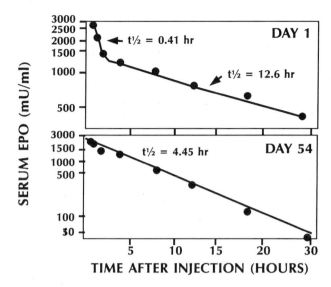

Figure 3. Serum concentration versus time curve of rhEpo. The serum concentration versus time profile on Day 1 (top), and following 54 days of dosing (bottom). (Reproduced with permission from Lim *et al.,* 1989.)

rhEpo given by intravenous or subcutaneous injection were also examined in a group of naive patients, and a group of patients on maintenance doses of rhEpo (Nielsen *et al.,* 1989; Nielsen, 1990). The plasma clearance in the naive population following intravenous administration was slower (7.4 ml/min) and the $t_{1/2}$ was longer (7.6 hr) than in the group on maintenance rhEpo (clearance 9.7 ml/min; $t_{1/2}$ 5.4 hr). Bioavailability of the subcutaneous dose in the naive group was estimated at 31.7%; the monitoring period was too short to allow estimations in the second group.

In contrast to the preceding studies, several studies have shown that multiple dosing regimens are not associated with significant changes in the pharmacokinetic profile of rhEpo. In one of these studies, rhEpo was given to hemodialysis patients intravenously two to three times a week for 2 months (Wikstrom *et al.,* 1988). There was no accumulation in the plasma during the dosing period and there were no significant differences in the estimated pharmacokinetic parameters between the start and the end of the study. The initial clearance, $t_{1/2}$, and volume of distribution were 11 ± 7 ml/min, 5.3 ± 1.3 hr, 4.7 ± 1.5 liters, respectively. After 2 months of treatment, these values were 8 ± 4 ml/min, 5.8 ± 1.2 hr, and 4.2 ± 0.7 liters. In a second study, hemodialysis and CAPD patients received rhEpo by intravenous injection; pharmacokinetic evaluations were performed on Day 1, and after 14–54 weeks of treatment (Hughes *et al.,* 1989). There were no significant differences in the plasma half-life on Day 1 or at later time points in either patient population.

The efficacy of rhEpo in the treatment of anemia in hemodialysis patients has been well established; similar information in CAPD patients is scarce. Because the intravenous route is impractical for chronic dosing, a number of clinical studies have been performed to evaluate subcutaneous and intraperitoneal dosing. Although intraperitoneal rhEpo shows low bioavailability, this method of administration is efficacious (Frenken *et al.,* 1988, 1989). In a study performed in renal dialysis patients (Bommer *et al.,* 1990), rhEpo administered subcutaneously was more efficacious than an equivalent intravenous dose, perhaps as a result of the sustained plasma levels seen with subcutaneous dosing. A dose reduction of 50% may be possible with subcutaneous administration to hemodialysis patients when compared with intravenous administration (Bommer *et al.,* 1988). The biologic effects of rhEpo administered by subcutaneous and intraperitoneal injection were compared in CAPD patients (Coppens *et al.,* 1990). Higher doses were required for the intraperitoneal route; the rate of hematocrit increase was lower after intraperitonial administration than after subcutaneous administration.

Boelaert *et al.* (1988) compared subcutaneous, intravenous, and intraperitoneal doses of rhEpo given to CAPD patients as a single dose of 300 U/kg. The $t_{1/2}$ of the intravenous dose was 11.2 ± 0.4 hr and the bioavailability of intraperitoneal and subcutaneous doses over 24 hr was 2.5 and 10.2%, respectively. These results, however, are not representative of total bioavailability due to the short evaluation period. Peak serum concentrations of Epo were seen 8 to 12 hr following subcutaneous administration in patients with end-stage renal disease (Egrie *et al.,* 1988). These levels were maintained for several hours. Following seven treatments, the peak concentrations were reduced 40 to 70%.

In a study in CAPD patients, a single dose of rhEpo was administered by the intravenous, subcutaneous, or intraperitoneal route (Macdougall *et al.,* 1989a,b,c, 1990). The average plasma $t_{1/2}$ of the intravenous dose was 8.2 hr (range 6.2–10.2) and clearance was 0.047 ml/min per kg (range 0.032–0.085). Bioavailability following intraperitoneal administration was only 2.9% (range 1.2–6.8), with maximum plasma concentrations seen at 12 hr. Bioavailability after a subcutaneous dose was higher (21.5%), and the time to maximum plasma concentration was 18 hr. The differences seen between clinical and preclinical studies may reflect differences in absorption kinetics. An alternative explanation is that peritoneal dialysate instillation and removal may affect the absorption of Epo. Although these studies showed low bioavailability following intraperitoneal administration, Epo levels in the serum continued to increase after the initial 8-hr dwell of the peritoneal dialysate, and the subsequent elimination half-life of the Epo was approximately 23 hr, as compared to approximately 8.2 hr following intravenous administration. The continued absorption of Epo may be due to pooling in the lymphatic system. In another study in CAPD patients, the bioavailability of a subcutaneous dose was 14% at 24 hr and 31% at 72 hr (Hughes *et al.,* 1989). The bioavailability of an equivalent intraperitoneal dose was 56 and 46% of the subcutaneous dose, respectively. The time to maximum plasma levels following the subcutaneous dose was 12 hr and a level above baseline was sustained for 72 hr.

Higher bioavailability for Epo has been reported in patients on chronic hemodialysis (Besarab *et al.,* 1990; Flaharty *et al.,* 1990b,c). The serum half-life was 6.9 ± 3.2 hr, with a clearance rate of 8 ± 2 ml/hr per kg, and a volume of distribution of 70 ± 20 ml/kg. The bioavailability of a subcutaneous dose was $49 \pm 20\%$, with a maximum serum concentration seen at 25.6 ± 13.1 hr. Approximately one-half of the total weekly intravenous dose could be administered by the subcutaneous route to achieve an equivalent biologic effect. Kromer *et al.* (1990) estimated Epo bioavailability following subcutaneous and intraperitoneal administration at 41 and 15%, respectively.

The low bioavailability of rhEpo following subcutaneous administration may be due to its molecular size (36 kDa). Studies by Emanuele and Fareed (1987) have demonstrated a negative correlation between molecular size and bioavailability for heparin.

3. GRANULOCYTE-COLONY STIMULATING FACTOR

G-CSF supports the growth and proliferation of relatively late granulocyte progenitor cells already committed to the neutrophil lineage and enhances the function and activation of mature neutrophils (Burgess and Metcalf, 1980; Nicola *et al.,* 1983). This factor has been cloned and expressed in both bacterial and mammalian cells (Souza *et al.,* 1986; Nagata *et al.,* 1986). Recombinant human G-CSF (rhG-CSF) shows broad species cross-reactivity with respect to its characteristic biologic response, a rapid and dose-dependent neutrophilia. The potential clinical applications for rhG-CSF include myelorestoration following bone marrow transplantation, mitigating chemotherapy-induced neutropenia, boosting host defense against infections, and augmenting effector cell function (Morstyn *et al.,* 1989a).

G-CSF Pharmacokinetics

Cohen *et al.* (1987) reported biexponential clearance of *E. coli*-derived rhG-CSF administered intravenously to hamsters; the α and β serum half-lives were 0.5 and 3.8 hr (all data were normalized to the amount recovered at 5 min after injection). Following subcutaneous administration, biologic effects were noted at 2 hr and were sustained for 36 hours, suggesting rapid absorption.

Tanaka *et al.* (1989) also examined the influence of dose level and route of administration on the pharmacokinetics of *E. coli*-derived G-CSF following intravenous and subcutaneous administration to Sprague–Dawley rats at doses of 10 or 100 μg/kg. Biexponential plasma clearance was noted following an intravenous dose of 100 μg/kg; the α and β half-lives were 25 and 102 min, respectively. Peak serum concentrations were seen at approximately 2 hr postdose (Fig. 4). The calculated bioavailability of the subcutaneous dose was approximately 78%.

In a second study (Tanaka and Tokiwa, 1990), the pharmacokinetics of rhG-CSF were studied in the rat after intravenous or subcutaneous doses of 1, 5, 10, and 100 μg/kg. There was an approximate 30% decrease in the rate of clearance from 80 to 56 ml/hr per kg as the dose was increased from 1 to 10 μg/kg; no further decreases in clearance were noted up to 100 μg/kg. The calculated volume of distribution was equivalent at each dose. Following subcutaneous administration, peak serum levels were observed at approximately 2–3 hr postdose. There were nonproportional increases in AUC; a

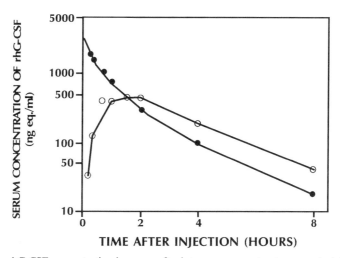

Figure 4. rhG-CSF concentration in serum after intravenous or subcutaneous administration to rats. rhG-CSF was administered to rats intravenously (●) or subcutaneously (○) at a dose level of 100 μg/kg, and the concentration of rhG-CSF in serum was assessed at subsequent time points. Each point is the mean of three animals. (Reproduced with permission from Tanaka *et al.*, 1989.)

100-fold increase in the dose resulted in a 164-fold increase in AUC. Bioavailability following subcutaneous dosing ranged from approximately 50 to 70%. The results of this study suggested that two mechanisms may be involved in the clearance of G-CSF. One pathway is saturated at low concentrations and represents high-affinity, low-capacity receptor-mediated uptake by target cells (neutrophils). The other mechanism represents clearance by nonspecific pathways such as renal excretion.

Layton and co-workers (1989) evaluated the pharmacokinetics of *E. coli*-derived rhG-CSF as part of a Phase I/II clinical trial. The protein was administered by intravenous or subcutaneous routes at dose levels of 3 to 30 µg/kg and serum G-CSF levels were determined by RIA (Fig. 5, right). The elimination half-life following intravenous dosing was 1.4 hr at a dose of 1 µg/kg, and increased to 4.2 hr at a dose of 60 µg/kg. These results indicated saturation of clearance mechanisms at doses greater than 10 µg/kg. When a continuous subcutaneous infusion was performed for 5 days, steady state was not achieved and there was a rapid reduction in G-CSF levels during the last 2 days of administration. Neutrophil levels increased during this period and these results were interpreted as neutrophil receptor-mediated regulation of G-CSF uptake and degradation. Peak levels of G-CSF were achieved at 6 hr after initiation of subcutaneous dosing and these levels were maintained for up to 16 hr. When high doses of melphalan were coadministered

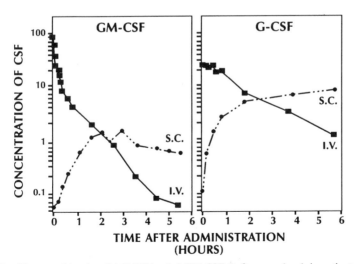

Figure 5. Pharmacokinetics of rhG-CSF and rhGM-CSF. Left: serum levels in patients receiving intravenous or subcutaneous rhGM-CSF (3 µg/kg). Right: serum levels in patients receiving intravenous (1.5 µg/kg) or subcutaneous rhG-CSF (3 µg/kg). Data are for individual patients. (Modified from Morstyn *et al.*, 1989c.)

with G-CSF, the neutrophil levels did not increase and serum G-CSF levels remained constant. The serum half-life results were similar to those reported by Gabrilove *et al.* (1988), who reported an average elimination half-life of 5.1 ± 0.5 hr for doses of 10–60 μg/kg administered to patients in a Phase I/II oncology trial.

rhG-CSF, produced by Amgen (Thousand Oaks, Calif.), has recently been granted marketing approval by the FDA. The following information on the pharmacokinetics of rhG-CSF in humans was contained in the approval documentation. The absorption and clearance of rhG-CSF follow first-order kinetics without any apparent concentration dependence. When rhG-CSF was administered by constant intravenous infusion for 24 hr at a dose level of 20 μg/kg the mean serum concentration was 48 ng/ml. Constant intravenous infusion for 11–20 days produced steady-state serum concentrations over the infusion period. Subcutaneous administration of doses of 3.45 and 11.5 μg/kg resulted in peak serum concentrations of 4 and 49 ng/ml, respectively. The calculated mean volume of distribution was 150 ml/kg. The elimination half-life was 3.5 hr following either intravenous or subcutaneous administration, with a clearance rate of 0.5–0.7 ml/min per kg. The administration of a daily dose for 14 consecutive days did not affect the half-life.

4. GRANULOCYTE MACROPHAGE-COLONY STIMULATING FACTOR

GM-CSF is a multipotential growth factor that effects the proliferation and differentiation of early lineage-specific progenitors that are capable of forming both granulocytes and macrophages (Metcalf, 1985, 1986). In addition to these activities, GM-CSF potentiates the function of the mature cells. GM-CSF has been cloned and expressed in bacteria (Burgess *et al.,* 1987), yeast (Moonen *et al.,* 1987), and mammalian cells (Metcalf *et al.,* 1986). The predominate *in vivo* effects of GM-CSF are a dose-related leukocytosis and monocytosis. The potential uses of GM-CSF include prevention of chemotherapy-induced leukopenia, myelorestoration, augmentation of effector cell functions, and stimulation of monocyte and macrophage function and activity (Morstyn *et al.,* 1988, 1989b,c).

4.1. GM-CSF Preclinical Pharmacokinetics

In an early report on the pharmacokinetics of GM-CSF, the plasma half-life of a partially purified human urinary-CSF was determined following intravenous administration to mice. The plasma bioactivity versus time curve was biphasic, with a calculated $t_{1/2}$ of 1.7 to 3.7 hr (Metcalf and Stanley, 1971). It is important to note that these preparations and the methods used for detection of the protein in the serum were at best imprecise. Burgess and Metcalf (1977) prepared GM-CSF from mouse lung conditioned medium and evaluated its pharmacokinetics and biodistribution in mice. The serum half-life of this preparation in mice was approximately 7 hr. There was some localization of the labeled protein to the spleen and liver; the majority of the label, however, was recovered in the kidney.

In studies with recombinant murine GM-CSF produced in *E. coli,* Metcalf *et al.* (1987) administered a single intraperitoneal dose to mice. Peak levels of GM-CSF were detected in the serum at 30 min after dosing, and the elimination half-life was estimated to be 35 min. In a study reported by Talmadge and co-workers (1989), yeast-derived GM-CSF was administered to mice by the intravenous and intraperitoneal routes. Peak serum concentrations of GM-CSF were observed 2 min after the intravenous injection. Donahue *et al.* (1986a) showed that ^{35}S-labeled rhGM-CSF produced in CHO cells showed a biexponential plasma radioactivity versus time profile following intravenous administration to primates. The calculated serum $t_{1/2}$ α and β were 7 and 90 min, respectively.

rhGM-CSF has been expressed in *E. coli* (Burgess *et al.,* 1987), yeast (Moonen *et al.,* 1987), and mammalian cell lines (Metcalf *et al.,* 1986). The glycosylation patterns of rhGM-CSF produced in these systems differ from nonglycosylated (bacterial), to glycosylated with nonhuman carbohydrate chains (yeast), to heavily glycosylated (mammalian cell lines). Receptor binding studies have shown the lowest receptor affinity for the heavily glycosylated species and the highest receptor affinity for the nonglycosylated species. There is significant size heterogeneity in rhGM-CSF produced in CHO cells due to the carbohydrate moiety. The effects of these carbohydrate modifications on the *in vivo* survival of GM-CSF were examined by Donahue and co-workers (1986b). Protein fractionation techniques were used to resolve three species of rhGM-CSF expressed by CHO cells: one with both *N*-linked glycosylation sites occupied, one with only one *N*-linked site occupied, and one with no occupied *N*-linked sites. The blood half-life of the protein was significantly increased by the addition of *N*-linked carbohydrate, primarily due to an increased $t_{1/2}$ α (Fig. 6). With no *N*-linked carbohydrate, the $t_{1/2}$ α was 2.5 min and the $t_{1/2}$ β was 16 min. With one *N*-linked carbohydrate, the $t_{1/2}$ α and β were 4 and 15 min. With both sites occupied, the $t_{1/2}$ α was 4.5 min and the $t_{1/2}$ β was 90 min. Biodistribution studies showed that the kidney was the major organ of localization at 30 min. Donahue and co-workers (1986b) also examined the effects of the extent of sialation on the clearance of the two *N*-linked species. The desialated form was rapidly cleared to the

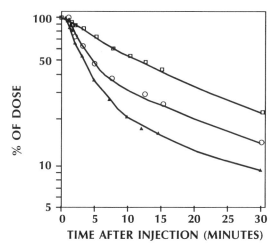

Figure 6. Effects of *N*-linked carbohydrate on the clearance of GM-CSF in the rat. Serum radioactivity versus time curves of preparations of radiolabeled rhGM-CSF having both *N*-linked sites (□), one *N*-linked site (○), or no *N*-linked sites (▲) modified by carbohydrate addition. (Reproduced with permission from Donahue *et al.*, 1986b.)

cleared to the liver and did not show biologic activity in the primate. These results are similar to the results reported for the clearance of sialated and desialated Epo.

Burchiel *et al.* (1989) examined the biodistribution and clearance of CHO-derived rhGM-CSF in the primate using ^{35}S-labeled protein for determinations of plasma pharmacokinetics, and ^{123}I-labeled protein with γ-scintigraphy for noninvasive determinations of biodistribution. Intravenous doses of 15 and 300 μg/kg and subcutaneous doses of 10 and 100 μg/kg were used. A serum $t_{1/2}$ of 4.2 hr was observed following intravenous dosing and a maximum plasma concentration was seen at 16 hr following the subcutaneous dose. No dose-dependent pharmacokinetics were observed. In the presence of an antibody response to the protein, there was rapid localization of radioactivity to the liver.

4.2. GM-CSF Clinical Pharmacokinetics

Several clinical trials have reported pharmacokinetic evaluations of rhGM-CSF. Cebon *et al.* (1988) reported the pharmacokinetics of *E. coli*-derived rhGM-CSF administered to patients by the intravenous or subcutaneous route. GM-CSF levels in the serum were assayed by an ELISA. Following an intravenous dose of 0.3 or 1 μg/kg, the serum concentration versus time curves were best described by a biexponential function, with α and β half-lives of 5 and 150 min, respectively. After a single subcutaneous dose of 10 μg/kg, the peak serum levels of GM-CSF were obtained within 1 hr and sustained levels above 1 ng/ml were seen for more than 12 hr. *In vitro* studies indicated that this serum concentration would produce the desired biologic effect *in vivo*, and that subcutaneous administration should be as efficacious as continuous intravenous infusion. Clinical trials using both intravenous and subcutaneous routes support this conclusion. Typical serum concentration versus time curves are shown in Fig. 5 (left panel) for rhGM-CSF administered by intravenous or subcutaneous injection. Further pharmacokinetic results from this study were published by Lieschke and co-workers (1989). A peak serum level of 15 ng/ml was seen between 4 and 6 hr after a subcutaneous dose of 15 μg/kg. Serum levels greater than 1 ng/ml were maintained for up to 16 hr after dosing. Cebon *et al.* (1990a) also tried to establish the relationship between efficacy and pharmacokinetic parameters. The highest correlations with responses were seen with dose level, length of time serum concentrations greater than 1 ng/ml were maintained, and area under the concentration–time curve. Peak serum concentrations did not show any correlation to response. Toxicity, however, was associated with maximum serum concentrations.

In another study, the pharmacokinetics of bacterially derived rhGM-CSF were evaluated in patients with myelodysplastic syndrome (Thompson *et al.*, 1989). The protein was administered as a daily subcutaneous dose at 0.3, 1, 3, or 10 μg/kg. At doses of 0.3 and 1 μg/kg, only one of three patients showed detectable serum GM-CSF (an ELISA assay was used). At doses of 3 and 10 g/kg, the peak serum levels were seen between 4 and 6 hr postdose, and measurable amounts were detectable out to 14–24 hr.

Short intravenous infusions (30 min) and continuous intravenous infusions were also used to administer bacterially derived rhGM-CSF to patients with advanced malignancies (Steward *et al.*, 1989, 1990a,b). The doses ranged from 0.3 to 60 μg/kg in the short infusion study, and from 3 to 10 μg/kg in the continuous intravenous infusion study. The continuous intravenous infusion was more effective than the short infusion in terms of reduced toxicity and in inducing a neutrophil increase. GM-CSF serum levels showed increases during the first 24 hr of the constant intravenous infusion. With the short infusion, higher peak serum levels were obtained than with constant intravenous infusion. At a dose of 10 μg/kg, the peak concentration at the end of a short infusion was approximately 40 ng/ml as compared to 4 ng/ml at the end of a continuous intravenous infusion at a dose level of 3 μg/kg.

In a Phase Ib trial, Herrmann *et al.* (1989) evaluated the hematopoietic responses and pharmacokinetics of yeast-derived rhGM-CSF administered to patients with advanced malignancy. The rhGM-CSF was administered at dose levels of 120 to 1000 $\mu g/m^2$ by daily intravenous bolus administration or continuous intravenous infusion. Continuous intravenous infusion was far more effective than intravenous bolus dosing in producing a hematopoietic response. This observation is consistent with a report (Cebon *et al.*, 1990b) that showed a correlation between biologic response and sustained serum levels of GM-CSF. The pharmacokinetics of rhGM-CSF after intravenous bolus administration at doses levels of 250, 500, and 1000 $\mu g/m^2$ have been reported. The serum concentration–time curves were biexponential, with α and β half-lives of approximately 10 and 85 min.

The pharmacokinetics of rhGM-CSF produced in yeast were evaluated in normal volunteers (Lee *et al.*, 1990). The doses were administered by subcutaneous injection at dose levels of 3, 10, or 20 $\mu g/kg$. The peak serum concentrations (3.5 hr) were independent of dose. The mean terminal phase half-life was dose related and ranged from 1.6 hr at the low dose, to 3.0 hr at the high dose. In another study using yeast-derived rhGM-CSF, Shadduck *et al.* (1990) reported peak GM-CSF serum levels 2 hr after a subcutaneous dose; these levels persisted up to 6 hr postdose.

5. MACROPHAGE-COLONY STIMULATING FACTOR

M-CSF, also known as CSF-1, is a homodimeric glycoprotein which promotes the proliferation, differentiation, activation, and survival of cells of the monocytic lineage (Tushinski *et al.*, 1982). This factor has been cloned and expressed in both bacterial and mammalian systems (Kawasaki *et al.*, 1985; Wong *et al.*, 1987). The predominate *in vivo* effects of rhM-CSF are dose-related increases in the number and functional activity of peripheral blood monocytes and macrophagelike cells. Human clinical trials with the mammalian and bacterially produced forms are in progress. The therapeutic indications include cancer, infectious disease, and immune system deficiencies.

M-CSF Pharmacokinetics

Shadduck *et al.* (1979) purified CSF from murine L-cell culture. The iodinated protein was administered to mice and showed a rapid plasma clearance with α and β half-lives of 24–40 min and 2–2.5 hr, respectively.

Nephrectomy decreased the clearance. In control animals, there were only trace amounts of the bioactive protein recovered in the urine. These results appeared to indicate that the kidney is involved in the degradation of M-CSF.

Bartocci *et al.* (1987) proposed that macrophages regulate the clearance of M-CSF. M-CSF was prepared from L-cell culture medium and labeled with ^{125}I. The labeled material was injected into mice and the pharmacokinetics and biodistribution were investigated. The labeled protein was cleared from the serum with an estimated $t_{1/2}$ of 10 min. At 5 min after injection, greater than 40% of the dose was recovered in the liver and approximately 6 and 2.6% of the dose was recovered in the spleen and kidney, respectively. The liver uptake of radioactivity was attributed to the action of Kupffer cells. The splenic uptake was attributed to adherent cells. The clearance of M-CSF was saturable at low concentrations. There was a decrease in the rate of clearance of the labeled sample by prior administration of unlabeled M-CSF. The depletion of the macrophage population with carrageenan resulted in a fivefold decrease in the uptake of the [^{125}I]M-CSF in the liver and spleen, and a threefold increase in its half-life (Fig. 7). The authors concluded that macrophages regulate M-CSF clearance by a saturable mechanism involving M-CSF receptor-mediated endocytosis and degradation. This interpretation is consistent with *in vitro* data showing that the receptor for M-CSF is expressed predominantly on monocytes, macrophages, and their committed bone marrow precursors (Byrne *et al.*, 1981). At 37°C, these receptors medi-

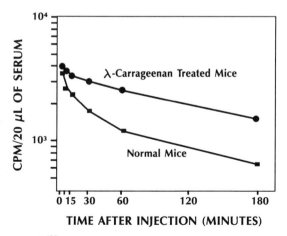

Figure 7. Clearance of [^{125}I]-CSF-1 from the serum of normal and carrageenen-treated C3H/ HeJ mice. Data are presented as means ± S.D.; three mice per point. (Reproduced with permission from Bartocci *et al.*, 1987.)

ate both M-CSF-induced proliferation and M-CSF destruction by the target cells (Tushinski *et al.*, 1982). *In vitro* studies have shown that iodinated M-CSF is degraded by macrophages after being internalized by receptor-mediated endocytosis. The M-CSF is rapidly utilized in a concentration-dependent and cell number-dependent manner over a range of M-CSF concentrations that result in both cell survival and proliferation (Tushinski *et al.*, 1982). Shadduck *et al.* (1989) indicated that after an intravenous injection of L-cell-derived M-CSF to mice, its $t_{1/2}$ was approximately 20 min. Increasing the dose led to an increase in the M-CSF half-life to 45 min, and a saturation and down-regulation of M-CSF receptors.

Pharmacokinetic evaluations of rhM-CSF have been performed in a variety of animal species, including man. The mammalian-produced rhM-CSF is fully glycosylated and has a molecular mass of 90 kDa. The bacterial form is nonglycosylated and is approximately 44 kDa. In studies with rhM-CSF expressed in *E. coli,* Young *et al.* (1989) reported a total body clearance of 0.6 ml/min per kg, with an initial half-life of 35 to 85 min, depending upon the animal species. The half-life in rats and mice was increased twofold following bilateral nephrectomy, although less than 0.01% of the injected dose was recovered as bioactive protein in the urine of control animals. Biodistribution studies in rats and mice were performed using ^{125}I-labeled rhM-CSF and whole body autoradiography. These studies demonstrated early (5 min) accumulation in the spleen red pulp, kidney cortex, bone marrow, and liver. At 20 min postdose, the label was recovered in the proximal tubules, the perimeter of the spleen periarteriolar lymphatic sheaths, and the lymphatic system. Macrophages in the spleen, liver, and kidneys were also labeled. These results suggest that clearance of the nonglycosylated form is due to a combination of renal filtration and metabolism, and an M-CSF receptor-mediated binding and clearance. The clearance rate was dose dependent, and decreased as the dose was increased. rhM-CSF was absorbed after both intraperitoneal and subcutaneous administration, with a reported bioavailability of approximately 50% (Aukerman *et al.*, 1990). The rapid clearance of the bacterially synthesized form requires the administration of a larger dose (up to 5 mg/kg per day) than the mammalian-derived species (where doses of less than 100 μg/kg are effective).

Stoudemire *et al.* (1989) and Stoudemire and Garnick (1991) reported the pharmacokinetics of rhM-CSF produced in mammalian cells. This recombinant form of M-CSF has a molecular mass of 90 kDa and is fully glycosylated; there are two *N*-glycosylation sites and numerous *O*-glycosylation sites. Following an intravenous bolus dose of 100 μg/kg to cynomologus monkeys or Sprague–Dawley rats, the plasma concentration–time curves were best described by a monoexponential function. The calculated clearances were 5–7 ml/hr per kg in the primate and 10 ml/hr per kg in the rat. The half-life in the primate and rat was 6.4 and 3.2 hr, respectively.

Dose ranging studies of rhM-CSF in the rat showed pronounced dose-dependent kinetics (Stoudemire and Garnick, 1991). As the dose increased from 5 to 1000 μg/kg, the clearance rate decreased from 60 to 2.4 ml/hr per kg; the half-life was increased from 23 to 578 min. A tenfold increase in peripheral blood monocytes is seen in Sprague–Dawley rats following five consecutive daily doses of rhM-CSF. The plasma half-life of rhM-CSF administered to these animals was decreased from 27 to less than 1 min. The rapid clearance of the rhM-CSF was associated with increased localization to the liver and spleen, with greater than 95% of the injected dose being recovered in these organs. In control animals, less than 10% of the injected dose was recovered in these organs over the same time interval.

In a study by Timony *et al.* (1990), cynomolgous monkeys were infused with rhM-CSF at a dose of 100 μg/kg per day for 7 days. Intravenous bolus doses of 100 μg/kg rhM-CSF were administered prior to the infusion, and at 72 and 168 hr. Serial blood samples were taken throughout the infusion for the determination of serum rhM-CSF levels by ELISA (Fig. 8). The observed serum rhM-CSF levels reached a plateau within 24 hr and remained at those levels through the next 48 hr. Over the next 24 hr, the rhM-CSF levels began to decrease steadily and returned to predosing levels by 100 hr. The total

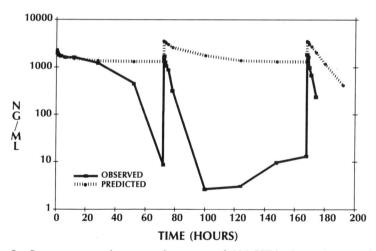

Figure 8. Serum concentration versus time curves of rhM-CSF in the nonhuman primate. rhM-CSF was administered by constant intravenous infusion for 7 days. An intravenous dose was administered on Days 1, 4, and 7. The solid line represents the actual levels and the dotted line represents predicted levels. (J. B. Stoudemire, unpublished.)

body clearance was increased from 35 ml/hr at baseline to 138 and 147 ml/hr at 72 and 168 hr, respectively. A peripheral monocytosis that peaked at Day 5 accompanied the increased rhM-CSF clearance. An antibody that is known to competitively inhibit rhM-CSF binding to its receptor *in vitro* was used to investigate the mechanism of accelerated clearance. The increased clearance of rhM-CSF could be decreased to baseline levels by the administration of this antibody. In biodistribution studies in the rat, administration of the antibody led to a fourfold decrease in liver localization of rhM-CSF. These results suggested that the observed change in clearance during rhM-CSF administration results from an increase in the population of mononuclear phagocytes that bear the rhM-CSF receptor.

Clinical trials are in progress with rhM-CSF. In a dose-escalation trial at Memorial Sloan–Kettering Hospital, patients with metastatic melanoma were treated with rhM-CSF administered as a continuous intravenous infusion for 7 days, followed by a 1-week rest period, and a second cycle of treatment (Bajorin *et al.,* 1990). In another Phase I clinical trial, rhM-CSF was administered by continuous intravenous infusion for 14 consecutive days to bone marrow transplant (BMT) patients following myeloid suppressive therapy (W. P. Peters, personal communication). Serum samples from the patients in these studies were evaluated for M-CSF levels by an ELISA assay. Based upon extrapolation from studies in nonhuman primates, the serum M-CSF concentrations of the melanoma patients were significantly lower than expected, while the M-CSF levels seen in the BMT patients were higher than expected (Fig. 9). Administration of rhM-CSF to melanoma patients resulted in an increase in the monocytes and macrophages expressing M-CSF receptors; these, in turn, mediate enhanced clearance of the protein from the serum. BMT patients, however, have very low populations of these cells, which leads to a reduced rate of receptor-mediated clearance.

6. CONCLUSIONS

The factors described in this chapter mediate quantifiable biologic effects on target cell populations. As clinical trials progress it should be possible to correlate the biologic responses with the pharmacokinetic profiles and use this information to define optimum therapeutic regimens and routes of administration. Unfortunately, preclinical pharmacokinetic studies are often regarded as an exercise to determine a plasma half-life rather than as a means of developing a pharmacokinetic model to assist in determining clinical dosing regimens. Although there have been numerous reports of preclinical and clinical trial results using CSFs, there has been little information

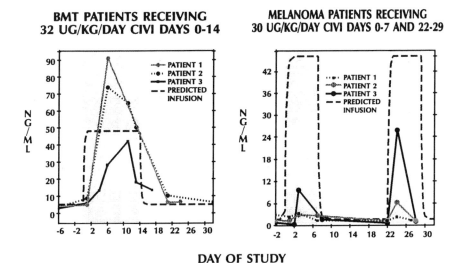

BMT PATIENTS RECEIVING 32 UG/KG/DAY CIVI DAYS 0-14

MELANOMA PATIENTS RECEIVING 30 UG/KG/DAY CIVI DAYS 0-7 AND 22-29

DAY OF STUDY

Figure 9. Serum concentration versus time curves of rhM-CSF in patients. rhM-CSF was administered to bone marrow transplant patients and patients with melanoma. (J. B. Stoudemire, D. Bajorin, and W. P. Peters, unpublished.)

reported about their pharmacokinetic profiles. The CSFs present a challenge to the pharmacokineticist. These factors affect a target cell population leading to proliferation and activation of that population. This in turn results in increased number and/or activity of receptors which are responsible in part for the clearance and metabolism of the factor. Under these conditions, a pharmacokinetic profile becomes a "moving target" which will change in response to the population dynamics of the target hematopoietic cells.

REFERENCES

Ashwell, G., and Kawasaki, T., 1978, A protein from mammalian liver that specifically binds galactose-terminated glycoproteins, *Methods Enzymol.* **50**:287–288.

Aukerman, S. L., Young, J. D., Chong, K. T., and Ralph, P., 1990, Preclinical pharmacology of recombinant human M-CSF, 27th Annual Meeting of the Society for Leukocyte Biology, October 14–18, Crete, Greece.

Bajorin, D. F., Jakubowski, A., Cody, B., Munn, D., Cheung, N.-K., Urmacher, C., Dantis, L., Templeton, M. A., Scheinberg, D. A., Chapman, P., Toner, G., Zakowski, M., Haines, C., Oettgen, H. F., Gabrilove, J., Garnick, M. B., and Houghton, A. N., 1990, Recombinant macrophage colony stimulating factor (rhM-CSF): A Phase I trial in patients with metastatic melanoma, in: *Proceed-*

ings of the 26th Annual Meeting of the American Society of Clinical Oncology, Saunders, Philadelphia, p. 183.

Bargman, J. M., Breborowicz, A., Rodela, H., Abraham, G., and Oreopoulos, D. G., 1990, Absorption of recombinant human erythropoietin (EPO) from the peritoneal cavity in rabbits, *Kidney Int.* **37**(1):243 (abstract).

Bartocci, A., Mastrogiannis, D. S., Migliorati, G., Stockert, R. J., Wolkoff, A. W., and Stanley, E. R., 1987, Macrophages specifically regulate the concentration of their own growth factor in the circulation, *Proc. Natl. Acad. Sci. USA* **84**:6179–6183.

Besarab, A., Vlasses, P., Caro, J., Flaherty, K., Medina, F., and Scalise, R., 1990, Subcutaneous (SC) administration of recombinant human erythropoietin (H-rEPO) for treatment of ESRD anemia, *Kidney Int.* **37**(1):236 (abstract).

Boelaert, J., Schurgers, M., Matthys, E., Daneels, M., DeCre, M., and Bogaert, M., 1988, Recombinant human erythropoietin pharmacokinetics in CAPD patients: Comparison of the intravenous, subcutaneous, and intraperitoneal routes, *Nephrol. Dial. Trans.* **3**:493 (abstract).

Bommer, J., Ritz, E., Weinreich, T., Bommer, G., and Ziegler, T., 1988, Subcutaneous erythropoietin, *Lancet* **2**:406 (letter).

Bommer, J., Schwobel, B., Ritz, E., Zeier, M., Barth, H. P., and Bommer, G., 1990, Efficacy of intravenous (iv) and subcutaneous (sc) recombinant human erythropoietin (rHuEPO) therapy in hemodialysis patients, *Kidney Int.* **37**(1):289 (abstract).

Bondurant, M. C., and Koury, M. J., 1986, Anemia induces accumulation of erythropoietin mRNA in the kidney and liver, *Mol. Cell. Biol.* **6**:2731–2733.

Briggs, D. W., Fisher, J. W., and George, W. J., 1974, Hepatic clearance of intact and desialylated erythropoietin, *Am. J. Physiol.* **227**(6):1385–1388.

Brown, J. H., Lappin, T. R. J., Elder, G. E., Bridges, J. M., and McGeown, M. G., 1990, Comparison of the metabolism of erythropoietin in normal and uremic rabbits, *Kidney Int.* **37**:526 (abstract).

Burchiel, S. W., Oette, D. H., and Stoll, R. E., 1989, Pharmacokinetic evaluation of radiolabeled CHO cell-derived human rGM-CSF in rhesus monkeys using gamma scintigraphy, *Fed. Am. Soc. Exp. Biol.* **3**(3):A501.

Burgess, A. W., and Metcalf, D., 1977, Serum half-life and organ distribution of radiolabeled colony stimulating factor in mice, *Exp. Hematol.* **5**:456–464.

Burgess, A. W., and Metcalf, D., 1980, Characterization of a serum factor stimulating the differentiation of myelomonocytic leukemia cells, *Int. J. Cancer* **26**:647–654.

Burgess, A. W., Begley, C. G., Johnson, G. R., Lopez, A. F., Williamson, D. J., Mermod, J. J., Simpson, R. J., Schmitz, A., and DeLarmarter, J. F., 1987, Purification and properties of bacterially synthesized human granulocyte-macrophage colony-stimulating factor, *Blood* **69**:43–57.

Burke, W. T., and Morse, B., 1962, Studies on the production and metabolism of erythropoietin in the rat liver and kidney, in: *Erythropoiesis* (L. O. Jacobson and M. A. Doyle, eds.), Grune & Stratton, New York, pp. 111–119.

Byrne, P. V., Guilbert, L. J., and Stanley, E. R., 1981, Distribution of cells bearing receptors for a colony stimulating factor (CSF-1) in murine tissues, *J. Cell Biol.* **91**:848.

Cebon, J., Dempsey, P., Fox, R., Kannourakis, G., Bonnem, E., Burgess, A. W., and Morstyn, G., 1988, Pharmacokinetics of human granulocyte-macrophage colony-stimulating factor using a sensitive immunoassay, *Blood* **72**(4): 1340–1347.

Cebon, J., Bury, R., Lieschke, G., O'Connor, M., Farrell, J., and Morstyn, G., 1990a, The pharmacokinetic correlates of GM-CSF efficacy and toxicity, *Proc. ASCO* **9**:701a.

Cebon, J., Nicola, N., Ward, M., Gardner, I., Dempsey, P., Layton, J., Duhrsen, U., Burgess, A. W., Nice, E., and Morstyn, G., 1990b, Granulocyte-macrophage colony stimulating factor from human lymphocytes, *J. Biol. Chem.* **265**:4483–4491.

Clark, S. C., and Kamen, R., 1987, The human hematopoietic colony-stimulating factors, *Science* **236**:1229–1237.

Cohen, A. M., Zsebo, K. M., Inoue, H., Hines, D., Boone, C., Chazin, V. R., Tsai, L., Ritch, T., and Souza, L. M., 1987, *In vivo* stimulation of granulopoiesis by recombinant human granulocyte colony-stimulating factor, *Proc. Natl. Acad. Sci. USA* **84**:2484–2488.

Coles, G. A., Macdougall, I. C., Jones, J., Davies, M., and Williams, J. D., 1990, The metabolism of ^{125}I-labelled recombinant human erythropoietin (rHuEPO) in man, *Kidney Int.* **37**(1):366 (abstract).

Coppens, P. J. W., Frenken, L. A. M., Struijk, D. G., Tiggeler, G. W. L., Krediet, R. T., and Koene, R. A. P., 1990, Comparative study of intraperitoneal and subcutaneous administration of recombinant-human erythropoietin (r-HuEPO) in patients treated with continuous ambulatory peritoneal dialysis (CAPD), *Kidney Int.* **37**(1):326 (abstract).

Cotes, P. M., Pippard, M. J., Reid, C. D. L., Winearls, C. G., Oliver, D. O., and Royston, J. P., 1989, Characterization of the anaemia of chronic renal failure and the mode of its correction by a preparation of human erythropoietin (r-HuEPO). An investigation of the pharmacokinetics of intravenous erythropoietin and its effects on erythrokinetics, *Q. J. Med.* **70**:113–137.

Davis, J. M., and Arakawa, T., 1987, Characterization of recombinant human erythropoietin produced in Chinese hamster ovary cells, *Biochemistry* **26**:2633–2638.

Dinkelaar, R. B., Hart, A. A. M., Engels, E. Y., Schoemaker, L. P., Chamuleau, R. A. F. M., and Bosch, E., 1981a, Metabolic studies on erythropoietin (Ep): I. A reliable micro assay method for Ep in rat plasma using fetal mouse liver cells (FMLC), *Exp. Hematol.* **7**(9):788–795.

Dinkelaar, R. B., Engels, E. Y., Hart, A. A. M., Schoemaker, L. P., Bosch, E., and Chamuleau, A. F. M., 1981b, Metabolic studies on erythropoietin (Ep): II. The role of liver and kidney in the metabolism of Ep, *Exp. Hematol.* **7**(9):796–803.

Donahue, R. E., Wang, E. A., Stone, D. K., Kamen, R., Wong, G. G., Sehgal, P. K., Nathan, D. G., and Clark, S. C., 1986a, Stimulation of haematopoiesis in primates by continuous infusion of recombinant human GM-CSF, *Nature* **321**:872–875.

Donahue, R. E., Wang, E. A., Kaufman, R. J., Foutch, L., Leary, A. C., Witek-Gian-netti, J. S., Metzger, M., Hewick, R. M., Steinbrink, D. R., Shaw, G., Kamen, R., and Clark, S. C., 1986b, Effects of N-linked carbohydrate on the *in vivo* proper-ties of human GM-CSF, *Cold Spring Harbor Symp. Quant. Biol.* **51**:685–692.

Dordal, M. S., Wang, F. F., and Goldwasser, E., 1985, The role of carbohydrate in erythropoietin action, *Endocrinology* **116**:2293–2299.

Egrie, J. C., Eschbach, J. W., McGuire, T., and Adamson, D. W., 1988, Pharmacoki-netics of recombinant human erythropoietin (rHuEpo) administered to hemodi-alysis (HD) patients, *Kidney Int.* **33**:262 (abstract).

Emanuele, R. M., and Fareed, J., 1987, The effect of molecular weight on the bioavail-ability of heparin, *Thromb. Res.* **48**:591–596.

Emmanouel, D. S., Goldwasser, E., and Katz, A., 1984, Metabolism of pure human erythropoietin in the rat, *Am. J. Physiol.* **247**:168–176.

Erslev, A. J., Lavietes, P. H., and Van Wagenen, G., 1953, Erythropoietic stimulation induced by "anemic" serum, *Proc. Soc. Exp. Biol. Med.* **83**:548–550.

Flaharty, K., Vlasses, P. H., Caro, J., Bjornsson, T. D., Whalen, J. J., Morris, E. M., and Erslev, A. J., 1988, Clinical pharmacology of single doses of human recombi-nant erythropoietin in healthy men, *Clin. Pharmacol. Ther.* **45**:135 (abstract).

Flaharty, K., Caro, J., Erslev, A. J., Whalen, J. J., Morris, E. M., Bjornsson, T. D., and Vlasses, P. H., 1990a, Pharmacokinetics and erythropoietic response to human recombinant erythropoietin in healthy men, *Clin. Pharmacol. Ther.* **47**(5):557–564.

Flaharty, K., Besarab, A., Caro, J., Medina, F., and Vlasses, P. H., 1990b, Intrave-nous and subcutaneous pharmacokinetics and pharmacodynamics of human recombinant erythropoietin (rh-EPO) for anemia in end-stage renal disease, *Clin. Pharmacol. Ther.* **47**(5):144 (abstract).

Flaharty, K. K., Besarab, A., Vlasses, P., and Caro, J., 1990c, Pharmacokinetics of human recombinant erythropoietin (H-rEPO, Marogen) in ESRD patients, *Kid-ney Int.* **37**(1):237 (abstract).

Frenken, L. A. M., Coppens, P. J. W., Tiggeler, R. G. W. L., and Koene, R. A. P., 1988, Intraperitoneal erythropoietin, *Lancet* **2**:1495 (letter).

Frenken, L. A. M., Coppens, P. J. W., Tiggeler, R. G. W. L., and Koene, R. A. P., 1989, Intraperitoneal erythropoietin, *Lancet* **1**:962.

Fu, J. S., Lertora, J. J. L., Brookins, J., Rice, J. C., and Fisher, J. W., 1988, Pharmaco-kinetics of erythropoietin in intact and anephric dogs, *J. Lab. Clin. Med.* **111**:669–676.

Fukuda, M. N., Sasaki, H., Lopez, L., and Fukuda, M., 1989a, Survival of recombi-nant erythropoietin in the circulation: The role of carbohydrates, *Blood* **73**:84–89.

Fukuda, M. N., Sasaki, H., and Fukuda, M., 1989b, Erythropoietin metabolism and the influence of carbohydrate structure, in: *Erythropoietin: From Molecular Structure to Clinical Application* (C. A. Baldamus, P. Scigalla, L. Wieczorek, and K. M. Koch, eds.), Karger, Basel, pp. 78–89.

Gabrilove, J. L., Jakubowski, A., Scjer, H., Sternberg, C., Wong, G., Grous, J., Yagoda, A., Fain, K., Moore, M. A. S., Clarkson, B., Oettgen, H. F., Alton, K., Welte, K., and Souza, L., 1988, Effect of granulocyte colony-stimulating factor on neutropenia and associated morbidity due to chemotherapy for transitional-cell carcinoma of the urothelium, *N. Engl. J. Med.* **318:**1414–1422.

Goldwasser, E., and Kung, C. K.-H., 1968, Progress in the purification of erythropoietin, *Ann. N.Y. Acad. Sci.* **149:**49–53.

Goldwasser, E., Kung, C. K.-H., and Eliason, J., 1974, On the mechanism of erythropoietin-induced differentiation, XIII, *J. Biol. Chem.* **249:**4202–4206.

Herrmann, F., Schulz, G., Lindemann, A., Meyenburg, W., Oster, W., Krumwieh, D., and Mertelsmann, R., 1989, Hematopoietic responses in patients with advanced malignancy treated with recombinant human granulocyte-macrophage colony-stimulating factor, *J. Clin. Oncol.* **2**(7):159–167.

Hughes, R. T., Cotes, P. M., Oliver, D. O., Pippard, M. J., Royston, P., Stevens, J. M., Strong, C. A., Tam, R. C., and Winearls, C. G., 1989, Correction of the anaemia of chronic renal failure with erythropoietin: Pharmacokinetic studies in patients on haemodialysis and CAPD, in: *Erythropoietin: From Molecular Structure to Clinical Application* (C. A. Baldamus, P. Scigalla, L. Wieczorek, and K. M. Koch, eds.), Karger, Basel, pp. 122–130.

Jacobs, K., Shoemaker, C., Rudersdorf, R., Neill, S. D., Kaufman, R. J., Mufson, A., Seehra, J., Jones, S. S., Hewick, R., Fritsch, E. F., Kawakita, M., Shimizu, T., and Miyake, T., 1985, Isolation and characterization of genomic and cDNA clones of human erythropoietin, *Nature* **313:**806–810.

Kampf, D., Kahl, A., Passlick, J., Pustelnik, A., Eckardt, E.-U., Ehmer, B., Jacobs, C., Baumelou, A., Grabensee, B., and Gahl, G. M., 1989, Single-dose kinetics of recombinant human erythropoietin after intravenous, subcutaneous and intraperitoneal administration, in: *Erythropoietin: From Molecular Structure to Clinical Application* (C. A. Baldamus, P. Scigalla, L. Wieczorek, and K. M. Koch, eds.), Karger, Basel, pp. 106–111.

Kawasaki, E. S., Ladner, M. B., Wang, A. M., Van Ardsdell, J., Warren, M. K., Coyne, M. Y., Schweickart, V. L., Lee, M. T., Wilson, K. J., Boosman, A., Stanley, E. R., Ralph, P., and Mark, D. F., 1985, Molecular cloning of a complementary DNA encoding human macrophage specific colony-stimulating factor (CSF-1), *Science* **230:**291–296.

Kindler, J., Eckardt, K. U., Ehmer, B., Jandeleit, K., Kurtz, A., Schreiber, A., Scigalla, P., and Sieberth, H. G., 1989, Single-dose pharmacokinetics of recombinant human erythropoietin in patients with various degrees of renal failure, *Nephrol. Dial. Trans.* **4:**345–349.

Koury, M. J., Bondurant, M. C., and Koury, M. J., 1988, Localization of erythropoietin synthesizing cells in murine kidneys by in situ hybridization, *Blood* **71:**524–527.

Kromer, G., Solf, B., Ehmer, B., Kaufmann, E., and Quelhorst, E., 1990, Single dose pharmacokinetics of recombinant human erythropoietin (rHuEpo) comparing intravenous (IV), subcutaneous (SC), and intraperitoneal (IP) administration in IPD patients, *Kidney Int.* **37**(1):386 (abstract).

Krumdieck, N., 1943, Erythropoietic substance in the serum of anemic animals, *Proc. Soc. Exp. Biol. Med.* **54**:14–17.

Layton, J. E., Hockman, H., Sheridan, W. P., and Morstyn, G., 1989, Evidence for a novel *in vivo* control mechanism of granulopoiesis: Mature cell-related control of a regulatory growth factor, *Blood* **4**(74):1303–1307.

Lee, L.-J., Affrime, M., Bordens, R., DeVries, J. K., Jacobs, S., Patrick, J., and Smith, D. E., 1990, Pharmacokinetics of recombinant human granulocyte-macrophage colony stimulating factor (rHu-GM-CSF) after subcutaneous administration to healthy male volunteers, *Pharm. Res. Suppl.* **7**(9):S-218.

Lieschke, G. J., Maher, D., Cebon, J., O'Connor, M., Green, M., Sheridan, W., Boyd, A., Rallings, M., Bonnem, E., Metcalf, D., Burgess, A. W., McGrath, K., Fox, R. M., and Morstyn, G., 1989, Effects of bacterially synthesized recombinant human granulocyte-macrophage colony-stimulating factor in patients with advanced malignancy, *Ann. Intern. Med.* **110**:357–364.

Lim, V. S., DeGowin, L., Zavala, D., Kirchner, P. T., Abels, R., Perry, P., and Fangman, J., 1989, Recombinant human erythropoietin treatment in pre-dialysis patients, *Ann. Intern. Med.* **110**:108–114.

Lin, F.-K., Suggs, S., Lin, C. K., Browne, J. K., Smalling, R., Egrie, J. C., Chen, K. K., Fox, G. M., Martin, F., Stabinsky, Z., Badrawi, S. M., Lau, P. H., and Goldwasser, E., 1985, Cloning and expression of the human erythropoietin gene, *Proc. Natl. Acad. Sci. USA* **82**:7580–7584.

Lukowsky, W. A., and Painter, R. H., 1972, Studies on the role of sialic acid in the physical and biological properties of erythropoietin, *Can. J. Biochem.* **50**:909–917.

Macdougall, I. C., Cavill, I., Davies, M. E., Hutton, R. D., Coles, G. A., and Williams, J. D., 1989a, Subcutaneous recombinant erythropoietin in the treatment of renal anaemia in CAPD patients, in: *Erythropoietin: From Molecular Structure to Clinical Application* (C. A. Baldamus, P. Scigalla, L. Wieczorek, and K. M. Koch, eds.), Karger, Basel, pp. 219–226.

Macdougall, I. C., Roberts, D. E., Neubert, P., Dharmasena, A. D., Coles, G. A., and Williams, J. D., 1989b, Pharmacokinetics of recombinant human erythropoietin in patients on continuous ambulatory peritoneal dialysis, *Lancet* **2**:425–427.

Macdougall, I. C., Roberts, D. E., Neubert, P., Dharmasena, A. D., Coles, G. A., and Williams, J. D., 1989c, Pharmacokinetics of intravenous, intraperitoneal, and subcutaneous recombinant erythropoietin in patients on CAPD, in: *Erythropoietin: From Molecular Structure to Clinical Application* (C. A. Baldamus, P. Scigalla, L. Wieczorek, and K. M. Koch, eds.), Karger, Basel, pp. 219–226.

Macdougall, I. C., Coles, G. A., and Williams, J. D., 1990, The pharmacokinetics of recombinant erythropoietin in CAPD patients, *Kidney Int.* **37**(1):273 (abstract).

McMahon, F. G., Vargas, R., Ryan, M., Jain, A. K., Abeis, R. I., Perry, B., and Smith, I. L., 1990, Pharmacokinetics and effects of recombinant human erythropoietin after intravenous and subcutaneous injections in healthy volunteers, *Blood* **76**(9):1718–1722.

Metcalf, D., 1985, The granulocyte-macrophage colony stimulating factors, *Science* **229**:16–22.

Metcalf, D., 1986, The molecular biology and functions of the granulocyte-macrophage colony stimulating factors, *Blood* **67**:257–267.

Metcalf, D., and Stanley, E. R., 1971, Serum half life in mice of colony stimulating factor prepared from human urine, *Br. J. Haematol.* **20**:549–556.

Metcalf, D., Begley, C. G., Johnson, G. R., Nicola, N. A., Vadas, M. A., Lopez, A. F., Williamson, D. J., Wong, G. G., Clark, S. C., and Wang, E. A., 1986, Biologic properties *in vitro* of a recombinant human granulocyte-macrophage-colony stimulating factor, *Blood* **67**:37–45.

Metcalf, D., Begley, C. G., Williamson, D. J., Nice, E. C., DeLamarter, J., Mermod, J. J., Thatcher, D., and Schmidt, A., 1987, Hemopoietic responses in mice injected with purified recombinant murine GM-CSF, *Exp. Hematol.* **15**:1–9.

Miller, M. E., 1985, Plasma clearance of erythropoietin in erythropoietically perturbed mice, *Hematopoietic Stem Cell Physiology,* Liss, New York, pp. 415–424.

Mladenovic, J., Eschbach, J. W., Koup, J. R., Garcia, J. F., and Adamson, J. W., 1985, Erythropoietin kinetics in normal and uremic sheep, *J. Lab. Clin. Med.* **105**:659–663.

Moonen, P., Mermod, J. J., Ernst, J. F., Hirschi, M., and DeLamarter, J. F., 1987, Increased biologic activity of deglycosylated recombinant human granulocyte/ macrophage colony-stimulating factor produced by yeast or animal cells, *Proc. Natl. Acad. Sci. USA* **84**:4428–4431.

Morell, A. G., Irvine, R. A., Sternlieb, I., Scheinberg, I. H., and Ashwell, G., 1968, Physical and chemical studies on ceruloplasmin: V. Metabolic studies on sialic acid-free ceruloplasmin *in vivo, J. Biol. Chem.* **243**:155–159.

Morell, A. G., Greroriadis, G., Scheinberg, I. H., Hickman, J., and Ashwell, G., 1971, The role of sialic acid in determining the survival of glycoproteins in the circulation, *J. Biol. Chem.* **246**:1461–1467.

Morstyn, G., Campbell, L., Souza, L. M., Alton, N. K., Keech, J., Green, M., Sheridan, W., and Metcalf, D., 1988, Effect of granulocyte colony stimulating factor on neutropenia induced by cytotoxic chemotherapy, *Lancet* **1**:667–672.

Morstyn, G., Lieschke, G. J., Sheridan, W., Layton, J., Cebon, J., and Fox, R. M., 1989a, Clinical experience with recombinant human granulocyte colony-stimulating factor and granulocyte macrophage colony-stimulating factor, *Semin. Hematol.* **26**:9–13.

Morstyn, G., Campbell, L., Lieschke, G., Layton, J. E., Maher, D., O'Connor, M., Green, M., Sheridan, W., Vincent, M., Alton, K., Souza, L., McGrath, K., and Fox, R. M., 1989b, Treatment of chemotherapy-induced neutropenia by subcutaneously administered granulocyte colony-stimulating factor with optimization of dose and duration of therapy, *J. Clin. Oncol.* **7**:1554–1562.

Morstyn, G., Lieschke, G. J., Sheridan, W., Layton, J., and Cebon, J., 1989c, Pharmacology of the colony-stimulating factors, *Trends Pharm. Sci.* **10**:154–159.

Muirhead, N., Keown, P. A., Slaughter, D., Mazaheri, R., Jevniker, A. M., Hollomby, D. J., Hodsman, A. B., Cordy, P. E., Lindsay, R. M., Clark, W. F., and Faye, W. P., 1988, Recombinant human erythropoietin in the anaemia of chronic renal failure: A pharmacokinetic study, *Nephrol. Dial. Trans.* **3**:499 (abstract).

Naets, J. P., and Wittek, M., 1968, Effect of erythroid hyperplasia on the disappearance rate of erythropoietin in the dog, *Acta Haematol.* **39**:42–50.

Naets, J. P., and Wittek, M., 1969, Erythropoietic activity of marrow and disappearance rate of erythropoietin in the rat, *Am. J. Physiol.* **217**:297–301.

Naets, J. P., and Wittek, M., 1974, A role of the kidney in the catabolism of erythropoietin in the rat, *J. Lab. Clin. Med.* **84**:99–106.

Nagata, S., Tsuchiya, M., Asano, S., Kaziro, Y., Yamakazi, T., Yamamoto, O., Hirata, Y., Kubota, N., Oheda, M., Nomura, H., and Ono, M., 1986, Molecular cloning and expression of cDNA for human granulocyte colony stimulating factor, *Nature* **319**:415–417.

Neumayer, H. H., Brockmoller, J., Fritschka, E., Roots, I., Scigalla, P., and Wattenberg, M., 1989, Pharmacokinetics of recombinant human erythropoietin after SC administration and in long-term IV treatment in patients on maintenance hemodialysis, in: *Erythropoietin: From Molecular Structure to Clinical Application* (C. A. Baldamus, P. Scigalla, L. Wieczorek, and K. M. Koch, eds.), Karger, Basel, pp. 131–142.

Nicola, N. A., Metcalf, D., Matsumoto, M., and Johnson, G. R., 1983, Purification of a factor inducing differentiation in murine myelomonocytic leukemia cells: Identification as granulocyte-colony stimulating factor (G-CSF), *J. Biol. Chem.* **258**:9017–9023.

Nielsen, O. J., 1990, Pharmacokinetics of recombinant human erythropoietin in chronic hemodialysis patients, *Pharmacol. Toxicol.* **66**:83–86.

Nielsen, O. J., Egfjord, M., and Hirth, P., 1989, Erythropoietin metabolism in the isolated perfused rat liver, in: *Erythropoietin: From Molecular Structure to Clinical Application* (C. A. Baldamus, P. Scigalla, L. Wieczorek, and K. M. Koch, eds.), Karger, Basel, pp. 90–97.

Paul, P., Rothmann, S. A., McMahon, J. T., and Gordon, A. S., 1984, Erythropoietin secretion by isolated rat Kupffer cells, *Exp. Hematol.* **12**:825–830.

Powell, J. S., Berkner, K. L., Lebo, R. U., and Adamson, J. W., 1986, Human erythropoietin gene: High level expression in stably transfected mammalian cells and chromosome localization, *Proc. Natl. Acad. Sci. USA* **83**:6465–6469.

Reissmann, K. R., Diederich, D. A., Ito, K., and Schmaus, J. W., 1965, Influence of disappearance rate and distribution space on plasma concentration of erythropoietin in normal rats, *J. Lab. Clin. Med.* **65**:967–975.

Roh, B. L., Paulo, L. G., Thompson, J., and Fisher, J. W., 1972a, Plasma disappearance of ^{125}I-labeled erythropoietin in anesthetized rabbits, *Proc. Soc. Exp. Biol. Med.* **141**:268–270.

Roh, B. L., Paulo, L. G., and Fisher, J. W., 1972b, Metabolism of erythropoietin by isolated perfused livers of dogs treated with SKF 525-A, *Am. J. Physiol.* **223**(6):1345–1348.

Sasaki, H. B., Bothner, B., Dell, A., and Fukuda, M., 1987, Carbohydrate structure of erythropoietin expressed in Chinese hamster ovary cells by a human erythropoietin cDNA, *J. Biol. Chem.* **262**:12059–12076.

Scigalla, P., Hoelk, G., and Pahlke, W., 1987, Pharmacokinetics of recombinant erythropoietin (rEpo) in normal and uraemic rats, *Nephrol. Dial. Trans.* **2**:389 (abstract).

Shadduck, R. K., Waheed, A., Porcellini, A., Rizzoli, V., and Pigoli, G., 1979, Physiologic distribution of colony-stimulating factor *in vivo*, *Blood* **54**:894–905.

Shadduck, R. K., Waheed, A., and Weed, E. J., 1989, Demonstration of a blood–bone marrow barrier to macrophage-colony stimulating factor, *Blood* **73**:68–73.

Shadduck, R. K., Waheed, A., Evans, C., Sulecki, M., and Rosenfeld, C. S., 1990, Serum and urinary levels of recombinant human granulocyte-macrophage colony-stimulating factor: Assessment after intravenous infusion and subcutaneous injection, *Exp. Hematol.* **18**:601 (abstract).

Sieff, C. A., 1987, Hematopoietic growth factors, *J. Clin. Invest.* **79**:1549–1557.

Souza, L. M., Boone, T. C., Gabrilove, J., Lai, P. H., Zsebo, K. M., Murdock, D. C., Chazin, V. R., Bruszewski, J., Lu, H., Chen, K. K., Barendt, J., Platzer, E., Moore, M. A. S., Mertelsmann, R., and Welte, K., 1986, Recombinant human granulocyte colony-stimulating factor: Effects on normal and leukemic myeloid cells, *Science* **232**:61–65.

Spivak, J. L., and Hogans, B. B., 1989, The *in vivo* metabolism of recombinant human erythropoietin in the rat, *Blood* **73**:90–99.

Steinberg, S. E., Garcia, J. F., Matzke, G. R., and Mladenovic, J., 1986, Erythropoietin kinetics in rats: Generation and clearance, *Blood* **67**:646–649.

Steward, W. P., Scarffe, J. H., Austin, R., Bonnem, E., Thatcher, N., Morgenstern, G., and Crowther, D., 1989, Recombinant human granulocyte macrophage colony stimulating factor (rhGM-CSF) given as daily short infusions—A Phase I dose-toxicity study, *Br. J. Cancer* **59**:142–145.

Steward, W. P., Scarffe, J. H., Bonnem, E., and Crowther, D., 1990a, Clinical studies with recombinant human granulocyte-macrophage colony-stimulating factor, *Int. J. Cell Cloning* **8**:335–346.

Steward, W. P., Scarffe, J. H., Dirix, L. Y., Chang, J., Radford, J. A., Bonnem, E., and Crowther, D., 1990b, Granulocyte-macrophage colony stimulating factor (GM-CSF) after high-dose melphalan in patients with advanced colon cancer, *Br. J. Cancer* **61**:749–754.

Stohlman, F., and Howard, D., 1962, Humoral regulation of erythropoiesis. IX. The rate of disappearance of erythropoietin from plasma, in: *Erythropoiesis* (L. O. Jacobson and M. A. Doyle, eds.), Grune & Stratton, New York, pp. 120–124.

Stoudemire, J. B., and Garnick, M. B., 1991, *In vivo* evaluations of recombinant human macrophage colony stimulating factor, in: *Cytokines in Hemopoiesis, Oncology, and AIDS* (M. Freund, H. Link, and K. Welte, eds.), Springer-Verlag, Berlin, pp. 459–468.

Stoudemire, J. B., Metzger, M., Timony, G., Bree, A., and Garnick, M. B., 1989, Pharmacokinetics of recombinant human macrophage-colony stimulating factor (rhM-CSF) in primates and rodents, *Proc. AACR* **30**:2139a.

Takeuchi, M., Inoue, N., Strickland, T. W., Kubota, M., Wada, M., Shimizu, R., Hoshi, S., Kozutsumi, H., Takasaki, S., and Kobata, A., 1989, Relationship between sugar chain structure and biological activity of recombinant human erythropoietin produced in Chinese hamster ovary cells, *Proc. Natl. Acad. Sci. USA* **86**:7819–7822.

Talmadge, J. E., Tribble, H., Pennington, R., Bowersox, O., Schneider, M. A., Castelli, P., Black, P. L., and Abe, F., 1989, Protective, restorative, and therapeutic properties of recombinant colony-stimulating factors, *Blood* **73**(8):2093–2103.

Tanaka, H., and Tokiwa, T., 1990, Pharmacokinetics of recombinant human granulocyte-colony stimulating factor studied in the rat by a sandwich enzyme-linked immunosorbent assay, *J. Pharmacol. Exp. Ther.* **256**(1):724–729.

Tanaka, H., Okada, Y., Kawagishi, M., and Tokiwa, T., 1989, Pharmacokinetics and pharmacodynamics of recombinant human granulocyte-colony stimulating factor after intravenous and subcutaneous administration in the rat, *J. Pharmacol. Exp. Ther.* **251**(3):1199–1203.

Thompson, J. A., Lee, D. L., Kidd, P., Rubin, E., Kaufman, J., Bonnem, E. M., and Fefer, A., 1989, Subcutaneous granulocyte-macrophage colony-stimulating factor in patients with myelodysplastic syndrome: Toxicity, pharmacokinetics, and hematological effects, *J. Clin. Oncol.* **7**:629–637.

Timony, G., Bree, A., Metzger, M., Horgan, P., and Stoudemire, J. B., 1990, Pharmacokinetics of recombinant human macrophage colony stimulating factor (rhM-CSF), *Blood* **76**(10)(Suppl. 1):665a.

Tushinski, R. J., Oliver, I. T., Guilbert, L. J., Tynan, P. W., Warner, J. R., and Stanley, E. R., 1982, Survival of mononuclear phagocytes depends on a lineage-specific growth factor that the differentiated cells selectively destroy, *Cell* **28**:71–81.

Wasley, L. C., Horgan, P., Timony, G., Stoudemire, J., Krieger, M., and Kaufman, R., 1991, The importance of N- and O-linked oligosaccharides in the biosynthesis and *in vitro* and *in vivo* biological activities of erythropoietin, *Blood* **77**(12):2624–2632.

Weintraub, A. H., Gordon, A. S., Becker, E. L., Camiscoli, J. F., and Contrera, J. F., 1964, Plasma and renal clearance of exogenous erythropoietin in the dog, *Am. J. Physiol.* **207**:523–529.

Wikstrom, B., Salmonson, T., Grahnen, A., and Danielson, B. G., 1988, Pharmacokinetics of recombinant human erythropoietin in haemodialysis patients, *Nephrol. Dial. Trans.* **3**:503 (abstract).

Wong, G. G., Temple, P. A., Leary, A. C., Witek, J. S., Yang, Y. C., Ciarletta, A. B., Chung, M., Murtha, P., Kriz, R., Kaufman, R. J., Ferlenz, C. R., Sibley, B. S., Turner, K. J., Hewick, R. M., Clark, S. C., Yanai, N., Yokata, H., Yamada, M., Saito, M., Motoyoshi, K., and Takaku, F., 1987, Human CSF-1: Molecular cloning and expression of 4-kb cDNA encoding the human urinary protein, *Science* **235**:1504–1509.

Young, J. D., Aukerman, S. L., Chong, K. T., Moyer, B. R., and Winkelhake, J. L., 1989, Preclinical pharmacology of recombinant human M-CSF, First International Symposium on Cytokines in Hemopoiesis, Oncology, and AIDS, Hanover, West Germany.

Zanjani, E. D., Poster, J., Burlington, H., Mann, L. I., and Wasserman, L. R., 1974, Liver as the site of erythropoietin formation in the fetus, *J. Lab. Clin. Med.* **89**:640–644.

Chapter 8

Pharmacokinetics and Metabolism of Therapeutic and Diagnostic Antibodies

John M. Trang

1. INTRODUCTION

1.1. Background

1.1.1. Historical Perspectives

Biotechnology has been described as the third great technological revolution of the 20th century, following atomic technology and computer technology (Montague, 1989). Over 100 years have passed since Mendel published his original work on genetics in 1866. During this time tremendous advances have been made in our understanding of the universal molecular alphabet which governs how all organisms store genetic information and replicate.

The work of Avery, Wilkins, Crick, Watson, and others during the 1940s and 1950s led to an understanding of the double-helical structure of DNA, the isolation of DNA polymerase, the discovery of RNA polymerase, the establishment of the complete genetic code, and the isolation of restriction endonucleases. These discoveries provided the necessary tools for the first successful DNA recombination experiments in 1973 and for the cre-

John M. Trang • Department of Pharmaceutical Sciences, Samford University Pharmacokinetics Center, School of Pharmacy, Samford University, Birmingham, Alabama 35229.

Protein Pharmacokinetics and Metabolism, edited by Bobbe L. Ferraiolo *et al.* Plenum Press, New York, 1992.

ation of the first recombinant DNA (rDNA) molecules containing mammalian DNA in 1977 (Larkin, 1988).

During the past decade, numerous companies have begun to develop and produce rDNA products with applications in human health care, agricultural productivity, and environmental improvement. Since the approval of the first rDNA human insulin in 1982 by the FDA, biotechnology has proven to be a critical factor in the search for pharmaceuticals to treat and perhaps cure some of the most devastating diseases facing society. Over one-half of the genetically engineered medicines and vaccines that are in human clinical testing or under review by the FDA are in development for cancer and cancer-related conditions.

Of the over 100 biotechnology pharmaceuticals currently in development, monoclonal antibodies comprise the largest category, with 37 products in development. Fifteen of these thirty-seven monoclonal antibodies are being tested for use in either the diagnosis or therapy of cancer. The remaining antibodies in development will have applications in the treatment of septic shock and sepsis, immune response modulation, and organ transplant therapy.

1.1.2. Hybridoma Technology

At the turn of the century Paul Ehrlich proposed the concept of "magic bullets" that would specifically bind infectious organisms and not bind to host tissue (Ehrlich, 1906). His work with Behring led to the discovery of a natural antitoxin to diphtheria toxin in the blood of animals that survived diphtheria infections. Their work and the recognition of antibodies as the immune system's "magic bullets" has been reviewed by Brodsky (1988).

Prior to the development of hybridoma technology in 1975 by Kohler and Milstein, the only source of antibodies for exogenous administration was the serum of immunized animals. Polyclonal antibodies obtained by these methods are usually present in low quantities, are usually heterogeneous, and are usually not reproducible (Kohler, 1986).

The method of lymphocyte fusion developed by Kohler and Milstein (1975) involves the fusion of mouse myeloma tumor cells to spleen cells derived from a mouse that has been previously immunized with an antigen. About 50% of the hybrid cells combine the desired traits of vigorous growth in tissue culture derived from the myeloma tumor cell and antibody production derived from the spleen B cell (Kohler, 1986). This technique has many advantages, including the development of antibodies with high specificity, the production of antibodies in unlimited supply, the development of pure antibody reagents from impure antigens, and the possibility of genetic manipulation.

1.1.3. Monoclonal Antibodies

The application of rDNA technology to the production of murine, human, and chimeric monoclonal antibodies has resulted in the development of hundreds of antibodies with potential applications in cancer, transplant rejection therapy, sepsis, septic shock, and immunomodulation. The administration of these antibodies to patients has prompted a renewed interest in the metabolism and pharmacokinetics of immunoglobulins. The factors that contribute to the variation in the metabolism of immunoglobulins have been reviewed by Waldmann *et al.* (1970) and will be discussed in Section 1.2.

The immunologic and pharmacologic characteristics of monoclonal antibodies have been reviewed (Zuckier *et al.*, 1989). The authors noted that three important variables must be considered when interpreting data: the origin and nature of the immunoglobulin studied; the recipient species; and the method of antibody production, purification, labeling, and assay. Heterogeneous polyclonal antibodies will exhibit different characteristics than homogeneous monoclonal antibodies. Equally important are the antibody isotype and whether the antibody is administered intact or as a fragment.

Species-related factors to be considered include the size of the species and the presence or absence of autogenic or xenografted tumors, since the half-lives of endogenous, homologous immunoglobulins in a given species are inversely proportional to the size of the species, and since the size of the species and the resulting dilution of the antibody are a factor in the rate of interaction of the antibody with the target. Finally, the methods of antibody production, purification, labeling, and assay can have a profound effect on the results. Harsh production, purification, labeling, and conjugation techniques can cause structural damage and lead to reduced antigen–antibody affinity and accelerated antibody catabolism. Nonspecific analytical methods which do not differentiate between biologically active parent antibodies and biologically inactive metabolites result in the generation of erroneous pharmacokinetic parameter estimates (Zuckier *et al.*, 1989).

1.2. Overview of Immunoglobulins

1.2.1. Immunoglobulin Types, Structure, Disposition, Production, and Distribution

Endogenous immunoglobulins have been classified into five major categories: IgG, IgA, IgM, IgD, and IgE. All of the major classes of immunoglobulins have unique rates of synthesis, distribution volumes, steady-state serum concentrations, and fractional catabolic rates (Waldmann *et al.*, 1970).

IgG is a Y-shaped protein composed of four polypeptides, two light (L) chains and two heavy chains, linked by four disulfide linkages. There are two functional regions for IgG antibodies. The arms are referred to as the Fab domain and contain variable regions that are responsible for antigen binding. The Fc domain contains content sequences that are involved in the immunological functions of the antibody. The Fc, Fab, and L chain fragments of intact IgG can be obtained by treating IgG with papain or pepsin. Each of these fragments has unique kinetic characteristics. The Fc fragment has a half-life of 10 to 20 days, but the Fab and L chain fragments have very short half-lives of 0.18 and 0.14 day, respectively (Waldmann *et al.*, 1970).

The primary factor controlling immunoglobulin synthesis is the presence of antigenic stimulation; there are no truly "normal" serum immunoglobulin levels. In addition to the presence of antigen, which is the primary controlling factor for the rate of synthesis of immunoglobulins, other factors include the age of the animal, the concentration of the antibody, the genetic characteristics of the animal, and numerous hormonal factors.

The total body pool size for immunoglobulins ranges from very small body loads of about 0.04 mg/kg body weight for IgE to 1100 mg/kg body weight for IgG (Waldmann *et al.*, 1970). This corresponds to the relatively low rate of synthesis for IgE of 0.016 mg/kg body weight per day to 33 mg/kg body weight per day for IgG. The percent of the total body pool present intravascularly ranges from 42% for IgA to 76% for IgM, with IgG being approximately evenly distributed between intravascular and extravascular fluid.

1.2.2. Immunoglobulin Metabolism and Catabolism

Fractional catabolic rates of immunoglobulins range from 6.7% of the intravascular pool per day for IgG to 89% of the intravascular pool per day for IgE. The catabolism of IgG subclasses varies; IgG1, IgG2, and IgG4 have half-lives of approximately 23 days, while IgG3 has a shorter half-life of 7.5 to 9 days. The half-lives for IgA, IgM, IgD, and IgE are 5.8, 5.1, 2.8, and 2.5 days, respectively (Waldmann *et al.*, 1970).

The catabolism of intact immunoglobulins is determined by factors that are specific for the immunoglobulin class (Waldmann *et al.*, 1970). The catabolic rate of IgM and IgA subclasses is independent of serum concentration. In contrast, the catabolic rate of IgG varies directly in proportion to its concentration in the plasma. The fractional catabolic rate is lowest in patients with reduced IgG concentrations. Half-lives for IgG as long as 70 days and as short as 11 days have been reported (Solomon *et al.*, 1963). The variation in IgG catabolism has been related to the serum IgG concentration and specifically the concentration of the Fc fragment (Fahey and Robinson, 1963).

Hypogammaglobulinemia may result from a variety of pathophysiologic conditions related to defective synthesis, abnormalities in the endogenous rate of immunoglobulin catabolism, and excessive loss of immunoglobulins into the urinary or gastrointestinal tracts. Hypercatabolism of immunoglobulins can occur as a result of familial idiopathic hypercatabolic hypoproteinemia, Wiskott–Aldrich syndrome, myotonic dystrophy, and abnormal immunoglobulin interactions (Waldmann *et al.*, 1970). The loss of immunoglobulins into the urinary tract has been observed in patients with nephrotic syndrome and with Fanconi syndrome. Wochner and co-workers (1967) have shown that the kidney plays an important role in the catabolism of L chains. Excessive loss of gastrointestinal protein is another major pathophysiologic disorder that can lead to hypogammaglobulinemia, and has been observed in association with over 40 diseases (Waldmann, 1966).

1.2.3. Human Anti-Mouse Antibody Response

The administration of exogenous human antibodies to humans normally results in distribution and elimination characteristics that are very similar to endogenous immunoglobulins of the same isotype. The administration of immunoglobulins from nonhuman sources such as murine, porcine, goat, sheep, or even other primates will usually result in the formation of an anti-antibody response. When a murine monoclonal antibody is administered, the response observed is termed a human anti-mouse antibody response (HAMA).

The development of a HAMA response will normally occur within 2 to 4 weeks after administration of a murine monoclonal antibody. It is thought that this antibody response may be involved in some of the adverse side effects observed with the administration of murine monoclonal antibodies. It has been shown that the development of HAMA responses in humans can dramatically alter the pharmacokinetic characteristics of the exogenously administered murine monoclonal antibodies. A comprehensive review of the development of HAMA and antiglobulin responses to monoclonal antibodies in man has been presented by Dillman (1990).

2. PHARMACOKINETICS AND METABOLISM OF MURINE ANTIBODIES

2.1. Murine Native and Radioimmunoconjugate Antibodies and Fragments

Monoclonal antibodies of murine origin are the most extensively studied group of monoclonal antibodies to date (Table I). Numerous native,

Table I

Pharmacokinetic Characteristics of Murine Native and Radioimmunoconjugate Antibodies and Fragments (Section 2.1)

Name / Form / Isotope / Isotope (method)[a]	Disease/target / Antigen / Species (n)[b]	Route,[c] dose / Pharmacokinetic characteristics and comments	References
9.2.27 / Whole / IgG2a / —	Melanoma / Glycoprotein / Man (8)	Single IV 1, 10, 50, 100, and 200 mg doses resulted in proportional increases in peak concentrations	Oldham et al. (1984)
9.2.27 / Whole / IgG2a / [111]In (DTPA)	Melanoma / Glycoprotein / Man (12)	Single IV 1, 50, and 100 mg doses indicated dose-dependent pharmacokinetics. A three-compartment model was used with plasma, nonsaturable, and saturable compartments	Eger et al. (1987)
9.2.27–F(ab')$_2$ / Fragment / IgG2a / [99m]Tc (TFPE)	Melanoma / Glycoprotein / Man (8)	Single IV 2.5 mg dose plus IV 7.5 mg of unlabeled fragment resulted in a mean serum $t_{1/2} = 11.2 \pm 3.8$ hr	Eary et al. (1989)
9.2.27–Fab / Fragment / IgG2a / [99m]Tc (TFPE)	Melanoma / Glycoprotein / Man (19)	Single IV 2.5 mg dose plus IV 7.5 mg of unlabeled fragment resulted in a mean serum $t_{1/2} = 2.1 \pm 0.7$ hr	Eary et al. (1989)
17-1A / Whole / IgG2a / —	Colorectal / Cell surface / Man (20)	Multiple IV 400 mg doses weekly (see sub-table below) HAMA response in all patients within 29 days	LoBuglio et al. (1986)
17-1A / Whole / IgG2a / —	Colorectal / Cell surface / Man (25)	Multiple IV 400 mg doses on days 1, 3, 5, and 8 (see sub-table below)	Khazaeli et al. (1988)

Multiple IV 400 mg doses weekly

	1st (n = 5)	2nd (n = 10)	3rd (n = 3)
Cmax (µg/ml):	139 ± 8	141 ± 5	108 ± 2
$t_{1/2}$ (hr):	15 ± 2	14 ± 1	24 ± 2

Multiple IV 400 mg doses on days 1, 3, 5, and 8

Dose number:	1 (n = 5)	2 (n = 10)	3 (n = 3)	4 (n = 5)
Cmax (µg/ml):	156	150	118	143
$t_{1/2}$ (hr):	15.0	15.1	25.3	14.4
Vd (ml/kg):	45 ± 6	39 ± 4	57 ± 8	44 ± 7
CL (ml/hr/kg):	2.41	2.21	1.92	2.57

Antibody	Parameters	Values			Reference
19-9–F(ab')₂ Fragment IgG1 ¹¹¹In (DTPA) Colorectal, pancreatic, ovarian, lung Oligosaccharide Man (14)	Single IV 1 mg dose $t_{1/2}\alpha$ (hr): $t_{1/2}\beta$ (hr): Maximal activity accumulation occurred at 20 hr in liver, spleen, and kidneys	1.9 19.3			Hnatowich et al. (1985a)
19-9–F(ab')₂ Fragment IgG1 ¹¹¹In (DTPA) Colorectal, pancreatic, Ovarian, lung Oligosaccharide Man (14)	Single IV 1 mg dose $t_{1/2}\alpha$ (hr): $t_{1/2}\beta$ (hr): Vc (L): Vdss (L): MRT (hr): CL (L/hr):	5.1 ± 3.8 27.0 ± 6.6 2.5 ± 1.0 3.5 ± 1.5 28 ± 10 0.13 ± 0.05			Hnatowich et al. (1987)
96.5–Fab Fragment IgG2a ¹²⁵I (CTM) Melanoma p97 glycoprotein Athymic mice (−)	IV 80 μg biodistribution in blood $t_{1/2}\alpha$ (days): $t_{1/2}\beta$ (days): Maximum tumor uptake was 8 to 15% of the dose	1 (70% of dose) 7 (30% of dose)			Larson et al. (1983)
96.5 Whole IgG2a ¹¹¹In (DTPA) Melanoma p97 glycoprotein Man (12)	IV elimination from plasma Dose (mg): $t_{1/2}$ (hr): Vd (L):	1 ($n=4$) 27 ± 9 7.8 ± 0.7	2 ($n=5$) 36 ± 3 7.5 ± 0.6	20 ($n=7$) 27.4 ± 2 3.0 ± 0.1	Rosenblum et al. (1985)
96.5 Whole IgG2a ¹¹¹I (DTPA) Melanoma p97 glycoprotein Man (10)	Single IV 1 mg dose $t_{1/2}$ (hr): Vd (L):	Without IFN 29.8 ± 3.2 3.15 ± 0.50	With IFN 39.7 ± 3.3 5.56 ± 0.67		Rosenblum et al. (1988)
96.5 Whole IgG2a ¹¹¹In (DTPA) Melanoma p97 glycoprotein Man (12)	Single IV 1 mg doses after preinfusion of: $t_{1/2}\alpha$ (min): $t_{1/2}\beta$ (min): Vd (L): CL (ml/min/kg):	NIR-MoAb 42 ± 26 2394 ± 401 6.9 ± 1.5 0.08 ± 0.02	96.5 37 1988 ± 287 3.5 ± 0.3 0.02 ± 0.005		Murray et al. (1988)

(continued)

Table I (*Continued*)

Antibody	Tumor / Antigen / Source	Pharmacokinetics		Reference
260F9 Whole IgG1 ^{111}In (DTPA)	Breast 55-kDa protein Man (7)	Single IV 1 mg doses of [^{111}In]-260F9:	Alone ($n = 6$) / After 20 mg cold 260F9 ($n = 1$)	Griffin et al. (1989)
		$t_{1/2}\alpha$ (hr):	3.1 ± 2.9 / —	
		$t_{1/2}\beta$ (hr):	22.9 ± 12.2 / 28.0	
		Vc (L):	2.11 ± 0.8 / 1.97	
		Vdss (L):	3.0 ± 1.7 / 1.99	
		CL (L/hr):	0.12 ± 0.03 / 0.05	
		MRT (hr):	23.7 ± 10.4 / 40.8	
791T/36 Whole IgG2b ^{131}I (DTPA)	Colorectal 72-kDa protein Man (9)	Single IV doses	200 µg ($n = 8$) / 1 mg ($n = 1$)	Pimm et al. (1985)
		$t_{1/2}\alpha$ (d):	0.62 / 0.5	
		$t_{1/2}\beta$ (d):	1.85 ± 0.30 / 2.0	
791T/36 Whole IgG2b ^{111}In (DTPA)	Colorectal 72-kDa protein Man (4)	IV pharmacokinetics	1 mg ($n = 4$)	Pimm et al. (1985)
		$t_{1/2}\alpha$ (d):	0.46 ± 0.06	
		$t_{1/2}\beta$ (d):	1.40 ± 0.23	
791T/36 Fab IgG2b ^{131}I (−)	Colorectal M_r 72,000 Man (3)	IV 200 µg pharmacokinetics	200 µg ($n = 3$)	Pimm et al. (1985)
		$t_{1/2}\alpha$ (d):	0.20 ± 0.05	
		$t_{1/2}\beta$ (d):	0.78 ± 0.16	
Anti-CEA Whole IgG1 ^{131}I (CTM)	Various cancers CEA Man (6)	Multiple IV 7.5 mg administration following immunosuppression with cyclosporin A Clearance of anti-CEA was accelerated and tumor localization absent with HAMA >30 µg/ml		Ledermann et al. (1988)
Anti-idiotype Whole Various IgG —	BCLL Various Man (14)	Multiple IV doses	HAMA (−) / HAMA (+)	Brown et al. (1987)
		Dose (mg):	500 / 900	
		Cmax (µg/ml):	150 / 350–400	
		$t_{1/2}$ (hr):	6 / 24 (estimated)	
AUA1 Whole IgG1 ^{131}I (Iodogen)	Ovarian 35-kDa protein Man (11)	Single IV 20–30 mg doses	HAMA (−) / HAMA (+)	Stewart et al. (1988)
		Cmax (%):	26 / 5	
		$t_{1/2}\alpha$ (hr):	— / 19	
		$t_{1/2}\beta$ (hr):	50 / 57	
		Du^{110} (%):	70 / 90	

6

Antibody	Target	Parameter	No EDTA	EDTA		Reference
Whole IgG1 ^{90}Y (DTPA)	35-kDa protein Man (19)	Cmax (%):	20	20		...et al. (1988)
		$t_{1/2}$ (hr):	50	50		
		Du^{110} (%):	11	31		
B72.3 Whole IgG1 ^{131}I (Iodogen) ^{125}I (−)	Colorectal TAG-72 Mice (−) −IV −IP	Single IV and IP doses; plasma levels as % of dose	^{131}I (IV)	^{125}I (IP)		Hand et al. (1989)
		C-3 to 5 hr:	—	45 (peak)		
		C-12 hr:	50	—		
		C-72 hr:	14–23	9–16		
B72.3 Whole IgG1 ^{131}I (−) ^{125}I (−)	Colorectal TAG-72 Rats (−)	Single IV and IP doses; plasma levels as % of dose	^{131}I (IV)	^{125}I (IP)		Hand et al. (1989)
		C-4 to 5 hr:	—	50 (peak)		
		C-9 hr:	50	—		
		C-72 hr:	10–12	10–20		
B72.3 Whole IgG1 ^{131}I (Iodogen) ^{125}I (Iodogen)	Colorectal TAG-72 Monkeys (−)	Single IV and IP doses; plasma levels as % of dose	^{131}I (IV)	^{125}I (IP)		Hand et al. (1989)
		C-24 hr:	—	40 (peak)		
		C-20 to 30 hr:	50	—		
		C-72 hr:	20	20		
B72.3 Whole IgG1 ^{131}I (Iodogen) ^{125}I (Iodogen)	Colorectal TAG-72 Man	IV and IP; peak plasma levels after IP injection at 48 hr				Hand et al. (1989)
		Time of peak following IP administration correlated to body surface area				
B72.3 Whole IgG1 ^{99m}Tc (MT)	Breast, colon TAG-72 Monkey (−)	Single IV doses				Burchiel et al. (1989)
		Dose (mg/kg):	0.03	0.3	1.0	
		$t_{1/2}\beta$ (hr):	24.6	27.6	26.8	
		Blood MRT (hr):	22.0	25.4	22.4	
		Systemic MRT (hr):	34.3	38.7	37.1	
HMFG1 Whole IgG1 ^{131}I (Iodogen)	Ovarian Mucinlike Man (11)	See AUA1				Stewart et al. (1988)

(continued)

Table I (*Continued*)

Antibody / Isotype	Tumor / Specificity / n[b]	Data	Reference
HMFG1 Whole IgG1 ^{90}Y (DTPA)	Ovarian Mucinlike Man (19)	see AUA1	Stewart *et al.* (1988)
Lym-1 Whole IgG2a ^{131}I (CTM)	Lymphoma B-cell specific Man (5)	IV 5–12 mg; elimination of ^{131}I and [^{131}I]-Lym-1 was characterized by rapid distribution (hours) and slow elimination (0.5–1.5 days)	DeNardo *et al.* (1988)
Lym-1 Whole IgG2a —	Lymphoma B-cell specific Man (10)	Single IV doses Dose (mg): 58.1 116.2 232.4 464.8 C-1 hr (µg/ml): 4–6 4–17 10–22 30–35 $t_{1/2}\alpha$ (hr): 2 to 12 $t_{1/2}$: days (not specified)	Hu *et al.* (1989)
OC-125-F(ab') Fragment IgG1 ^{111}In (DTPA)	Ovarian Epithelial Man (18)	Single IV 1 mg doses $t_{1/2}\beta$ (hr): 21 ± 8.6 Vc (L): 2.3 ± 0.7 Vdss (L): 2.2 ± 0.6 CL (L/hr): 0.056 ± 0.025 MRT (hr): 44 ± 14	Hnatowich *et al.* (1987)
OC-125-F(ab)$_2$ Fragment IgG1 ^{90}Y (DTPA)	Ovarian Epithelial Man (5)	IP biodistribution during second-look surgery indicated that the mean tumor/normal tissue radioactivity ratio varied between 3:1 and 25:1	Hnatowich *et al.* (1987)
T101 Whole IgG2a ^{131}I (CTM)	CTCL T65 Man (5)	Single IV doses Dose (mg): 9.6–10.5 9.9–16.9 $t_{1/2}\alpha$ (hr): 0.2–1.5 0.5–2.7 $t_{1/2}\beta$ (hr): 11.0–29.0 2.0–27.5 Lesion/normal: 0.9–2.1 1.1–6.8	Rosen *et al.* (1987)
T101 Whole IgG2a ^{131}I (CTM)	CLL T65 Man (4)	Single IV 10 mg dose Plasma $t_{1/2}\alpha/t_{1/2}\beta$ (hr): 0.02–2.7/15–177 Whole body $t_{1/2}\beta$ (days): 0.9–1.2 Urinary excretion $t_{1/2}$ (days): 1.2–4.0	Zimmer *et al.* (1988)

[a] Isotope labeling method if provided: CTM, chloramine T method; DTPA, diethylenetriaminepentaacetic acid method; MT, metallothionein method; TFPE, tetrafluorophenol ester method; PTN, pertechnetate method.

[b] n, number of animals/subjects in study.

radioimmunoconjugate, chemoimmunoconjugate, immunotoxin conjugate whole antibody and antibody fragments have been developed and tested in animal models and man. A number of the murine native and radioimmuno-conjugate monoclonal antibodies and fragments have been developed for use in the diagnosis and treatment of cancer. Biodistribution and/or pharma-cokinetic data have been reported for many of these agents.

2.1.1. Disposition

The pharmacokinetics of murine native and radioimmunoconjugate antibodies and fragments have been characterized using one-, two-, and three-compartment pharmacokinetic models (LoBuglio *et al.,* 1986; Hnato-wich *et al.,* 1987; Eger *et al.,* 1987). Single-dose administration of the IgG1 and IgG2 isotype murine monoclonal antibodies to man has generally re-sulted in biphasic elimination of the antibody from plasma. The rapid initial phase half-life has been characterized to be as short as 0.2 to 2.5 hr (e.g., T101; Rosen *et al.,* 1987) and as long as 12 hr (e.g., 791T/36; Pimm *et al.,* 1985). The mean elimination half-lives for intact murine native monoclonal antibodies have been reported to range from 24 to 48 hr (e.g., T101; Rosen *et al.,* 1987; and 791T/36; Pimm *et al.,* 1985, respectively). These ranges for IgG2 antibodies were also characteristic of IgG1 isotype antibodies (e.g., 260F9; Griffin *et al.,* 1989).

Many murine intact native and radioimmunoconjugate IgG1 and IgG2 isotype antibodies have been characterized using a one-compartment phar-macokinetic model. The elimination half-lives in man generally ranged from a minimum of 12 hr up to a maximum of 48 hr as reported for Lym-1 (DeNardo *et al.,* 1988) and 17-1A (LoBuglio *et al.,* 1986). These same mu-rine antibodies, if administered to mice or other closely related rodent ani-mal models, exhibited two-compartment pharmacokinetic characteristics with significantly longer distributive half-lives of approximately 1 day and elimination half-lives of approximately 7 days (Larson *et al.,* 1983). Other nonmurine species such as monkeys have exhibited distributive and elimina-tion half-lives for murine monoclonal antibodies very similar to those ob-served in man (Hand *et al.,* 1989; Burchiel *et al.,* 1989).

The distribution and elimination of bispecific F(ab')$_2$ antibody frag-ments have been reported to be slightly more rapid than for intact monoclo-nal antibodies (Hnatowich *et al.,* 1985a). For example, 19-9-F(ab')$_2$ exhib-ited two-compartment pharmacokinetics in man with a mean distributive half-life of 5 h and a 27-h mean elimination half-life (Hnatowich *et al.,* 1987). In contrast, monospecific Fab fragments were cleared very rapidly from blood in man, with elimination half-lives reported to be as short as 2 hr

(Eary *et al.*, 1989). The longest half-life reported for a Fab fragment was 0.8 day (Pimm *et al.*, 1985). When antibody fragments were administered to the species of origin (i.e., murine to mice), the distributive and elimination half-lives were considerably longer than when they were administered to an alternate species (i.e., murine to human).

2.1.2. Distribution and Clearance

Regardless of the pharmacokinetic model used to characterize the disposition of murine monoclonal antibodies and fragments, the reagents appeared to distribute throughout the vascular space and into limited extracellular fluids. When estimated, the volume of distribution of the central compartment has normally been in the range of 2 to 3 liters, which is very consistent with plasma volume. The apparent volume of distribution and the steady-state volume of distribution estimates have been reported to be 3.5 to 4 liters (Hnatowich *et al.*, 1987; LoBuglio *et al.*, 1986; Khazaeli *et al.*, 1988; Rosenblum *et al.*, 1988). These volume estimates were very consistent with the distribution volume of 6 to 7 liters reported for IgG, which distributes between the vascular (55%) and extravascular (45%) spaces (Waldmann *et al.*, 1970).

Clearance of murine antibodies and fragments has been reported for only a few antibodies (Hnatowich *et al.*, 1987; Khazaeli *et al.*, 1988; Murray *et al.*, 1988; Griffin *et al.*, 1989). The total body clearance of 17-1A ranged from 1.9 to 2.6 ml/hr per kg following multiple dose intravenous administration of 400 mg to man. The clearance of a 1-mg intravenous dose of 19-9-$(Fab')_2$ fragment in man was similar, with a value of 0.13 liter/hr reported (Hnatowich *et al.*, 1987).

2.1.3. Dose-Proportionality and Route of Administration

With two exceptions, all of the murine monoclonal antibodies and fragments reviewed that have been administered in increasing doses have exhibited linear pharmacokinetic characteristics. Increasing the dose has resulted in proportional increases in maximum plasma concentration and area under the curve (Oldham *et al.*, 1984; Hu *et al.*, 1989). The two exceptions are monoclonal antibody 9.2.27, which showed dose-dependent pharmacokinetics in the dose range of 1 to 100 mg (Eger *et al.*, 1987), and monoclonal antibody 96.5, which exhibited a decrease in the apparent volume of distribution as the dose was increased from 2 to 20 mg (Rosenblum *et al.*, 1985).

Several different routes of administration have been utilized, particularly for those antibodies directed at abdominal carcinomas. Comparison of intravenous and intraperitoneal administration of B72.3 resulted in a very interesting observation (Hand *et al.*, 1989). The authors noted that the time of the peak plasma antibody level correlated to the body surface area of the species tested. Specifically, the studied mice, rats, monkeys, and man and found that species with larger body surface area exhibited a longer the time to peak concentrations. The peak concentration in man occurred as long as 48 hr after administration, whereas in mice the peak concentration occurred within 3 to 5 hr. It was suggested that this was a result of the different surface areas of the peritoneal cavities across species.

2.1.4. Factors Influencing Pharmacokinetics

While most of the antibodies reviewed were administered as radioimmunoconjugates, several groups of investigators utilized both radioactivity and ELISA to characterize the disposition of the monoclonal antibodies (Griffin *et al.*, 1989; Rosenblum *et al.*, 1988). Very similar pharmacokinetic estimates were obtained regardless of the analytical method utilized, indicating that characterization of antibody pharmacokinetics using radiolabels appeared to be a reliable method.

Several investigators have studied the influence of the coadministration of nonimmunoreactive proteins (Murray *et al.*, 1988), immunomodulating agents such as α-interferon (Rosenblum *et al.*, 1988), and immunosuppressive agents such as cyclosporin A (Ledermann *et al.*, 1988) on antibody disposition. The coadministration of [^{111}In]-96.5 with α-interferon resulted in a prolongation of the half-life of the radioimmunoconjugate from 29.8 hr to 39.7 hr, and an increase in the apparent volume of distribution from 3.2 liters to 5.6 liters. The authors concluded that the administration of human α-interferon along with monoclonal antibody therapy significantly enhanced the uptake of the antibody by tumors.

The coadministration of either nonlabeled 96.5 or a nonimmunoreactive monoclonal antibody with ^{111}In-labeled 96.5 also resulted in alterations in the apparent volume of distribution, clearance, and elimination half-life (Murray *et al.*, 1988). The authors suggested that the difference between the two volume terms, 6.9 liters with nonimmunoreactive monoclonal antibody and 3.6 liters without, or 3.3 liters, was the volume of the saturable antigen compartment. The difference in clearance with the nonimmunoreactive monoclonal antibody (0.08 ml/min per kg) versus 0.02 ml/min per kg with coadministration of nonlabeled 96.5, was the result of saturation of antigen binding sites by the immunoreactive antibody and not merely saturation of clearance mechanisms by the nonimmunoreactive antibody. Similar results were obtained with 260F9 (Griffin *et al.*, 1989).

Coadministration of cyclosporin A and [^{131}I]anti-CEA resulted in a reduction in HAMA response (Ledermann *et al.*, 1988). The mean HAMA level in patients receiving cyclosporin A was 3.5 μg/ml. In three other patients who did not receive cyclosporin A, the HAMA level was nearly 2000 μg/ml. Patients receiving cyclosporin A showed evidence of antibody accumulation in the tumors after each consecutive dose, whereas those who did not receive cyclosporin A had an accelerated clearance of anti-CEA and no tumor localization (Ledermann *et al.*, 1988). The development of HAMA responses in patients receiving multiple-dose administration of murine monoclonal antibodies 17-1A and AUA1 was virtually 100% (LoBuglio *et al.*, 1986; Khazaeli *et al.*, 1988; Stewart *et al.*, 1988). The development of HAMA responses in patients receiving multiple-dose administration did not appear to influence the pharmacokinetics of either 17-1A or AUA1 significantly (Khazaeli *et al.*, 1988; Stewart *et al.*, 1988).

2.2. Murine Chemoimmunoconjugate Antibodies and Fragments

The potential application of monoclonal antibody-directed chemotherapy has been a focus of considerable investigation (Takahashi *et al.*, 1988; Dillman *et al.*, 1988; Pimm *et al.*, 1988; Yang and Reisfeld, 1988; Pietersz *et al.*, 1990). Doxorubicin (DXR), daunomycin (DNM), daunorubicin (DNR), desacetylvinblastine (DAVLB), methotrexate, mitomycin C, *N*-acetylmelphalan (N-AcMEL), neocarzinostatin (NCS), and others have been conjugated to various murine monoclonal antibodies using a variety of linkage technologies. In many instances the results from experimental animal models have not been encouraging. It has been suggested that the efficacy of monoclonal antibody drug conjugates may be profoundly affected by *in vivo* parameters (Yang and Reisfeld, 1988). The pharmacokinetics of several of these monoclonal antibody drug conjugates are summarized in Table II.

2.2.1. Disposition

The pharmacokinetics of several chemoimmunoconjugates have been investigated in mice and man (Yang and Reisfeld, 1988; Pimm *et al.*, 1988; Takahashi *et al.*, 1988; Pietersz *et al.*, 1990; Schneck *et al.*, 1990). Two chemoimmunoconjugates, 9.2.27–DXR and 791T/36–DNM, exhibited two-compartment pharmacokinetic characteristics in tumor-bearing mice models. These were very similar to the pharmacokinetics of unconjugated 9.2.27 and 791T/36, respectively (Yang and Reisfeld, 1988; Pimm *et al.*,

1988). The mean distributive half-lives ranged from 0.5 to 5 hr and the mean elimination half-lives ranged from 50 to 100 hr.

Similarly, the pharmacokinetics of anti-Ly-2.1–N-AcMEL exhibited two-compartment pharmacokinetic characteristics in mice, with mean distributive half-lives for the antibody and the chemoimmunoconjugate of approximately 5 hr and elimination half-lives arranging from 60 to 80 hr (Pietersz et al., 1990). The half-lives for the chemoimmunoconjugates were very similar to those of the unconjugated antibody and significantly longer than the two-compartment pharmacokinetic half-lives for unconjugated doxorubicin, daunomycin, and N-acetylmelphalan.

2.2.2. Distribution, Clearance, and Dose-Proportionality

The intravenous administration of KS1/4–DAVLB in man resulted in pharmacokinetic characteristics that were described using one- and two-compartment models which were very similar to those reported for nonconjugated murine monoclonal antibody 17-1A (Schneck et al., 1990). A one-compartment pharmacokinetic model was adequate to describe 4 of the 13 patients who received the KS1/4–DAVLB. In these patients, the mean elimination half-life was 31.5 hr, the mean total body clearance was 0.09 liter/hr, and the mean apparent volume of distribution was 4.43 liters.

In those patients (6 of 13) exhibiting two-compartment characteristics, the mean distributive half-life ranged from approximately 1.4 to 5.7 hr and the mean elimination half-life was 31.5 hr. The chemoimmunoconjugates appeared to exhibit very similar pharmacokinetic characteristics when compared to the unconjugated antibody following administrations to the same species (Schneck et al., 1990). Evaluation of AUC versus weight-normalized KS1/4–DAVLB dose resulted in a linear relationship, suggesting that the elimination of the chemoimmunoconjugate was not dose-dependent in the range of doses administered (40–250 mg/m^2). The mean total urinary and fecal recovery of the radiolabeled dose was 30% over 5 days. Two-thirds of this was recovered in feces and one-third was present in the urine (Schneck et al., 1990).

2.2.3. Route of Administration

The route of administration for chemoimmunoconjugates appeared to be a very important factor for the biodistribution and efficacy of the reagents. Intravenous administration produced distribution of the chemoimmunoconjugate throughout the entire systemic circulation. This resulted in

Table II
Pharmacokinetic Characteristics of Murine Chemoimmunoconjugate Antibodies (Section 2.2)

Name / Form / Isotype / Conjugate[a] / Linkage[a]	Disease/target Antigen Species (n)[b] Ratio[d]	Route[c], dose Pharmacokinetic characteristics and comments	References
9.2.27 Whole IgG2a Doxorubicin CAA	Melanoma Proteoglycan Mice (−) 8–10:1	Single IV doses of [^{125}I]-9.2.27–DXR and [^{125}I]-9.2.27 　　　　　　　　9.2.27–DXR　　9.2.27 $t_{1/2}\alpha$ (hr):　　　5　　　4 $t_{1/2}\beta$ (hr):　　　102　　97 Less than 10% of the injected IP [^{125}I]-9.2.27–DXR and [^{125}I]-9.2.27 appeared in serum with a peak at days 2 and 3	Yang and Reisfeld (1988)
791T/36 Whole IgG2b Daunomycin 14BD	Colorectal — Mice (−) 50:1	Single IV 4–100 µg doses resulted in biexponential decline Rapid distribution (3–5 h) was followed by slow elimination (50 h)	Pimm et al. (1988)
A7 Whole IgG1 Neocarzinostatin SPDP	Colorectal — Man (41) 2–3:1	Single IV, IA, or IP 15–90 mg doses were well tolerated NCS concentration in tumors was higher than in normal tissue	Takashashi et al. (1988)

			[³H]-N-AcMEL	[³H]-MoAb	[³H]-MoAb-N-AcMEL	
Anti-Ly-2.1	Thyoma	Single IV doses				Pietersz et al. (1990)
Whole	—					
IgG2a	Mice (−)					
N-Acetylmelphalan	10–16:1					
AE		$t_{1/2}\alpha$ (hr):	0.5	5	5.5	
		$t_{1/2}\beta$ (hr):	9.5	80	60	
		C-48 hr (%):	1.3	—	—	
		C-96 hr (%):	—	—	10	
KS1/4	Adenocarcinoma 40-K	Single IV 40–250 mg/m² doses				Schneck et al. (1990)
Whole	Man (13)	$t_{1/2}\beta$ (hr):	31.5			
IgG2a	4–6:1	Vd (L):	4.43			
Desacetylvinblastine		CL (L/hr):	0.09			
HSN		Du-5 days (%):	30			

[a] Linkage methods: CAA, cis aconitic anhydride; 14 BD, 14 = bromodaunomycin; SPDP, N-succinimidyl 3-(2-pyridyldithio)-propionate; AE, active ester; HSN, hemisuccinate.
[b] Number of subjects/animals in study.
[c] Route of administration: IV, intravenous; IP, intraperitoneal; IA, intraarterial.
[d] Drug-to-antibody ratio.

limited exposure of the chemotherapeutic agent to the tumor, even though the antibody targeted efficiently to the tumor site.

Intraperitoneal administration has been evaluated as well and has been shown to be a very effective route for the treatment of intraperitoneal tumors. Intraarterial administration to sites proximal to the tumor may be the most effective route of administration for chemoimmunoconjugates, since it results in exposure of the tumor to high concentrations of the reagent with minimal exposure in the systemic circulation.

2.2.4. Obstacles and Applications

The potential application of chemoimmunoconjugates in cancer therapy is promising. Biodistribution studies have indicated that the conjugation of a chemotherapeutic agent to a monoclonal antibody causes localization of the drug to the target antigen site. The evaluation of the efficacy of certain chemoimmunoconjugates in animal models has not been encouraging (Yang and Reisfeld, 1988; Pimm et al., 1988). In contrast, the efficacy of AZ7-NCS in man following intraarterial administration was very encouraging.

Obstacles to be overcome include the ability to conjugate sufficient quantities of the chemotherapeutic drug to the antibody without reducing the antigen binding affinity of the antibody.

In addition, the type of conjugation bonding must be such that once the antibody chemoimmunoconjugate complex reaches the tumor site the drug will be released in a sufficiently high concentration to effect cell killing. The longer half-lives of chemoimmunoconjugates when compared to the unconjugated drug are a clear advantage. The conjugation of chemotherapeutic agents with either human or chimeric monoclonal antibodies may lead to greater efficacy and reduced overall toxicity, since they will allow the readministration of the chemoimmunoconjugate on a multiple-dose basis without the development of HAMA responses.

2.3. Murine Immunotoxin Antibodies and Fragments

Immunotoxins are a new class of agents that have recently been used to treat cancer in a number of Phase I clinical trials. The immunotoxin conjugate consists of two components, usually a carrier protein such as a monoclonal antibody, and a cytotoxic agent such as diphtheria toxin, pseudomonas exotoxin, ricin toxin A-chain (RTA), or ribosome-inactivating toxin (RIT) (Gould et al., 1989). The rationale underlying immunotoxin therapy involves transport of the toxin to the tumor cell surface by an antibody which has specific activity for a tumor antigen; antibody binding to the cell surface; internalization of the toxin and release into the cytosol; and toxin-mediated inhibition of protein synthesis. The pharmacokinetic characteristics of several immunotoxins have been investigated in animal models and patients with cancer (Bourrie *et al.*, 1986; Laurent *et al.*, 1986; Letvin *et al.*, 1986a,b; Blakely *et al.*, 1987, 1988; Byers *et al.*, 1987; LoBuglio *et al.*, 1988, 1990; Manske *et al.*, 1988; Weiner *et al.*, 1989; Gould *et al.*, 1989). The results of these investigations are summarized in Table III.

2.3.1. Disposition

The administration of antibody conjugates with RTA or recombinant RTA (rRTA), such as 260F9–rRTA (Weiner *et al.*, 1989; Gould *et al.*, 1989), 791T/36–RTA (Byers *et al.*, 1987, 1989), anti-Thy 1.1–RTA (Blakely *et al.*, 1987, 1988), J12.5–RTA (Fulton *et al.*, 1988), and T101–RTA (Bourrie *et al.*, 1986) in mice and other rodent models has resulted in plasma concentration versus time data requiring a two-compartment pharmacokinetic model for accurate characterization.

The mean distributive half-lives for these RTA immunotoxin conjugates have been reported to be as short as 5 to 7 min (Bourrie *et al.*, 1986) and as long as 30 min (Blakely *et al.*, 1988) in mice and other rodent models. The mean elimination half-lives reported for these antibodies in mice and other rodent models have ranged from 8 to 12 hr. The disposition of immunotoxin monoclonal antibody conjugates in mice and other rodent models reflects a much more rapid clearance of the immunotoxin from blood than the corresponding unconjugated monoclonal antibodies. The presence of the RTA dramatically increased the clearance of the antibody and resulted in distributive and elimination half-lives that were very similar to the RTA molecule.

The disposition of anti-T11 murine monoclonal antibody conjugated with the ribosome-inactivating toxins gelonin and saporin has been investigated in monkeys (Letvin *et al.*, 1986a,b). Although the pharmacokinetics were not clearly characterized in these studies, the authors indicated that the immunotoxins were cleared much more rapidly in monkeys than was the unconjugated monoclonal antibody.

The plasma disposition of RTA monoclonal antibody immunotoxins in man has also been characterized using a two-compartment model (Gould *et al.*, 1989; Weiner *et al.*, 1989; LoBuglio *et al.*, 1990). Distributive half-lives as short as 14 min have been reported for XMMME-001–RTA (LoBuglio *et al.*, 1990) and as long as 1 to 3 hr for 260F9–rRTA (Weiner *et al.*, 1989). The serum disposition of the immunotoxin–RTA conjugates reflected much

Table III
Pharmacokinetic Characteristics of Murine Immunotoxin Antibodies and Fragments (Section 2.3)

Name / Form / Isotype / Conjugate / Linkage[a]	Disease/target Antigen Species (n)[b] Ratio[d]	Route[c], dose / Pharmacokinetic characteristics and comments				References
260F9 Whole, IgG1, Ricin A, Disulfide	Breast, 55-kDa protein, Man (4), —	Multiple IV doses for 6–8 days;				Weiner et al. (1989)
		Dose (μg/kg/day): 10-1	10-2	50-3	50-4	
		Cmax (μg/ml): 260	170	850	500	
		$t_{1/2}\alpha$ (hr): 1.1	2.6	3.0	2.0	
		$t_{1/2}\beta$ (hr): 7.9	8.7	10.9	9.9	
		Vc (ml/kg): 39.1	69.1	63.7	90.7	
		Vss (ml/kg): 62.7	95.5	91.9	139.0	
		CL (ml/hr/kg): 6.2	10.4	10.4	11.8	
260F9 Whole, IgG1, Ricin A, Disulfide	Breast, 55-kDa protein, Man (5), —	Single IV infusions				Gould et al. (1989)
		Infusion (μg/kg/day): 50 (n = 3)	100 (n = 2)			
		Tss (hr): 24	24			
		Css (μg/ml): 0.23	0.91			
		Vd (L) assumed: 3	3			
		$t_{1/2}$ (hr) estimate: 4–6	4–6			
		Anti-260F9 (μg/ml): 10–100	0			
		Anti-ricin A (μg/ml): 49–54	0–71			
791T/36 Whole, IgG2b, Ricin A, Disulfide	Colon, ovarian, 72-k antigen, Mice (-), 4.2:1	Single IV doses				Byers et al. (1987)
		$t_{1/2}\alpha$ (min): 9				
		$t_{1/2}\beta$ (hr): 9.8				
		Presence of ricin a-chain accelerated clearance				
Anti-T11 Whole, IgG1, Gelonin, Saporin, Disulfide	T cell, Erythrocyte-rosette receptor, Monkey (-), 2.3:1	Single IV 2–5 mg/kg doses				Letvin et al. (1986a)
			Anti-T11	Anti-T11–IT		
		Cmax (μg/ml):	65–90	70–90		
		C-1 hr (%):		60		
		C-24 hr (%):		13		

Anti-T11
T cell
Erythrocyte-rosette receptor
Monkey
2.3:1
Whole
IgG1
Gelonin
Saporin
Thioether
Disulfide

Single IV 1–5 mg/kg doses
Cmax (µg/ml): Anti-T11 57–106 Gelonin/saporin 12–35
Devitalized antibody was cleared more rapidly than unconjugated
Ribosome-inactivating protein was present on lymph node and spleen T cells at 16 hr

Letvin et al. (1986b)

Anti-Thy 1.1
Lymphoma
—
Mice (15)
—
Whole
IgG1
Ricin A
Disulfide

Single IV 10 µg doses of anti-Thy 1.1–RTA and anti-Thy 1.1–RTA.dg

	IT-A	IT-dg.A
$t_{1/2}\alpha$ (hr):	0.25	1.5
$t_{1/2}\beta$ (hr):	9.5	9.5
D_B^{24} (%):	<20	40

Blakely et al. (1987)

Anti-Thy 1.1
Lymphoma
—
Mice (—)
—
Whole
IgG1
Ricin A
Saporin
Disulfide

IV doses of anti-Thy 1.1–RTA and anti-Thy 1.1–SAP

	IT-RTA	IT-SAP	Anti-Thy 1.1
$t_{1/2}\alpha$ (hr):	0.52 ± 0.13	1.1 ± 0.2	3.0 ± 0.3
$t_{1/2}\beta$ (hr):	9.7 ± 1.3	17.1 ± 0.6	114 ± 3
VD (ml/kg):	211	104	98
MRT (hr):	9.4	21.8	158

Blakey et al. (1988)

JA12.5
Spleen
—
Mice (—)
—
Whole and fragment
IgG2b
Ricin A
Disulfide

Single IV 5–10 µg doses

	$t_{1/2}\alpha$ (min)	$t_{1/2}\beta$ (hr)
JA12.5	13	12
JA12.5-RTA	11	5
JA12.5-RTA.dg	12	6
JA12.5-Fab'	10	7
JA12.5-Fab'-RTA	11	4
JA12.5-Fab'-FTA.dg	10	4

Fulton et al. (1988)

T101
T cell
65-kDa glycoprotein
Rabbits (—)
—
Whole
IgG2a
Ricin A
Disulfide

Single IV 1.25 mg/kg doses

	T101-RTA	T101	RTA
$t_{1/2}\alpha$ (min):	5–7	—	5–7
$t_{1/2}\beta$ (hr):	7–8	13	5–7

Bourrie et al. (1986)

Table III (*Continued*)

			Pharmacokinetics		Reference
T101	T-cell lymphoma	Man (2)	Single IV 13.5–25 mg doses		Laurent et al. (1986)
Whole	65-kDa glycoprotein			13.5 mg × 2	25 mg × 3
IgG2a			Cmax (μg/ml):	115–450	600
Ricin A			C-24 hr (mg/ml):	5	1
Disulfide					
T101	Leukemia	Mice (5)	Single IV 5 μg doses		Manske et al. (1988)
Whole	65-kDa glycoprotein			T101-RIT	[125I]T101
IgG2a			$t_{1/2}\beta$ (hr):	25.7 ± 1.5	91.3 ± 13.3
Ricin A					
Disulfide					
XMMME-001	Melanoma	Man (10)	Single IV 0.4 mg/kg doses		LoBuglio et al. (1988)
Whole	220-k antigen		Cmax (μg/ml):	6.8 ± 2.2	
IgG2a			$t_{1/2}\beta$ (min):	24	
Ricin A					
Disulfide					
XMMME-001	Melanoma	Man (10)	Single IV 0.4 mg/kg doses		LoBuglio et al. (1990)
Whole	220-k antigen		Cmax (μg/ml):	6.8 ± 2.2	
IgG2a			$t_{1/2}\alpha$ (min):	14	
Ricin A			$t_{1/2}\beta$ (hr):	2.7	
Disulfide			Vd (L/kg):	0.103	
			CL (ml/hr/kg):	41	

[a] Linkage method.

[b] Number of animals/subjects in study.

[c] Route of administration: IV, intravenous.

[d] Drug-to-antibody ratio.

more rapid elimination of the immunotoxin than for the corresponding unconjugated monoclonal antibody in man.

2.3.2. Distribution and Clearance

The distribution volume for RTA immunotoxins has been reported to be as large as 200 ml/kg in mice (Blakely *et al.,* 1988) and approximately 90 to 140 ml/kg in man (Weiner *et al.,* 1989; LoBuglio *et al.,* 1990). These apparent volume of distribution estimates indicated that immunotoxins tend to distribute into a larger volume than the corresponding unconjugated monoclonal antibodies. Volume estimates in man of approximately 0.1 liter/kg indicated that immunotoxins appear to distribute throughout plasma water and limited extravascular space; this volume was approximately twofold larger than the 3-liter volume of distribution reported for unconjugated murine monoclonal antibodies.

Several investigators have examined the influence of deglycosylation of the ricin A-chain on both the distribution and clearance of immunotoxins in mice (Blakely *et al.,* 1987, 1988; Fulton *et al.,* 1988). The elimination of both anti-Thy 1.1–RTA and anti-Thy 1.1–RTA.dg was characterized as biexponential with an initial rapid distributive phase followed by a slower elimination phase. After 8 hr, when the distributive phase appeared to be complete, only 4.3% of the anti-Thy 1.1–RTA remained in the bloodstream compared with 18.4% of the anti-Thy 1.1–RTA.dg (Blakely *et al.,* 1987). Thereafter, both of the immunotoxins were eliminated from the blood at similar rates with elimination half-lives of approximately 9.5 hr. These were both significantly shorter than the 190-hr elimination half-life of unconjugated anti-Thy 1.1 in mice (Blakely *et al.,* 1987).

The authors indicated that approximately one-half of the accelerated disappearance of anti-Thy 1.1–RTA.dg and one-fifth of the disappearance of anti-Thy 1.1–RTA were accounted for by breakdown of the immunotoxins to free antibody and RTA. Clearance of the antibody from the liver and spleen was accompanied by its accumulation in the stomach, thyroid, and salivary gland. The presence of RTA and its glycosylation influenced the tissue distribution of the molecules; accumulation in the stomach, thyroid, and salivary gland reached a maximum at 4 hr. The rank order of accumulation was: anti-Thy 1.1–RTA > anti-Thy 1.1–RTA.dg > anti-Thy 1.1. Anti-Thy 1.1–RTA was metabolized and excreted more rapidly than the deglycosylated immunotoxin. After 24 hr, less than 20% of anti-Thy 1.1–RTA was present in the mouse, while 40% of the deglycosylated immunotoxin remained in the animal (Blakely *et al.,* 1987).

The authors noted that the chemical deglycosylation of RTA retarded the clearance of anti-Thy 1.1–RTA substantially, indicating that clearance of anti-Thy 1.1–RTA.dg was similar to that of an anti-Thy 1.1 immunotoxin prepared using Abrin A-chain, which does not have oligosaccharide side chains. They commented that the deglycosylation appeared to have eliminated the contribution that the A-chain-associated sugar residues make to the immunotoxin clearance. It also appeared that the changes in distribution influence the plasma concentration versus time profile, since anti-Thy 1.1–RTA achieved greater concentrations at 4 hr in stomach, thyroid, and salivary tissues than either the deglycosylated or the unconjugated anti-Thy 1.1. They noted that the slower clearance of anti-Thy 1.1–RTA.dg was primarily due to its decreased entrapment by the liver. Chemical deglycosylation destroyed all of the terminal and most of the remaining mannose residues and all of the fructose residues; thus, hepatic uptake of anti-Thy 1.1–RTA.dg by either of these pathways should be eliminated (Blakely et al., 1987).

The administration of T101 as the intact murine monoclonal and as the F(ab′)$_2$ fragment conjugated with RTA to rabbits, rats, mice, and rhesus monkeys resulted in rapid distribution half-lives and comparable elimination phase half-lives (Bourrie et al., 1986). It appeared that the elimination of both the immunotoxin conjugated to intact T101 and the immunotoxin conjugated to T101 F(ab′)$_2$ fragment was very similar to the elimination of RTA alone, indicating that the presence of the RTA on the monoclonal antibody had a significant influence on its clearance.

2.3.3. Dose-Proportionality and Multiple-Dose Administration

Two studies have reported the administration of different doses of immunotoxins (Weiner et al., 1989; Gould et al., 1989). The immunotoxin 260F9–rRTA was administered to man by intravenous bolus and continuous intravenous infusion. In these studies, maximum serum concentrations after intravenous administration were proportional to the increase in dose (Weiner et al., 1989). In contrast, continuous intravenous infusion of both 50 and 100 µg/kg per day resulted in roughly a fourfold increase in steady-state serum concentrations (Gould et al., 1989).

Multiple infusions of XMMME-001–RTA resulted in an IgM response to the immunotoxin by day 28 in all patients. The level of IgM activity was not significantly different in the groups receiving immunosuppressive therapy (LoBuglio et al., 1988). In addition, all patients exhibited an IgG anti-antibody response. Patients receiving high-dose cytoxan/prednisone therapy had a significantly greater IgG response than those who did not receive immunosuppressive therapy or who received moderate-dose cytoxan and low-dose continuous azathioprine/prednisone. The degree of IgM response after multiple infusions of XMMME-001–RTA did not influence its pharmacokinetics. In contrast, the clearance of the immunotoxin upon reinfusion correlated with the degree of IgG response present at the time of reinfusion (Lo-Buglio *et al.*, 1988).

2.3.4. Summary

In contrast to the chemoimmunoconjugate monoclonal antibodies which appear to exhibit very similar pharmacokinetic characteristics in terms of their volume and clearance as the unconjugated monoclonal antibody, the combination of RTA or either of the ribosome-inactivating proteins with the murine monoclonal antibodies dramatically alters both their distribution and their clearance. The distribution volume of RTA immunotoxins appeared to be slightly larger than the corresponding unconjugated monoclonal antibodies, and the clearance of immunotoxins was much more similar to the clearance of the toxin itself rather than the monoclonal antibody. In other words, the presence of the RTA or the ribosome-inactivating protein significantly influenced the antibodies' uptake and clearance by the liver.

3. PHARMACOKINETICS AND METABOLISM OF NONMURINE, HUMAN, AND CHIMERIC ANTIBODIES

3.1. Nonmurine Polyclonal Antibodies

While the development and testing of murine monoclonal antibodies have been extensive, the potential for polyclonal nonmurine antibodies as diagnostic and therapeutic agents has also been investigated (Goldenberg *et al.*, 1981; Leichner *et al.*, 1983, 1988; Rostock *et al.*, 1984; Vaughn *et al.*, 1985; Order *et al.*, 1986; Sharkey *et al.*, 1988). The development and application of several polyclonal antibodies have resulted in a significant increase in our understanding of the many factors which influence the disposition of exogenously administered antibodies in man. The results of these investigations are summarized in Table IV.

3.1.1. Disposition

The administration of anti-CEA polyclonal antibody of sheep and porcine origin to hamsters, rats, and mice has resulted in serum disposition that was characterized using a one-compartment pharmacokinetic model and elimination half-lives ranging between 2.5 and 4.5 days (Goldenberg *et al.*, 1981; Vaughn *et al.*, 1985). The administration of radiolabeled antiferritin

Table IV

Pharmacokinetic Characteristics of Nonmurine Polyclonal Antibodies (Section 3.1)

Name Form Isotype Source Isotope (method)[a]	Disease/target Antigen Species (n)[b]	Route[c], dose Pharmacokinetic characteristics and comments	References
Anti-CEA Whole IgG Goat ^{131}I (CTM)	Colon Carcinoembryonic antigen Hamster (–)	Single IC 25–105 µg doses resulted in elimination half-lives of 2.5 to 4.5 days Uptake of label by tumors continued up to day 25 Single dose resulted in marked inhibition of tumor growth and increase in survival time	Goldenberg et al. (1981)
Anti-CEA Whole IgG Sheep ^{90}Y (DTPA) ^{125}I (CTM)	Colon Carcinoembryonic antigen Rat (–)	Single IV 30 µg doses of [^{90}Y]anti-CEA and [^{125}I]anti-CEA resulted in similar elimination of ^{90}Y and ^{125}I labels from plasma The elimination half-life was approximately 50 hr Substantial uptake of ^{90}Y occurred in the liver	Vaughan et al. (1985)
Anti-CEA Whole IgG Goat ^{131}I (CTM)	Colon Carcinoembryonic antigen Hamster (–)	Single IC 15 µg doses of [^{131}I]anti-CEA primary antibody (PA) and donkey anti-goat IgG (SA) PA alone resulted in biphasic elimination with $t_{1/2}\alpha$ = 24 hr and $t_{1/2}\beta$ = days SA resulted in a rapid drop in circulating PA and an increase in tumor/blood ratios	Sharkey et al. (1988)

Antiferritin
Whole
IgG
Rabbit/pig
[131]I (LPO)

Hepatoma
Ferritin
Man (22)

Single IV 10 mg doses

	Total body	Liver and tumor
$t_{1/2}\alpha$ (days):	0.28	0.48
$t_{1/2}\beta$ (days):	2.7	3.9

Leichner et al. (1983)

Antiferritin
Whole
IgG
Rabbit
[131]I (LPO)

Hepatoma
Ferritin
Rat (–)

Single IC 180–200 μg doses

	Blood	Liver	Tumor
Tmax (hr):	0.5	0.5	8
Cmax (μCi/g):	5	1	1.5
$t_{1/2}\alpha$ (hr):	0.15	–	–
$t_{1/2}\beta$ (hr):	16	17	12

Rostock et al. (1984)

Antiferritin
Whole
IgG
Rabbit/pig
[90]Y (chelation)

Hepatoma
Ferritin
Man (6)

Single IV 10–20 mCi/mg doses

	Rabbit	Procine
Origin:		
$t_{1/2}$ (days):	1.5–2.0	1.5–1.7

Order et al. (1986)

Antiferritin
Whole
IgG
[111]In (–)

Hepatoma
Ferritin
Man (11)

Single IV 7 mCi/doses

	Blood	Liver	Tumor
$t_{1/2}\alpha$ (days):	0.38	–	–
$t_{1/2}\beta$ (days):	2.7	2.9	3.1

Leichner et al. (1988)

[a] Isotope labeling method if provided: CTM, chloramine T method; DTPA, diethylene pentaacetic acid; LPO, lactoperoxidase.
[b] Number of animals/subjects in study.
[c] Route of administration: IC, intracardiac; IV, intravenous.

to rodent animal models and patients with hepatocellular carcinoma has resulted in serum disposition that has been characterized using one- and two-compartment models in both the animal models and man. When antiferritin of rabbit origin was administered to rats and man, a rapid mean distributive half-life of 6 hr was observed and the mean elimination half-lives ranged between 16 and 36 hr (Leichner *et al.*, 1983, 1988; Rostock *et al.*, 1984; Order *et al.*, 1986).

3.1.2. Distribution

The biodistribution of both anti-CEA and antiferritin has been characterized in animal models particularly with respect to the tumor/blood ratios and tumor/normal tissue ratios that were achieved. Ratios from 1 to 1 to 6 to 1 have been reported (Leichner *et al.*, 1983; Rostock *et al.*, 1984; Sharkey *et al.*, 1988). The factors influencing the biodistribution of both of these polyclonal antibodies include tumor volume, tumor vascularity, tumor necrosis, external beam irradiation, and the subsequent administration of a second antibody.

Leichner and co-workers (1983) reported that there was a linear correlation of the activity reached in the tumors of patients given antiferritin with tumor volume up to about 2300 cm^3. Tumors larger than 2300 cm^3 exhibited lower activity per tumor volume than those below 2300 cm^3. Rostock *et al.* (1984) studied the biodistribution of antiferritin in rats and found a clear correlation between tumor size and the ratio established, with smaller tumors reaching higher ratios than larger tumors. The authors attributed this to the greater vascularity of smaller tumors. Other factors that they noted as important in the biodistribution of antiferritin were necrosis in the tumor and the antigen concentration itself.

3.1.3. Factors Influencing Pharmacokinetics

Sharkey and co-workers (1988) conducted an interesting experiment involving the administration of a second donkey anti-goat IgG antibody in hamsters 24 hr after the administration of a goat polyclonal anti-CEA. Administration of the second donkey anti-goat IgG resulted in a precipitous drop in the primary antibody blood levels, which in turn resulted in a very large increase in the primary antibody tumor/blood ratio. Leichner *et al.* (1988) demonstrated that external beam irradiation of patients with hepatocellular carcinoma who had received antiferritin also produced increased tumor uptake. They were able to correlate this with the increased vascularity that occurs in tumors 24 to 48 hr after external beam irradiation.

3.1.4. Applications and Limitations

Polyclonal antibodies, while less specific than monoclonal antibodies, may have a place in the therapy of certain cancers. Investigation of the disposition of polyclonal antibodies in species other than the species of origin has shown elimination half-lives of 2 to 4 days; these are much shorter than the elimination half-life of endogenous IgG. Limitations to therapy with polyclonal antibodies will be similar to those for murine monoclonal antibodies, since the potential for the development of an anti-antibody response exists. Nevertheless, the administration of polyclonal antiferritin to patients with hepatocellular carcinoma has produced very encouraging clinical responses (Order *et al.*, 1986).

3.2. Human Antibodies and Fragments

The development of methods to produce human monoclonal antibodies in sufficiently large quantities to enable their administration to man has resulted in major advances in antibody therapy. The frequent production of HAMA responses following the administration of murine monoclonal and polyclonal antibodies to patients has led to the development of human monoclonal antibodies to human tumor tissues (Hnatowich *et al.*, 1985b; Ehrlich *et al.*, 1987; McCabe *et al.*, 1988; Fisher *et al.*, 1990; Khazaeli *et al.*, 1990; Steis *et al.*, 1990).

Theoretically, human monoclonal antibodies should be less immunogenic than murine monoclonal antibodies when administered to humans. Technical limitations in the production of human monoclonal antibodies prevented the development of sufficient quantities until the late 1980s. With the development of large scale production methods for human monoclonal antibodies, it has become possible to study these reagents in patients with cancer and other diseases. The results of several investigations using human monoclonal antibodies are summarized in Table V.

3.2.1. Disposition

The administration of two human IgM monoclonal antibodies, 16.88 and 28A32, to mice resulted in distributive and elimination half-lives in the range of 6 to 8 and 8 to 12 hr, respectively (McCabe *et al.*, 1988). In contrast, the administration of these same two human IgM monoclonal antibodies to man resulted in distributive and elimination half-lives of 16 and 40 to 57 hr, respectively (Steis *et al.*, 1990).

Table V
Pharmacokinetic Characteristics of Human Antibodies and Fragments (Section 3.2)

Name Type Isotope Isotope (method)[a]	Disease/target Antigen Species (n)[b]	Route[c], dose Pharmacokinetic characteristics and comments	References
16-88 Whole IgM ^{125}I (Iodogen)	Colon Epstein–Barr Mice (−)	Single IV 50 µg doses $t_{1/2}\alpha$ (hr) $\quad$ $t_{1/2}\beta$ (hr) Serum: $\quad$ 7–8 $\quad$ 12 Spleen: $\quad$ 8 $\quad$ 19 Liver: $\quad$ 5 $\quad$ 17 Kidney: $\quad$ 12 $\quad$ 24 Muscle: $\quad$ 9 $\quad$ 16 Tumor: $\quad$ 48–72 Tumor/blood ratio: 12:1	McCabe et al. (1988)
16-88 Whole IGM ^{131}I (Iodogen)	Colon Epstein–Barr Man (12)	Single IV 8–208 mg doses Plasma $\quad$ Whole body $t_{1/2}\alpha$ (hr): $\quad$ 16.0 $\quad$ — $t_{1/2}\beta$ (hr): $\quad$ 56.9 $\quad$ 51 No significant differences over entire dose range No HAMA response	Steis et al. (1990)
28A32 Whole IgM ^{125}I (Iodogen)	Colon Surface and cytoplasm Mice (−)	IV 50 µg pharmacokinetics in non-tumor-bearing mice were similar to 16-88 The initial $t_{1/2}\alpha$ (hr) = 6–8 (plasma), 5–12 (tissues) Tumor/blood ratio = 10.1 at 4–7 days	McCabe et al. (1988)
28A32 Whole IgM ^{131}I (Iodogen)	Colon Surface and cytoplasm Man (16)	Single IV 8–208 mg doses resulted in 63% present in serum at end of infusion The mean $t_{1/2}\beta$ was 40 hr Similar disposition from 8–108 mg but slower clearance at 208 mg dose	Steis et al. (1990)
EV1-15 Whole IgG1 ^{125}I (Iodogen)	AIDS Cytomegalovirus Monkeys (2)	Single and multiple IV 0.5 mg/kg doses Single dose $\quad$ Multiple dose $t_{1/2}$ (days) ELISA: $\quad$ 14.4–19.4 $\quad$ 19.1–23.5 $t_{1/2}$ (days) label: $\quad$ 13.5–15.0 $\quad$ — Immune response in one monkey after first dose	Ehrlich et al. (1987)

HA-1A
Whole IgM
—
Sepsis
Bacterial endotoxin
Man (9)

Khazaeli et al. (1990)

Single IV doses

Dose (number):	10 mg (2)	25 mg (3)	100 mg (2)
Cmax (μg/ml):	3.0	10.4	41.6
C-24 hr (μg/ml):	1.4	4.5	20.1
$t_{1/2}$ (hr):	25.8	28.4	35.0
AUC (μg/ml · hr):	101	410	1955
Vdss (ml/kg):	50.8	47.3	39.5
CL (ml/hr/kg):	1.4	1.1	0.9
AUC/D (μg/ml · hr/mg):	10.1	16.4	19.6

No immune responses observed in 28 days

HA-1A
Whole IgM
Sepsis
Bacterail endotoxin
Man (34)

Fisher et al. (1990)

Single IV doses of HA-1A

Dose (number):	25 mg (13)	100 mg (16)	250 mg (5)
Cmax (μg/ml):	9.4	33.2	83.8
C-24 hr (μg/ml):	2.7	9.1	33.9
$t_{1/2}$ (hr):	14.1	16.2	19.7
AUC (μg/ml · hr):	199	705	2367
Vdss (ml/kg):	48.3	48.2	49.6
CL (ml/hr/kg):	3.0	2.9	2.1
AUC/D (μg/ml · hr/mg):	7.96	7.05	9.47

No adverse reactions or immune response

IgG
Whole IgG
^{90}Y (DTPA)
^{111}In (DTPA)
CA 19-9 antigen
Mice (–)

Hnatowich et al. (1985b)

Single IV (0.3–1.2 μCi/μg) doses

(% of dose) at:	1 hr (n = 4)		24 hr (n = 5)	
	^{111}In	^{90}Y	^{111}In	^{90}Y
Blood:	22.4	24.5	8.7	6.8
Liver:	20.6	17.4	19.1	13.5
Kidney:	6.3	6.6	17.8	5.0
Muscle:	0.5	0.5	0.8	0.6

[a] Isotope labeling method if provided: DTPA, diethylene pentaacetic acid.
[b] Number of animals/subjects in study.
[c] Route of administration: IV, intravenous.

The elimination half-life for the human IgM monoclonal antibody HA-1A in nonsepsis patients with cancer ranged from 26.8 to 35 hr at doses of 10 to 100 mg, respectively (Khazaeli *et al.*, 1990). Patients with sepsis exhibited more rapid elimination of HA-1A, with half-lives ranging from 14.1 to 19.7 hr following doses of 25 to 250 mg, respectively (Fisher *et al.*, 1990). Ehrlich and co-workers (1987) reported the pharmacokinetics of two IgG1 human monoclonal antibodies, EV1-15 and EV2-7, following administration to monkeys. Extremely long half-lives following both single-dose and multiple-dose administration were observed.

3.2.2. Distribution and Clearance

The distribution of several human monoclonal antibodies has been characterized in mice and man. The IgG human monoclonal antibody 19-9 radiolabeled with ^{111}In and ^{90}Y exhibited similar biodistribution at 1 hr, and significantly different distribution characteristics at 24 hr in mice (Hnatowich *et al.*, 1985b). McCabe and co-workers (1988) indicated that the IgM human monoclonal antibody 28A32 showed extremely high tumor/blood ratios of 10 to 1 following administration to mice.

The apparent volume of distribution of HA-1A following administration of 10, 25, and 100 mg to man was reported to be 50.8, 47.3, and 39.5 ml/kg in patients without sepsis. Patients with sepsis exhibited mean apparent distribution volumes of 48.3, 48.2, and 49.6 ml/kg following administration of 25, 100, and 250 mg, respectively (Fisher *et al.*, 1990). The mean apparent volume of distribution in patients with and without sepsis was approximately 3.5 liters. This is consistent with the distribution volume reported for endogenous IgM, which has been shown to reside predominantly (76% of the total body pool) in the intravascular space (Waldmann *et al.*, 1970).

The total body clearance of HA-1A in man following doses of 10 to 100 mg ranged from 0.9 to 1.4 ml/hr per kg; clearance decreased with increasing dose (Khazaeli *et al.*, 1990). Following doses of 25 to 250 mg, the clearance of HA-1A in man ranged from 2.1 to 3.0 ml/hr per kg (Fisher *et al.*, 1990). This study in patients with sepsis also indicated that clearance decreased with increasing dose.

3.2.3. Factors Influencing Pharmacokinetics

Human monoclonal antibody 16.88 was reported to exhibit linear pharmacokinetic characteristics over the dose range of 8 to 208 mg (Steis *et al.*, 1990). This has also been shown for HA-1A in the dose range of 25 to 250 mg

(Fisher *et al.,* 1990). Ehrlich and co-workers (1987) reported that the disposition characteristics of EV1-15 and EV2-17 determined by ELISA and radio-label monitoring were comparable. Finally, Fisher and co-workers (1990) noted that the more rapid clearance of HA-1A in patients with sepsis ($\sim$2.5 ml/hr per kg) compared to patients without sepsis ($\sim$1.1 ml/hr per kg) was probably due to the overall catabolic state of patients with sepsis and their enhanced reticuloendothelial system activity.

3.2.4. Summary

Overall, the administration of human IgM antibodies to man indicates that exogenous human monoclonal antibodies appear to distribute into approximately the same distribution volume as endogenous IgM human antibodies. They appear to be cleared at a more rapid rate than endogenous IgM, resulting in half-lives that are approximately 2.5 days compared to 5 days for endogenous IgM. The reasons for the enhanced clearance and reduced half-life of exogenously administered IgM are unknown.

3.3. Chimeric Antibodies and Fragments

The development of methods to clone genetic constructs composed of the variable regions of mouse monoclonal antibodies and the constant region of human immunoglobulins has resulted in chimeric mouse/human monoclonal antibodies with desirable specificity and pharmacological activity (Morrison *et al.,* 1984). Some of the chimeric monoclonal antibodies investigated in animals and man include chimeric monoclonal antibody B6.2 (Brown *et al.,* 1987), chimeric monoclonal antibody anti-idiotype IgG (Hamblin *et al.,* 1987), chimeric aglycosylated IgG (Tao and Morrison, 1989), chimeric monoclonal antibody C-17-1A (LoBuglio *et al.,* 1989; Trang *et al.,* 1990), and chimeric monoclonal antibody B72.3 (Begent *et al.,* 1990). The results of these investigations are summarized in Table VI.

3.3.1. Disposition

Following administration to one human, the elimination half-lives of chimeric anti-idiotype IgG intact and Fab IgG fragment were greater than 24 hr (Hamblin *et al.,* 1987). The mean elimination half-life for chimeric B72.3 was reported to be 116 hr, with a range of 37 to 166 hr (Begent *et al.,* 1990). The serum disposition of chimeric C-17-1A was characterized using a two-

Table VI
Pharmacokinetics of Chimeric Murine/Human Monoclonal Antibodies and Fragments (Section 3.3)

Name Form Isotype Isotope (method)[a]	Disease/target Antigen Species (n)[b]	Route[c], dose Pharmacokinetic characteristics and comments	References
Aglycosylated IgG Whole IgG1&3 [35S]-Met (−)	— DNS chloride Mice (5)	Single IV doses of glycosylated and aglycosylated (a) [35S]-IgG1 and [35S]-IgG3 $\quad$ IgG1 $\quad$ 1-IgG1 $\quad$ IgG3 $\quad$ a-IgG3 $t_{1/2}\alpha$ (hr): distribution complete in 48 hr $t_{1/2}\beta$ (days): $\quad$ 6.5 ± 0.5 $\quad$ 6.5 ± 0.5 $\quad$ 5.1 ± 0.4 $\quad$ 3.52 ± 0.2	Tao and Morrison (1989)
Anti-idio IgG Fragment IgG ELISA	Lymphoma Surface IgM Man (1)	Single IV doses of anti-Id FabIgG and anti-Id IgG1 $\qquad$ IgG1 $\qquad$ FabIgG $t_{1/2}$ (hr): $\quad$ <24 $\qquad$ <24 An anti-Id FabIgG control with Fab' activity against guinea pig L_2 cells and normal human IgG resulted in a $t_{1/2} > 10$ days	Hamblin et al. (1987)
B6.2 Whole IgG1 125/131I (Iodogen)	Cancer Surface Mice (−)	Single IV 5 µCi doses Biodistribution in human tumor-bearing mice was identical in blood, tumor, liver, lung, and other tissue	Brown et al. (1987)
B72.3 Whole IgG1/IgG4 131I (−)	Colon TAG-72 Man (6)	Single IV 10–20 mg doses $\qquad$ Mean $t_{1/2}$ $\qquad$ (hr) Blood: $\qquad$ 116 $\qquad$ (37–166) Tumor: $\qquad$ 226 $\qquad$ (162–256) Liver: $\qquad$ 155 $\qquad$ (141–173) Lung: $\qquad$ 166 $\qquad$ (135–182) Human-antichimeric antibody and HAMA in 2/6 after first treatment	Begent et al. (1990)

C-17-1A	Colon	Single IV doses			LoBuglio et al. (1989)
Whole	41-kDa glycoprotein	Dose (number);	10 mg ($n = 5$)	40 mg ($n = 5$)	
IgG1	Man (10)	$t_{1/2}\alpha$ (hr):	16 ± 2	19 ± 3	
—		$t_{1/2}\beta$ (hr):	95 ± 24	107 ± 30	
		AUC (mg/L · hr):	298 ± 134	1036 ± 363	
		Vdss (ml/kg):	59 ± 9	73 ± 19	
		CL (ml/hr/kg):	0.52 ± 0.16	0.61 ± 0.31	
		The multiple-dose pharmacokinetics were similar to single doses			
		1/6 exhibited a modest human-antichimeric antibody response			
C-17-1A	Colon	Single IV doses			Trang et al. (1990)
Whole	41-kDa glycoprotein	Dose (number):	10 mg ($n = 5$)	40 mg ($n = 5$)	
IgG1	Man (10)	$t_{1/2}\alpha$ (hr):	15.8	18.5	
—		$t_{1/2}\beta$ (hr):	90.0	97.6	
		Cmax/D (μg/L/mg):	398 ± 101	333 ± 85	
		AUC/D (μg/ml · hr/mg):	29.8 ± 13.4	25.9 ± 9.1	
		Vc (L):	2.65 ± 0.65	3.26 ± 0.58	
		Vβ (L):	37 ± 14	40 ± 12	
		Significant correlations between AUC and dose and between CL and tumor load			

[a] Isotope labeling method if provided.
[b] Number of animals/subjects in study.
[c] Route of administration: IV, intravenous.

compartment pharmacokinetic model with mean distributive half-lives of 15.8 and 18.5 hr for the 10- and 40-mg doses, respectively (Trang *et al.,* 1990). The mean elimination half-lives in these patients were 90.0 and 97.6 hr for the 10- and 40-mg doses, respectively.

3.3.2. Distribution and Clearance

A mean elimination half-life from tumor of 226 hr (range 162–256 hr) was reported for chimeric B72.3 (Begent *et al.,* 1990). The apparent distribution volume of the central compartment for chimeric C-17-1A was reported to be approximately 2.7 and 3.3 liters at doses of 10 and 40 mg, respectively. The mean apparent volume of distribution for chimeric C-17-1A was reported to be 4.8 and 6.1 liters for doses of 10 and 40 mg, respectively (Trang *et al.,* 1990).

The total body clearance of chimeric C-17-1A was estimated to be 37 and 40 ml/hr for doses of 10 and 40 mg, respectively (Trang *et al.,* 1990). The apparent volume of distribution for chimeric C-17-1A of approximately 5 to 6 liters (0.7 to 0.8 liter/kg) indicates that chimeric C-17-1A appears to distribute throughout the vascular space and into limited extravascular space. The volume terms are slightly larger than those observed for endogenous IgG1. The mean total body clearances for chimeric C-17-1A at doses of 10 and 40 mg were 0.52 and 0.61 ml/hr per kg, respectively (Trang *et al.,* 1990).

3.3.3. Factors Influencing Pharmacokinetics

In 1989, Tao and Morrison presented the results of their investigation of the structural and functional roles of the carbohydrates present in the C_H2 domain of human IgG molecules using a chimeric mouse/human IgG. The chimeric mouse/human monoclonal antibody used in their studies consisted of the variable region from the mouse hybridoma 27–44 with specificity for the hapten, DNS chloride, which was joined to human IgG1 or IgG3 constant regions. The structure responsible for retaining IgG in the circulation was thought to be related to the Fc region and has been suggested to be in the C_H2 domain. The presence of terminal sialic acid on the oligosaccharide chains protects many serum glycoproteins from clearance by asialoglycoprotein receptors on hepatocytes that recognize terminal galactose. The authors noted that the glycosylated and aglycosylated chimeric IgG1 gave the same elimination half-life in mice (6.5 ± 0.5 days), which was similar to the range reported for the half-lives of monoclonal mouse IgG1, IgG2a, and IgG3 (Tao and Morrison, 1989).

These results suggested that the asialoglycoprotein receptor did not play an important role in the clearance of human IgG1 in mice and that increased sensitivity of the aglycosylated IgG1 to proteases did not result in a shorter serum half-life. The shorter elimination half-life of aglycosylated chimeric IgG3 in mice (3.5 ± 0.2 days) compared to the wild-type chimeric IgG3 elimination half-life (5.1 ± 0.4 days) suggested that carbohydrates may not play a major role in IgG catabolism (Tao and Morrison, 1989).

In a clinical study, ten patients received chimeric C-17-1A; six received three infusions of either a 10- or 40-mg dose at 2-week intervals. One of these six individuals exhibited a modest human antichimeric antibody response (LoBuglio *et al.*, 1989). Interestingly, the multiple-dose pharmacokinetics of chimeric C-17-1A were very similar to those observed after single-dose administration (LoBuglio *et al.*, 1989).

The authors noted that the pharmacokinetics of chimeric C-17-1A differed substantially from the pharmacokinetics of murine 17-1A. The circulating half-life of murine 17-1A was reported to be approximately 15 hr in five patients receiving a single infusion and 16.2 hr in 23 patients receiving from one to four infusions. In contrast, chimeric C-17-1A showed a mean distributive half-life of around 18 hr and the mean elimination half-life was approximately 100 hr (Trang *et al.*, 1990). The distributive half-life for chimeric C-17-1A was suggested to be reflective of the time required for equilibration (Solomon *et al.*, 1963). The elimination half-life of chimeric C-17-1A (100 hr) appeared to be intermediate between murine 17-1A and human IgG immunoglobulin (Trang *et al.*, 1990).

In addition, the authors noted interesting correlations between chimeric C-17-1A total body clearance and circulating IgG levels. These findings were consistent with those of Morell *et al.* (1970), who reported that elevated serum concentrations of any IgG subclass in patients with multiple myeloma were associated with shortened biological half-lives and increased catabolic rates of all the subclasses. The authors also pointed out a correlation between chimeric C-17-1A total body clearance and tumor size, which was thought to be due to the accelerated catabolism of the protein associated with the elevation of total serum proteins that occurs with increasing tumor burden (Trang *et al.*, 1990).

3.3.4. Summary

The therapeutic advantages of chimeric antibodies result from their potential to exhibit reduced immunogenicity, longer circulating half-lives, and enhanced pharmacologic activity (Morrison *et al.*, 1984; Shaw *et al.*,

1987). The utility of native and conjugated murine and nonmurine monoclonal and polyclonal antibodies and antibody fragments has been limited by the natural activity of the antibody and by the development of side effects including anaphylaxis and HAMA production (Brodsky, 1988). Cancer therapy with murine monoclonal antibodies has been limited by moderate pharmacologic activity, immunogenicity, and relatively short circulating half-lives. The immunogenicity of monoclonal antibodies has restricted their use to single- and/or multiple-dose administration over a short time period.

4. CONCLUSIONS

4.1. Summary of Antibody Pharmacokinetics

Generally, the administration of murine monoclonal antibodies to man results in a disposition similar to that of the immunoglobulin subtype from which the murine monoclonal antibody was derived. In contrast, the elimination of a murine monoclonal antibody in man is much more rapid than the endogenous human immunoglobulin of the same isotype, and is also faster than the same murine monoclonal antibody when administered to the species of origin or a related species.

Generally, the administration of native and radioimmunoconjugate murine monoclonal antibodies in man has resulted in linear pharmacokinetic characteristics with proportional increases in maximum plasma concentration and area under the curve for increasing doses. The shorter half-lives in man are a result of the more rapid clearance of the murine monoclonal antibody when administered to man, and may also be influenced by the development of HAMA responses.

Native and radioimmunoconjugate polyclonal antibodies of both murine and nonhuman origin also appear to exhibit many of the same characteristics discussed above. Polyclonal antibodies exhibit comparable distribution volumes as the endogenous immunoglobulin of similar isotype and dramatically shorter elimination half-lives in man than those observed when the same murine or nonmurine polyclonal antibody was administered to the species of origin.

The administration of human monoclonal antibodies to man results in a significantly different pharmacokinetic profile. The biodistribution and distribution volumes of human monoclonal antibodies in man are comparable to the distribution volumes of the endogenous antibody from which the antibody was derived. The elimination half-life observed for human monoclonal antibodies is significantly shorter than that observed for the endogenous immunoglobulin of the same isotype. Murine/human chimeric monoclonal antibodies when administered to man exhibit distribution volumes and distributive half-lives that are comparable to the endogenous immunoglobulin of the same isotype. In contrast, chimeric antibodies appear to have much shorter elimination half-lives than human immunoglobulins of the same isotype, and significantly longer half-lives than the corresponding murine immunoglobulins of the same isotype.

The administration of a murine monoclonal antibody to a mouse or other rodent model generally results in an elimination half-life that is very similar to that of the endogenous native immunoglobulin of the corresponding isotype. Across species, generally, as the size of the animal increases the half-life of the monoclonal antibody increases also. These relationships have been described for small molecules by Mordenti (1986), who proposed a power function to describe the disposition of compounds in various species.

4.2. Factors Influencing Antibody Pharmacokinetics

For endogenous circulating immunoglobulins, the presence of antigen clearly dictates the increased synthesis and production of the immunoglobulin and correspondingly influences the circulating level. The role of the quantity of the antigen in the catabolism and pharmacokinetics of exogenously administered antibodies is unclear. The amount of antibody required to saturate an antigenic tumor site is unique to the individual tumor under investigation and the individual antibody being utilized for either imaging or therapy of that particular tumor.

Tumor size and tumor volume have been shown to influence the biodistribution of murine monoclonal antibodies with affinities for tumor-specific antigens. Generally, the larger the tumor, both in dimension and in volume, the higher is the ratio of drug in tumor versus drug in blood achieved, although extremely large tumors have resulted in decreased tumor/blood ratios. These decreases with extremely large tumors are thought to be the result of the decreasing vascularity and possible necrosis that occurs in larger tumors. The use of external beam irradiation, which produces an increased vascularization of tumors, has been shown to result in increased tumor/blood ratios.

The immunogenicity of murine monoclonal antibodies is an important factor in their more rapid clearance in man than in corresponding mouse or rodent systems. The administration or coadministration of immunomodulating agents such as α-interferon or a second antibody has been shown to dramatically influence the pharmacokinetic characteristics of murine monoclonal antibodies. These interactions can be understood on an antibody-by-antibody basis when one examines the relationships between the immunomodulator or coadministered antibody and the immune response to the primary antibody.

4.3. Limitations of Monoclonal Antibody Applications in Diagnosis and Therapy

The utility of a monoclonal antibody, regardless of its origin, in either diagnosis or therapy is dependent on the specific affinity of the antibody for the antigenic site, the unique population of antigenic sites relative to nonspecific sites in the body, the ability of the monoclonal antibody to produce a cytotoxic effect, and the efficacy or cytotoxicity of any chemotoxic agent or toxin associated with the antibody.

To date, the major problem associated with the application of chemoimmunoconjugates has been the inability to conjugate a sufficiently large quantity of chemotoxic agent to the antibody so that the chemotoxic drug is presented and released at the tumor site in sufficient quantity to result in cytotoxic effects. The primary disadvantages or obstacles related to the use of immunotoxins are related to the rapid clearance of the immunotoxin, toxin dissociation from the antibody, and potentially toxic side effects due to free toxin. As discussed above, the development of an immune response to the antibody, whether to a murine, chimeric, or human monoclonal antibody, limits the potential utility of the antibody, particularly if multiple-dose administration would be required to produce the desired effect.

Methodologic concerns in the study of the pharmacokinetics and metabolism of antibodies for diagnostic and therapeutic use include limitations in analytical methodology, compromised study design, and limited pharmacokinetic evaluation and interpretation. Most studies characterizing the biodistribution and/or pharmacokinetics of murine monoclonal antibodies have utilized radioimmunoconjugates to characterize the distribution and disposition of the radionuclide as a function of time in blood and various tissue/tumor sites. Early studies utilizing this approach were flawed because the radiolabeling methods employed were at times harsh and resulted in a radioimmunoconjugate which did not accurately reflect the distribution and pharmacokinetic characteristics of the unconjugated native monoclonal antibody. Recently, milder conjugating methods have produced radiolabeled antibodies that are very reflective of the native unconjugated monoclonal antibody. A major concern with respect to the analytical aspects of pharmacokinetic studies is that the method must evaluate the intact, unchanged therapeutic moiety. It has been suggested that in addition to radiolabeled studies of disposition and distribution, immunoassay or biologic activity assays should be employed.

Many of the clinical trials utilizing monoclonal antibodies have been compromised with respect to the study design simply because very limited quantities of the antibody have been available. In many instances it has been necessary to design a protocol around a finite amount of antibody. This has resulted in limitations in the study design that have not permitted comprehensive evaluation of the antibodies' pharmacokinetic characteristics. Relatively few studies have undertaken the characterization of the parent monoclonal antibody and its catabolic or metabolic breakdown products. Very little is known regarding the actual mechanisms of catabolism.

Finally, many of the studies utilizing monoclonal antibodies in man have evaluated serum or plasma concentration versus time data utilizing a one- or two-compartment model. In the absence of comprehensive data delineating the actual mechanisms of clearance and routes of loss for the antibody and its catabolic or metabolic breakdown products, it is often not possible to state definitively that the "distributive phase" truly reflects antibody distribution or that the "elimination phase" truly reflects the catabolism or metabolism of the molecule. Relatively few studies have presented comprehensive evaluations of monoclonal antibody pharmacokinetics including determination of area under the curve, the apparent volume of distribution, and total body clearance.

4.4. Recommendations for Future Study

Many approaches are being taken to overcome some of the limitations and obstacles discussed above. The development of methods to construct mouse/human chimeric monoclonal antibodies and the development of methods to facilitate the large-scale production of human monoclonal antibodies have been very effective in addressing the primary limitation of murine monoclonal antibodies, the immune responses observed upon multiple dosing. Significant progress has been made in the development of more gentle conjugation techniques for the attachment of radionuclides, chemotoxic agents, and toxins to monoclonal antibodies. These have resulted in the generation of radioimmunoconjugates, chemoimmunoconjugates, and immunotoxins that have high specificity and antigen binding characteristics that are comparable to the native antibody.

Another significant trend involves the utilization of Fab and $F(ab')_2$ fragments, since these fragments have the potential of targeting directly to an antigenic site, yet lack antigenic determinants, which have been implicated in the development of anti-antibody responses. Other approaches are being taken to enhance the efficacy of chemoimmunoconjugates and immunotoxins by increasing the ratio of drug/toxin:antibody and by exploring different conjugation techniques to enhance stability and control release of the chemotoxic drug or toxin once the antibody has bound to the cell. Much additional work needs to be done in the area of coadministration of immunomodulators such as α-interferon and other antibodies to modify the pharmacokinetic characteristics of the primary antibody administered.

At present, monoclonal antibodies appear to be a viable therapeutic modality for both the diagnosis and therapy of a number of diseases that have not been successfully treated with other methods. These include selected cancers, drug toxicity, immunological disorders, sepsis and septic shock, and organ transplants. Additional applications of monoclonal antibodies will include their use in establishing a more complete understanding of the immune system and the mechanisms of many diseases that may have immune or autoimmune components such as AIDS, diabetes, atherosclerosis, Alzheimer's disease, Guillain–Barre syndrome, and arthritis.

ACKNOWLEDGMENTS

The author wishes to thank Drs. M. B. Khazaeli and A. F. LoBuglio for their assistance in the review of literature; Ms. Deborah A. Camel for her skillful and patient preparation of the manuscript; and Judy DiMeglio Trang for her unfailing love, support, and encouragement.

REFERENCES

Begent, R. H. J., Ledermann, J. A., Bagshawe, K. D., Green, A. J., Kelly, A. M. B., Lane, D., Glaser, M. G., Dewji, M. R., Baker, T. S., and Secher, D. S., 1990, Chimeric B72.3 antibody for repeated radio-immunotherapy of colorectal carcinoma, Fifth International Conference on Monoclonal Antibody Immunoconjugates for Cancer, March 15–17, San Diego, p. 73.

Blakely, D. C., Watson, G. J., Knowles, P. P., and Thorpe, P. E., 1987, Effect of chemical deglycosylation of ricin A chain on the in vivo fate and cytotoxic activity of an immunotoxin composed of ricin A chain and anti-Thy 1.1 antibody, *Cancer Res.* **47**:947–952.

Blakely, D. C., Skilleter, D. N., Price, R. J., Watson, G. J., Hart, L. I., Newell, D. R., and Thorpe, P. E., 1988, Comparison of the pharmacokinetics and hepatotoxic effects of saporin and ricin A-chain immunotoxins on murine liver parenchymal cells, *Cancer Res.* **48**:7072–7078.

Bourrie, B. J. P., Casellas, P., Blythman, H. E., and Jansen, F. K., 1986, Study of the plasma clearance of antibody–ricin-A-chain immunotoxins, *Eur. J. Biochem.* **155**:1–10.

Brodsky, F. M., 1988, Monoclonal antibodies as magic bullets, *Pharm. Res.* **5**:1–9.

Brown, B. A., Davis, G. L., Saltzgaber-Muller, J., Simon, P., Ho, M.-K., Shaw, P. S., Stone, B. A., Sands, H., and Moore, G. P., 1987, Tumor specific genetically engineered murine/human chimeric monoclonal antibody, *J. Cancer Res.* **47**:3577–3583.

Burchiel, S. W., Hadjian, R. A., Hladik, W. B., Drozynski, C. A., Tolman, G. L., Haber, S. B., and Gallagher, B. M., 1989, Pharmacokinetic evaluation of technetium-99-metallothionein-conjugated mouse monoclonal antibody B72.3 in rhesus monkeys, *J. Nucl. Med.* **30**:1351–1357.

Byers, V. S., Pimm, M. V., Pawluczyk, H. M., Lee, I. Z. A., Scannon, P. J., and Baldwin, R. W., 1987, Biodistribution of ricin toxin A chain–monoclonal antibody 791T/36 immunotoxin and influence of hepatic blocking agents, *Cancer Res.* **47**:5277–5283.

Byers, V. S., Rodvien, R., Grant, K., Durrant, L. G., Hudson, K. H., Baldwin, R. W., and Scannon, P. J., 1989, Phase I study of monoclonal antibody-ricin A chain immunotoxin ZomaZyme-791 in patients with metastatic colon cancer, *Cancer Res.* **49**:6153–6160.

DeNardo, S. J., DeNardo, G. L., O'Grady, L. F., Levy, N. B., Mills, S. L., Macey, D. J., McGahan, J. P., Miller, C. H., and Epstein, A. L., 1988, Pilot studies of radioimmunotherapy of B cell lymphoma and leukemia using I-131 Lym-1 monoclonal antibody, *Antibody Immunoconj. Radiopharm.* **1**:17–32.

Dillman, R. O., 1990, Human antimouse and antiglobulin responses to monoclonal antibodies, *Antibody Immunoconj. Radiopharm.* **3**:1–15.

Dillman, R. O., Johnson, D. E., Shawler, D. L., and Koziol, J. A., 1988, Superiority of an acid-labile daunorubicin–monoclonal antibody immunoconjugate compared to free drug, *Cancer Res.* **48**:6097–6102.

Eary, J. F., Schroff, R. W., Abrams, P. G., Fritzberg, A. R., Morgan, A. C., Kasina, S., Reno, J. M., Srinivasan, A., Woodhouse, C. S., Wilbur, D. S., Natale, R. B., Collins, C., Stehlin, J. S., Mitchell, M., and Nelp, W. B., 1989, Successful imaging of malignant melanoma with technetium-99m-labeled monoclonal antibodies, *J. Nucl. Med.* **30**:25–32.

Eger, R. R., Covell, D. G., Carrasquillo, J. A., Abrams, P. G., Foon, K. A., Reynolds, J. C., Schroff, R. W., Morgan, A. C., Larson, S. M., and Weinstein, J. N., 1987, Kinetic model for the biodistribution of an ^{111}In-labeled monoclonal antibody in humans, *Cancer Res.* **47**:3328–3336.

Ehrlich, P., 1906, *Collected Studies on Immunity*, Wiley, New York.

Ehrlich, P. H., Harfeldt, E., Justice, J. C., Moustafa, Z. A., and Ostberg, L., 1987, Rhesus monkey responses to multiple injections of human monoclonal antibodies, *Hybridoma* **6**:151–160.

Fahey, J. L., and Robinson, A. G., 1963, Factors controlling serum γ-globulin concentration, *J. Exp. Med.* **118**:845–868.

Fisher, C. J., Zimmerman, J., Khazaeli, M. B., Albertson, T. E., Dellinger, R. P., Panacek, E. A., Foulke, G. E., Dating, C., Smith, C. R., and LoBuglio, A. F., 1990, Initial evaluation of human monoclonal anti-lipid A antibody (HA-1A) in patients with sepsis syndrome, *Crit. Care Med.* **18**:1311–1315.

Fulton, R. J., Tucker, T. F., Vitetta, E. S., and Uhr, J. W., 1988, Pharmacokinetics of tumor-reactive immunotoxins in tumor-bearing mice: Effect of antibody valency and deglycosylation of the ricin A chain on clearance and tumor localization, *Cancer Res.* **48**:2618–2625.

Goldenberg, D. M., Gaffar, S. A., Bennett, S. J., and Beach, J. L., 1981, Experimental radioimmunotherapy of a xenografted human colonic tumor (GW-39) producing carcinoembryonic antigen, *Cancer Res.* **41**:4354–4360.

Gould, B. J., Borowitz, M. J., Groves, E. S., Carter, P. W., Anthony, D., Weiner, L. M., and Frankel, A. E., 1989, Phase I study of an anti-breast cancer immunotoxin by continuous infusion: Report of a targeted toxic effect not predicted by animal studies, *J. Natl. Cancer Inst.* **81**:775–781.

Griffin, T. W., Bokhari, F., Collins, J., Stochl, M., Bernier, M., Gionet, M., Siebecker, D., Wertheimer, M., Giroves, E. S., Greenfield, L., Houston, L. L., Doherty, P. W., and Wilson, J., 1989, A preliminary pharmacokinetic study of ^{111}In-labeled 260F9 anti-(breast cancer) antibody in patients, *Cancer Immunol. Immunother.* **29**:43–50.

Hamblin, T. J., Catta, A. R., Glennie, M. J., MacKenzie, M. R., Stevenson, F. K., Watts, H. F., and Stevenson, G. T., 1987, Initial experience in treating human lymphoma with a chimeric univalent derivative of monoclonal anti-idiotype antibody, *Blood* **69**:790–797.

Hand, P. H., Shrivastav, S., Colcher, D., Snoy, P., and Schlom, J., 1989, Pharmacokinetics of radiolabeled monoclonal antibodies following intraperitoneal and intravenous administration in rodents, monkeys, and humans, *Antibody Immunoconj. Radiopharm.* **2**:241–255.

Hnatowich, D. J., Griffin, T. W., Koscinczyk, C., Rusckowski, M., Childs, R. L., Mattis, J. A., Shealy, D., and Doherty, P. W., 1985a, Pharmacokinetics of an indium-111-labeled monoclonal antibody in cancer patients, *J. Nucl. Med.* **26**:849–858.

Hnatowich, D. J., Virzi, F., and Doherty, P. W., 1985b, DTPA-coupled antibodies labeled with yttrium-90, *J. Nucl. Med.* **26**:503–509.

Hnatowich, D. J., Gionet, M., Rusckowski, M., Siebecker, D. A., Roche, J., Shealy, D., Mattis, J. A., Wilson, J., McGann, J., Hunter, R. E., Griffin, T., and Doherty, W. P., 1987, Pharmacokinetics of ^{111}In-labeled OC-125 antibody in cancer patients compared with the 19-9 antibody, *Cancer Res.* **47**:6111–6117.

Hu, E., Epstein, A. L., Naeve, G. S., Gill, I., Martin, S., Sherrod, A., Nichols, P., Chen, D., Mazumder, A., and Levin, A. M., 1989, A Phase Ia clinical trial of Lym-1 monoclonal antibody serotherapy in patients with refractory B cell malignancies, *Hematol. Oncol.* **7**:155–166.

Khazaeli, M. B., Saleh, M. N., Wheeler, R. H., Huster, W. J., Holden, H., Carrano, R., and LoBuglio, A. F., 1988, Phase I trial of multiple large doses of murine monoclonal antibody CO17-1A. II. Pharmacokinetics and immune response, *J. Natl. Cancer Inst.* **80**:937–942.

Khazaeli, M. B., Wheeler, R., Rogers, K., Teng, N., Ziegler, E., Haynes, A., Saleh, M. N., Hardin, J. M., Bolmer, S., Cornett, J., Berger, H., and LoBuglio, A. F., 1990, Initial evaluation of a human immunoglobulin M monoclonal antibody (HA-1A) in humans, *J. Biol. Respir. Mod.* **9**:178–184.

Kohler, G., 1986, Derivation and diversification of monoclonal antibodies, *Science* **233**:1281–1286.

Kohler, G., and Milstein, C., 1975, Continuous cultures of fused cells secreting antibody of pre-defined specificity, *Nature* **256**:495–497.

Larkin, 1988, Biotechnology and human health care symposium: A summary, Pharmaceutical Manufacturer's Association, Washington, D.C.

Larson, S. M., Brown, J. P., Wright, P. W., Carrasquillo, J. A., Hellstrom, I., and Hellstrom, K. E., 1983, Imaging of melanoma with 1-131-labeled monoclonal antibodies, *J. Nucl. Med.* **24**:123–129.

Laurent, G., Pris, J., Farcet, J.-P., Carayon, P., Blythman, H., Casellas, P., Poncelet, P., and Jansen, F. K., 1986, Effects of therapy with T101 ricin A-chain immunotoxin in two leukemia patients, *Blood* **67**:1680–1687.

Ledermann, J. A., Begent, R. H. J., Bagshawe, K. D., Riggs, S. J., Searle, F., Glaser, M. G., Green, A. J., and Dale, R. G., 1988, Repeated antitumour antibody therapy in man with suppression of the host response by cyclosporin A, *Br. J. Cancer* **58**:654–657.

Leichner, P. K., Klein, J. L., Siegelman, S. S., Ettinger, D. S., and Order, S. E., 1983, Dosimetry of ^{131}I-labeled antiferritin in hepatoma: Specific activities in the tumor and liver, *Cancer Treat. Rep.* **67**:647–658.

Leichner, P. K., Yang, N.-C., Frenkel, T. L., Loudenslager, D. M., Hawkins, W. G., Klein, J. L., and Order, S. E., 1988, Dosimetry and treatment planning for ^{90}Y-labeled antiferritin in hepatoma, *Int. J. Radiat. Oncol. Biol. Phys.* **14**:1033–1042.

Letvin, N. L., Goldmacher, V. S., Ritz, J., Yetz, J. M., Schlossman, S. F., and Lambert, J. M., 1986a, In vivo administration of lymphocyte-specific monoclonal antibodies in nonhuman primates—In vivo stability of disulfide-linked immunotoxin conjugates, *J. Clin. Invest.* **77**:977–984.

Letvin, N. L., Chalifoux, L. V., Reimann, K. A., Ritz, J., Schlossman, S. F., and Lambert, J. M., 1986b, In vivo administration of lymphocyte-specific monoclonal antibodies in nonhuman primates—Delivery of ribosome-inactivating proteins to spleen and lymph node T cells, *J. Clin. Invest.* **78**:666–673.

LoBuglio, A. F., Saleh, M., Peterson, L., Wheeler, R., Carrano, R., Huster, W., and Khazaeli, M. B., 1986, Phase I clinical trial of CO17-1A monoclonal antibody, *Hybridoma* **5**:S117–S123.

LoBuglio, A. F., Khazaeli, M. B., Lee, J., Haynes, A., Sumerel, L., Mischak, R., and Spitler, L., 1988, Pharmacokinetics and immune response to XomaZyme-Mel in melanoma patients, *Antibody Immunoconj. Radiopharm.* **1**:305–310.

LoBuglio, A. F., Wheeler, R. H., Trang, J., Haynes, A., Rogers, K., Harvey, E. B., Sun, L., Ghrayeb, J., and Khazaeli, M. B., 1989, Mouse/human chimeric monoclonal antibody in man: Kinetics and immune response, *Proc. Natl. Acad. Sci. USA* **86**:4220–4224.

LoBuglio, A. F., Khazaeli, M. B., Trang, J. M., Lee, J., Haynes, A., Sumerel, L., Mischak, R., and Spitler, L., 1990, Pharmacokinetics of XomaZyme®-Mel in melanoma patients, *Pharm. Res.* **7**:S–40.

McCabe, R. P., Peters, L. C., Haspel, M. V., Pomato, N., Carrasquillo, J. A., and Hanna, M. G., Jr., 1988, Preclinical studies on the pharmacokinetic properties of human monoclonal antibodies to colorectal cancer and their use for detection of tumors, *Cancer Res.* **48**:4348–4353.

Manske, J. M., Buchsbaum, D. J., Hanna, D. E., and Vallera, D. A., 1988, Cytotoxic effects of anti-CD5 radioimmunotoxins on human tumors in vitro and in a nude mouse model, *Cancer Res.* **48**:7107–7114.

Montague, M. J., 1989, The impact of biotechnology on the practice of pharmacy in the year 2000, *Am. J. Pharm. Ed.* **53**:21S–26S.

Mordenti, J., 1986, Dosage regimen design for pharmaceutical studies conducted in animals, *J. Pharm. Sci.* **75**:852–857.

Morell, A., Terry, W. D., and Waldmann, T. A., 1970, Metabolic properties of IgG subclasses in man, *J. Clin. Invest.* **49**:673–680.

Morrison, S. L., Johnson, M. J., Herzenberg, L. A., and Oj, V. T., 1984, Chimeric human antibody molecules: Mouse antigen-binding domains with human constant region domains, *Proc. Natl. Acad. Sci. USA* **81**:6851–6855.

Murray, J. L., Lamki, L. M., Shanken, L. J., Blake, M. E., Plager, C. E., Benjamin, R. S., Schweighardt, S., Unger, M. W., and Rosenblum, M. G., 1988, Immunospecific saturable clearance mechanisms for indium-111-labeled anti-melanoma monoclonal antibody 96.5 in humans, *Cancer Res.* **48**:4417–4422.

Oldham, R. K., Foon, K. A., Morgan, A. C., Woodhouse, C. S., Schroff, R. W., Abrams, P. G., Fer, M., Schoenberger, C. S., Farrell, M., Kimball, E., and Sherwin, S. A., 1984, Monoclonal antibody therapy of malignant melanoma: In vivo localization in cutaneous metastasis after intravenous administration, *J. Clin. Oncol.* **2**:1235–1244.

Order, S. E., Klein, J. L., Leichner, P. K., Frincke, J., Lollo, C., and Carlo, D. J., 1986, ^{90}Yttrium antiferritin—A new therapeutic radiolabeled antibody, *Int. J. Radiat. Oncol. Biol. Phys.* **12**:277–281.

Pietersz, G. A., Krauer, K., Toohey, B., Smyth, M. J., and McKenzie, I. F. C., 1990, Biodistribution of N-acetylmelphalan–monoclonal antibody conjugates in mice, *Antibody Immunoconj. Radiopharm.* **3**:27–35.

Pimm, M. V., Perkins, A. C., Armitage, N. C., and Baldwin, R. W., 1985, The characteristics of blood-borne radiolabels and the effect of anti-mouse IgG antibodies on localization of radiolabeled monoclonal antibody in cancer patients, *J. Nucl. Med.* **26**:1011–1023.

Pimm, M. V., Paul, M. A., Ogumuyiwa, Y., and Baldwin, R. W., 1988, Biodistribution and tumour localization of a daunomycin–monoclonal antibody conjugate in nude mice with human tumour xenografts, *Cancer Immunol. Immunother.* 27:267–271.

Rosen, S. T., Zimmer, A. M., Goldman-Leikin, R., Gordon, L. I., Kazikiewicz, J. M., Kaplan, E. H., Variakojis, D., Marder, R. J., Dykewicz, M. S., Piergies, A., Silverstein, E. A., Roenigk, H. H., Jr., and Spies, S. M., 1987, Radioimmunodetection and radioimmunotherapy of cutaneous T cell lymphomas using an ^{131}I-labeled monoclonal antibody: An Illinois Cancer Council study, *J. Clin. Oncol.* 5:562–573.

Rosenblum, M. G., Murray, J. L., Haynie, T. P., Glenn, H. J., Jahns, M. F., Benjamin, R. S., Frincke, J. M., Carlo, D. J., and Hersh, E. M., 1985, Pharmacokinetics of ^{111}In-labeled anti-p97 monoclonal antibody in patients with metastatic malignant melanoma, *Cancer Res.* 45:2382–2386.

Rosenblum, M. G., Lamki, L. M., Murray, J. L., Carlo, D. J., and Gutterman, J. U., 1988, Interferon-induced changes in pharmacokinetics and tumor uptake of ^{111}In-labeled antimelanoma antibody 96.5 in melanoma patients, *J. Natl. Cancer Inst.* 80:160–165.

Rostock, R. A., Klein, J. L., Leichner, P. K., and Order, S. E., 1984, Distribution of and physiologic factors that affect ^{131}I-antiferritin tumor localization in experimental hepatoma, *Int. J. Radiat. Oncol. Biol. Phys.* 10:1135–1141.

Schneck, D., Butler, F., Dugan, W., Littrell, D., Petersen, B., Bowsher, R., DeLong, A., and Dorrbecker, S., 1990, Disposition of a murine monoclonal antibody vinca conjugate (KS1/4-DAVLB) in patients with adenocarcinomas, *Clin. Pharmacol. Ther.* 47:36–41.

Sharkey, R. M., Mabus, J., and Goldenberg, D. M., 1988, Factors influencing anti-antibody enhancement of tumor targeting with antibodies in hamsters with human colonic tumor xenografts, *Cancer Res.* 48:2005–2009.

Shaw, D. R., Khazaeli, M. B., Sun, L. K., Ghrayeb, J., Daddona, P. E., McKinney, S., and LoBuglio, A. F., 1987, Characterization of a mouse/human chimeric monoclonal antibody (17-1A) to a colon cancer tumor-associated antigen, *J. Immunol.* 138:4534–4538.

Solomon, A., Waldmann, T. A., and Fahey, J. L., 1963, Clinical and experimental metabolism of normal 6.6S γ-globulin in normal subjects and in patients with macroglobulinemia and multiple myeloma, *J. Lab. Clin. Med.* 62:1–17.

Steis, R. G., Carrasquillo, J. A., McCabe, R., Bookman, M. A., Reynolds, J. C., Larson, S. M., Smith, J. W., II, Clark, J. W., Dailey, V., Del Vecchio, S., Shuke, N., Pinsky, C. M., Urba, W. J., Haspel, M., Perentesis, P., Paris, B., Longo, D. L., and Hanna, M. G., Jr., 1990, Toxicity, immunogenicity, and tumor radioimmunodetecting ability of two human monoclonal antibodies in patients with metastatic colorectal carcinoma, *J. Clin. Oncol.* 8:476–490.

Stewart, J. S. W., Hird, V., Snook, D., Sullivan, M., Myers, M. J., and Epenetos, A. A., 1988, Intraperitoneal [131]I- and ^{90}Y-labelled monoclonal antibodies for ovarian cancer: Pharmacokinetics and normal tissue dosimetry, *Int. J. Cancer Suppl.* 3:71–76.

Takahashi, T., Yamaguchi, T., Kitamura, K., Suzuyama, H., Honda, M., Yokota, T., Kotanagi, H., Takahashi, M., and Hashimoto, Y., 1988, Clinical application of monoclonal antibody–drug conjugates for immunotargeting chemotherapy of colorectal carcinoma, *Cancer* **61**:881–888.

Tao, M.-H., and Morrison, S. L., 1989, Studies of aglycosylated chimeric mouse-human IgG, *J. Immunol.* **143**:2595–2601.

Trang, J. M., LoBuglio, A. F., Wheeler, R. H., Harvey, E. B., Sun, L., Ghrayeb, J., and Khazaeli, M. B., 1990, Pharmacokinetics of a mouse/human chimeric monoclonal antibody (C-17-1A) in metastatic adenocarcinoma patients, *Pharm. Res.* **7**:587–592.

Vaughan, A. T. M., Keeling, A., and Yankuba, S. C. S., 1985, The production and biological distribution of yttrium-90 labelled antibodies, *Br. J. Appl. Radiat. Isot.* **36**:803–806.

Waldmann, T. A., 1966, Protein-losing enteropathy, *Gastroenterology* **50**:422–443.

Waldmann, T. A., Strober, W., and Blaese, R. M., 1970, Variations in the metabolism of immunoglobulins measured by turnover rates, in: *Immunoglobulins. Biological Aspects and Clinical Uses* (E. Merler, ed.), National Academy of Sciences, Washington, D.C., pp. 33–51.

Weiner, L. M., O'Dwyer, J., Kitson, J., Comis, R. L., Frankel, A. E., Bauer, R. J., Konrad, M. S., and Groves, E. S., 1989, Phase I evaluation of an anti-breast carcinoma monoclonal antibody 260F9–recombinant ricin A chain immunoconjugate, *Cancer Res.* **49**:4062–4067.

Wochner, R. D., Strober, W., and Waldmann, T. A., 1967, The role of the kidney in the catabolism of Bence Jones proteins and immunoglobulin fragments, *J. Exp. Med.* **126**:207–221.

Yang, H. M., and Reisfeld, R. A., 1988, Pharmacokinetics and mechanism of action of a doxorubicin–monoclonal antibody 9.2.27 conjugate directed to a human melanoma proteoglycan, *J. Natl. Cancer Inst.* **80**:1154–1159.

Zimmer, A. M., Kaplan, E. H., Kazikiewicz, J. M., Goldman-Leikin, R., Gilyon, K. A., Dykewicz, M. S., Spies, W. G., Silverstein, E. A., Spies, S. M., and Rosen, S. T., 1988, Pharmacokinetics of I-131 T101 monoclonal antibody in patients with chronic lymphocytic leukemia, *Antibody Immunoconj. Radiopharm.* **1**:291–303.

Zuckier, L. S., Rodriguez, L. D., and Scharff, M. D., 1989, Immunologic and pharmacologic concepts of monoclonal antibodies, *Semin. Nucl. Med.* **19**:166–186.

Index